Arctic Flora and Fauna

Arctic Flora and Fauna

STATUS AND CONSERVATION

EDITA • HELSINKI

EDITORIAL TEAM:

Project lead: Paula Kankaanpää (Arctic Centre, University of Lapland/Ministry of Environment, Finland)
Lead author: Henry P. Huntington (Huntington Consulting, USA),
Project management: Snorri Baldursson (CAFF Secretariat, Iceland),
Project coordination and technical layout: Anna-Liisa Sippola (Arctic Centre, University of Lapland, Finland)
Graphics coordination: Seppo Kaitala (University of Helsinki, Finland)
Species information and graphics coordination: Christoph Zöckler (UNEP-World Conservation Monitoring Centre, UK)
Other members: Anne Gunn (Government of Northwest Territories, Canada), Marina Mirutenko (All-Russian Institute of Nature Protection, Russia), Peter Prokosch (WWF Arctic Programme, Norway), Arkady Tishkov (Russian Academy of Sciences, Russia).

CONTRIBUTORS:

Preparation of maps and figures: S. Blyth, N. Cox, K. Kiviaho, I. Lysenko, L. Malm, J. Nieminen
Main review: O. W. Heal
Chapter reviewers: T. Fenge, C. Freese, E. Helander, B. Stonehouse, J. Svoboda, S. Tuhkanen, W. Vincent

Box authors:

Adrésson, Ó.S.	Einarsson, Á.	Iljashenko, V.	Mulders, R.	Schliebe, S.	Tishkov, A.
Arnalds, Ó.	Fedorova, N.	Ingimundarson, J.H.	Ohenoja, E.	Sippola, A-L.	Vasander, H.
Baldursson, S.	Gau, R.J.	Itämies, J.	Oksanen, T.	Skúlason, S.	Vigfússon, O.
Belikov, S.	Gulden, G.	Kalxdorff, S.	Pagnan, J.	Smith, D.	Vincent, W.
Burek, K.A.	Gunn, A.	Leggat, A.	Petersen, A.	Snorrason, S.	Virtanen, R.
Byrd, V.	Hamre, J.	Magnússon, B.	Pluzhnikov, N.	Soppela, P.	Vogt, P.
Chardine, J.	Heal, B.	Mann, D. H.	Pospelova, E.	Sowls, A.	Vongraven, D.
Chernov, Y.U.	Helle, T.	Marteinsson, V.	Prokosch, P.	Springer, A.M.	Wielgolaski, F.E.
Cluff, H.D.	Henttonen, H.	McGovern, T.H.	Ragnarsson, S.A.	Svoboda, J.	Zöckler, C.
Danks, H.V.	Huberth-Hansen, J-P.	Meltofte, H.	Rasmus, S.	Talbot, S.	
Dyke, A.	Huntington, H. P.	Miller, F.L.	Reist, J.	Thorpe, N.	
Eggertsson, Ó.	Ikävalko, J.	Miller, P.A.	Rätti, O.	Timonen, M.	

Contributing experts:

Aamlid, D.	Estes, J.	Hebert, P. N. D.	Kojola, I.	Neuvonen, S.	Stirling, I.
Aiken, S.	Ewins, P.	Hik, D.	Kovacs, K.	Nilsson, A.	Strøm, H.
Alisauskas, R.T.	Falk, K.	Hines, J.	Kullerud, L.	Ollila, T.	Sulyandziga, P.
Anderson, M.E.	Feder, H. M.	Hodkinson, I. D.	Kurvits, T.	Orlov, V.	Sutherland, P.
Anker-Nilssen, T.	Feldman, K.	Hoefs, M.	Legare, G.	Pike, D.	Svensen, E.
Arnbom, T.	Forbes, B.	Hummel, H.	Lunn, N.	Raillard, M.	Touborg, K.
Bakken, V.	Fossum, P.	Høegh, K.	Machtans, C.	Reid, J.	Tyler, N.
Bogardus, D.	Fosså, J.H.	Jefferies, R. L.	Markon, C.	Rothe, T. C.	Tynan, C.
Bogstad, B.	Fritts, E.	Johnson, L.	McCormick, K.	Rugh, D. J.	Valkenburg, P.
Burch, E.S.	Gilchrist, G.	Johnson, V.	McFarland, F.	Russell, D.	Weller, G.
Burn, C.	Gilg, O.	Johnston, J.D.	McNair, M.	Rydahl, K.	Whiting, A.
Carriere, S.	Grady, S.	Kaissl, T.	Mela, M.	Semyonova, T.	Williams, M.
De Groot, P.	Grebmeier, J. M.	Kampp, K.	Meltofte, H. D.	Simon, P.	Yurtsev, B.
DeCicco, F.	Hario, M.	Kephart, J.	Merkel, F.	Sittler, B.	
Dickson, L.	Heal, O.W.	Klein, D. R.	Moore, S. E.	Smol, J. P.	
Douglas, M.	Heath, M.	Kline, T.	Mortensen, P.B.	Solovieva, D.	
Duffy, D.	Hebert, P.	Kolpashnikov, L.	Muir, M.	Stephenson, B.	

Contributing institutions: Arctic Centre, University of Lapland; BirdLife International; Norwegian Polar Institute; Institute of Biogeography, Russian Academy of Sciences; UNEP-GRID Arendal; UNEP-World Conservation Monitoring Centre.
Financial and other support: Nordic Council of Ministers; National Science Foundation (U.S.); Ministry of the Environment Finland; Canadian Wildlife Service – Environment Canada, Fisheries and Oceans Canada, Indian and Northern Affairs Canada; Ministry of Foreign Affairs Iceland; Swedish Environmental Protection Agency; Greenland Homerule Government; Directorate for Nature Management, Norway.

Photographs:

Antikainen, P.	Falk, K.	Kaikusalo, A.	Mikkola, K.	Talbot, S.L.	Vogt, P.
Arnalds, O.	Gilg, O.	Kassens, H.	Miller, P.	Tolvanen, P.	Väre, H.
Bakke, T.A.	Gislason, B.	Kristinsson, H.	Nicklen, P.	Tuominen, O.	Zöckler, C.
Bangjord, G.	Heikkinen, O.	Leinonen, A.	Overholt, J.	Tynys, T.	
Chardine, J.	Hillmarsson, J.Ó.	Luhta, J.	Prokosch, P.	Vasama, V.	
Dau, C.	Ikävalko, J.	Løfaldli, L.	Sabard, B.	Vasander, H.	
Einarsson, A.	Institute of Marine Research, Norway	Magnusson, B.	Sippola, A-L.	Virtanen, R.	
		Mattsson, J.	Svensen, E.	Virtanen, T.	

Paintings and drawings: M. Rapeli (pages 15, 110-111, 134, 162, 182, 216-218), Gerald Kuehl (page 46)
Cover photo: Snowy owls *Nyctea scandiaca*. Gilg & Sabard/GREA.
Back cover photo: Roseroot, *Rhodiola rosea*. Gilg & Sabard/GREA.

Artistic layout: P. Kivekäs
Printing and binding: Edita Plc

ISBN 9979-9476-5-9

Contents

Preface 9

1. INTRODUCTION — 11

The Report 11
Defining the Arctic 11
    Box 1. The mission of CAFF 12
    Box 2. The Arctic Council 13

2. ECOLOGY — 17

Climate 17
    Shaping the climate 17
    Characteristics of the Arctic climate 18
    Box 3. Snow: physics, ecology, and climate change 20
    Spatial climatic variation 21
    Climate variation with time 21
    Box 4. The effects of past climates in the North American Arctic 22
Life in the Arctic 24
    The physical setting 24
    Box 5. The North Atlantic Oscillation and reindeer husbandry 25
    Responding to climate 29
    Finding favorable habitats 29
    Box 6. Variations in snow cover affect plants 30
    Box 7. How land animals adapt to cold and snow 32
    Box 8. Insect adaptations in the High Arctic 36
    Box 9. Staying warm in a cold climate 38
    The importance of movement 39
    Ecological relationships 39
    Box 10. The volcanic island Surtsey: a natural experiment in colonization 40
    Box 11. The role of fungi 41
    Ecosystem relationships 42
    Box 12. The Arctic as a theater of evolution 44
The Concept of Biodiversity in the Arctic 44
    Box 13. Microbes everywhere 45
    Box 14. The diversity of arctic char 46
    Box 15. North-south trends in terrestrial species diversity 48

HUMANS — 51

The First People 51
    Box 16. Driftwood in the Arctic 52
    Box 17. Reindeer and the peopling of the Yamal 54

Later Migrations 57
Modern Development 59
   Box 18. Northern farming in a long-term perspective 60
Political Development 61
Current Settlement Patterns 63
   Box 19. Co-management in the Inuvialuit Settlement Region 64
   Box 20. Environmental impacts of diamond mining in the Canadian Arctic 66
Use of Living Resources 67
   Box 21. The Tłı̨chǫ and the Detsı̨̀ı̨̀làà ɂekwǫ̀: Showing respect for caribou 69
   Box 22. Why pine trees have different names: a Khanty folk tale 72
Cultural Significance of the Environment 73
   Box 23. The use of wild plants in northern Russia 74

4. CONSERVATION   77

Habitat 77
   The role of smaller areas 77
   Box 24. The Circumpolar Protected Areas Network (CPAN) 78
Wildlife 81
   Box 25. Protected areas in northern Fennoscandia: an important corridor for taiga species 82
   Box 26. Protecting the Lena Delta-an example of the rapid growth
             of protected areas in Russia 84
   Box 27. Threats to Arctic biodiversity – overview 86
   Economic importance of harvests 87
Industrial and Commercial Development 87
   Box 28. Unsustainable take of murres in Greenland 88
   Box 29. CAFF International Murre Conservation Strategy and Action Plan 90
   Box 30. Success with geese: proof that conservation is working? 93
Diffuse Threats 95
   Box 31. Oil development impacts on the North Slope of Alaska 96
   Box 32. Habitat fragmentation: a threat to Arctic biodiversity and wilderness 98
   Box 33. Impacts of infrastructure: the Arctic 2050 scenario 99
Understanding 100
   Box 34. Preserving sacred sites in the Russian Arctic 100
   Box 35. Climate and caribou: Inuit knowledge of the impacts of climate change 101
   Box 36. Tourism – threat or benefit to conservation? 102
   Box 37. Alien invasive species: a threat to global and Arctic biodiversity 104
   Box 38. International legal instruments for conservation 106
   Box 39. U.S./Russia bilateral agreement on polar bears
             breaks new ground in international conservation 108

5. FROM FOREST TO TUNDRA   110

The region 110
Key characteristics 111
Functionality 115
Interactions with other regions 116
Box 40. Mountain birch and the Nordic treeline 116
Box 41. Paludification, fire, and topography structure the northern forest 117
Natural variability 118

  Human uses and impacts 118
  Box 42. Ary-Mas: an island of remnant forest 121
  Box 43. The introduced Nootka lupine in Iceland: benefit or threat? 122
  Box 44. Management and biodiversity in Fennoscandian timberline forests 124
  Climate change 126
  Conservation 128

6. THE TUNDRA AND THE POLAR DESERT 131

  The region 131
  Box 45. The tundra and polar desert: dissimilarities between North America and Eurasia 132
  Key characteristics 133
  Box 46. Decomposition ensures recycling of nutrients 134
  Box 47. Mires and wetlands 136
  Box 48. Mosquitoes 139
  Functionality 139
  Box 49. Lemmings: key actors in the tundra food web 142
  Interactions with other regions 144
  Box 50. Parasitism: an underestimated stressor of Arctic fauna? 145
  Box 51. Sandpipers: birds of the tundra 146
  Natural variability 147
  Human uses and impacts 150
  Box 52. Erosion in Iceland 153
  Climate change 154
  Box 53. Peary Caribou: extreme demands on an endangered caribou 156
  Box 54. Rare endemic vascular plants of the Arctic 158
  Conservation 160

7. RIVERS, LAKES, AND WETLANDS 163

  The region 163
  Key characteristics 164
  Box 55. Conserving arctic char 166
  Functionality 167
  Box 56. Freshwater food webs 168
  Interactions with other regions 170
  Natural variability 171
  Box 57. The Atlantic salmon: a species in trouble 172
  Human uses and impacts 172
  Box 58. Food web contaminants in Arctic freshwaters 177
  Box 59. The Lake Mývatn ecosystem: many uses, many threats 178
  Climate change 180
  Conservation 180

8. THE OCEANS AND SEAS 183

  The region 183
  Key characteristics 184
  Box 60. The BIOICE Research Program maps benthic biodiversity 184
  Box 61. Vent and seep communities on the Arctic seafloor 186

Functionality	187
Interactions with other regions	190
Box 62. Life within sea ice	190
Box 63. Living water in the frozen Arctic: the Great Siberian Polynya	192
Natural variability	193
Box 64. The Anadyr Current and ocean productivity in the Bering Strait area	194
Human uses and impacts	198
Box 65. Beluga whale hunting in Russia	200
Box 66. The interaction of climate and fish stocks in the Barents Sea	202
Box 67. Bycatch: a threat to seabirds in Arctic waters	203
Box 68. Impacts of trawling on seabed animals and habitats	204
Box 69. Seabirds, foxes, and rats: conservation challenges on Bering Sea islands	206
Climate change	207
Conservation	208

9. STATUS AND TRENDS IN SPECIES AND POPULATIONS — 211

Box 70. Black guillemots in Iceland: a case-history of population changes	212
Population Levels and Trends	214
Marine mammals	215
Seabirds	215
Marine fishes	226
Terrestrial mammals	231
Terrestrial birds	232
Box 71. Population fluctuations in Svalbard reindeer	232
Box 72. Wolves	234
Box 73. Wolverine	236
Box 74. Brown bear	238
Shorebirds	239
Passerines	239
Box 75. Muskoxen	240
Globally Threatened Species	247

10. CONCLUSIONS — 253

Ecology	253
Humans	253
Overexploitation	253
Habitat disturbance	254
Biological disturbance	254
Contamination and pollution	254
Climate change	255
Knowledge gaps	255
Conservation	255
List of English and Russian common names and Latin scientific names of species mentioned in this book	257
List of Figures	265
List of Tables	267

Preface

■ It is with great pleasure that we present *Arctic Flora and Fauna: Status and Conservation*. This report draws on the diversity of projects that the Program for the Conservation of Arctic Flora and Fauna (CAFF) has undertaken in its first decade, underscoring the need to address conservation on a circumpolar basis. We thus find it highly appropriate that its publication coincides with the tenth anniversary of circumpolar Arctic environmental cooperation, initiated in Rovaniemi, Finland, with the creation of the Arctic Environmental Protection Strategy, now a central part of the Arctic Council.

The Arctic Council, at its meeting in Iqaluit, Canada, in 1998, requested that CAFF prepare an overview of Arctic ecosystems, habitats, and species. The resulting report is not a comprehensive scientific document, but a scientifically-based overview of the main features of the Arctic ecoregion, its ecology, and the status of its flora and fauna. Written for the non-specialist, the report highlights the importance of Arctic resources and processes for Arctic residents and for the world as a whole. General conclusions address the current status of Arctic biodiversity and the main conservation challenges that face us. This report is a major step in CAFF's efforts to look at conservation issues from a circumpolar perspective. We will use it as the basis for developing, in the months and years to come, specific assessments, recommendations, and actions to conserve the unique natural heritage of the Arctic region.

This report is a collective product of the member countries, permanent participants, and observers of the CAFF Program. Without their commitment and support, the project could not have been undertaken, much less successfully completed. In addition, we thank all those who provided material for the report, reviewed drafts, and contributed in other ways. In particular, however, we commend the Editorial team for the admirable job it has done.

Projects of this kind do not happen solely as a result of good intentions and hard work. They also require financial and other support. We are grateful to the Nordic Council of Ministers, the U.S. National Science Foundation, WWF-Arctic Programme, and the governments of the eight Arctic nations, all of which have provided funding support for the report.

It is our sincere hope that readers of this report will learn more about a fascinating region of the planet, better understand its conservation challenges, and renew their commitment to meeting those challenges so that future generations will continue to enjoy Arctic flora and fauna.

Sune Sohlberg   Chair of CAFF
Esko Jaakkola   National Representative of Finland

Gyrfalcon, **Falco rusticolus**

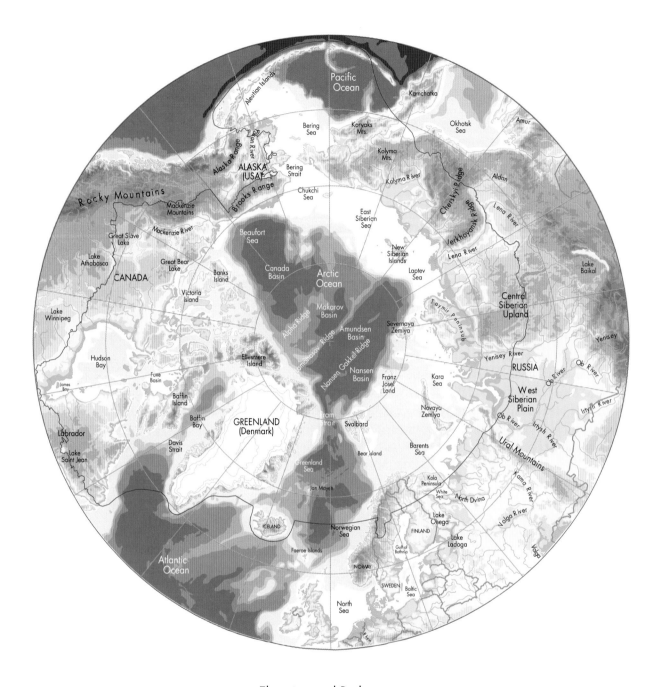

Elevation and Bathymetry

— CAFF Boundary

Figure 1. The Arctic.

1. Introduction

The Report

What is the overall state of the Arctic's natural environment? The aim of this report is to answer the many aspects of this seemingly straightforward question. Although several national and international efforts have looked at parts of the Arctic, this is the first attempt to assess the state of Arctic flora and fauna as a whole. Clearly, a brief report cannot cover all the relevant aspects and areas of the Arctic in detail. The chief criterion used to determine what to include has been international interest or significance. While much has of necessity been left out, we hope that the report nonetheless gives a thoughtful and compelling reply to our opening question.

Many volumes have been written, and will continue to be written, about the ecology, conservation, indigenous peoples, environmental threats, economics, oceanography, politics, and other aspects of the Arctic region. Adding this report has a threefold purpose:

- To summarize, in one place, the information required to assess the state of the Arctic's natural environment today, providing a marker from which to measure progress in conservation
- To provide a useful reference to a wide audience of policy makers, Arctic residents, researchers, and others active in the conservation of Arctic flora and fauna
- To point the way to improving our collective understanding and facilitating the international cooperative action required to conserve the Arctic's natural environment

The report is structured to strike a balance between covering many aspects of the Arctic in general terms and covering a few in more detail. This Introduction gives an overview of the Arctic region and its boundaries. Chapter 2 describes the characteristics of the Arctic and basic principles of ecology and biodiversity as they apply to the Arctic. Chapter 3 outlines the history and role of humans in the Arctic environment. Chapter 4 reviews the various approaches to conservation as they have been applied in the Arctic. Chapters 5 through 8 give an overview of the major biological regions of the Arctic. Chapter 9 reports on the status and trends of certain species and groups. Chapter 10 offers brief conclusions. Throughout, graphics and boxes examine specific topics in detail to illuminate the general points made in the main text.

Defining the Arctic

The word Arctic comes from the Greek word for bear, *arktos*, after the constellations Ursa major and Ursa minor that are visible year round in the northern night sky. Astronomically, the boundary of the Arctic is the latitude at which the sun does not set on the summer solstice, about 66°33' N, known as the Arctic or Polar Circle. This definition has been used for many things, among them identifying the eight Arctic nations that comprise the Arctic Council. Nonetheless, in biological terms, the Arctic Circle is merely an abstraction.

One botanical feature used to delineate the Arctic is the treeline. This marker is attractive because it seems easy to define, but in practice the exact location of the

BOX 1. THE MISSION OF CAFF

▶ The Program for the Conservation of Arctic Flora and Fauna is a distinct forum to discuss and address Arctic conservation issues. As one of the working groups of the Arctic Council, its primary role is to advise the Arctic governments on conservation matters of international significance and common concern.

Since its first meeting in 1992, CAFF has sponsored a variety of research projects and analyses and launched several circumpolar conservation strategies:

- CAFF's conservation strategies and action plans for murres (guillemots) and the four eider species inhabiting the Arctic are being implemented and have greatly improved the conservation status of these seabirds.
- CAFF has created the unique Circumpolar Protected Areas Network, which aims to maintain in perpetuity the diverse habitats and biodiversity of the circumpolar Arctic.
- A CAFF/Inuit Circumpolar Conference project on traditional ecological knowledge of the beluga whale has paved the way for using indigenous knowledge in conservation work.
- The Atlas of Rare Endemic Vascular Plants is a major step in protecting a vital class of species unique to the Arctic.
- The Circumpolar Vegetation Map, soon to be finished, represents a step towards a harmonized habitat classification system for the circumpolar region on which to base future monitoring activities and assessments.

Several major new initiatives are underway:

- CAFF is developing a program to monitor circumpolar biodiversity. Expert networks have already been established to harmonize monitoring of several key species and species groups, including reindeer/caribou, Arctic char, ringed seals, seabirds, polar bears, waders/shorebirds, geese, and vascular plants.
- In cooperation with its sister working group AMAP, CAFF is involved in the Arctic Climate Impact Assessment, which will deliver in 2004 a major scientific assessment of the impacts of climate variability and change and UV-B radiation on ecosystems and societies in the circumpolar region.
- Together with the Russian Association of Indigenous Peoples of the North (RAIPON), CAFF is assessing the conservation value and status of sacred sites of indigenous people in two regions of Arctic Russia as preparation for a nationwide assessment.
- In collaboration with the United Nations Environment Programme (UNEP) and the Global Environment Facility (GEF), CAFF is working to reduce disturbance and fragmentation of Russia's extensive undisturbed tundra and taiga ecosystems.

In the 10 years of its existence, CAFF has made significant progress. Perhaps CAFF's greatest achievement has been to create working links among Arctic scientists, managers, and indigenous peoples who share similar challenges and interests. These links have changed the face of conservation work in the Arctic. They represent the seeds from which future successes will grow.

---

treeline is harder to pin down. For instance, in northern Russia, the transition zone from boreal forest to open tundra covers up to 300 kilometers. On the other hand, in mountainous terrain, altitude can be the main factor in determining where trees grow, and indeed Arctic-like conditions are found on mountaintops far to the south. Furthermore, isolated stands of trees exist well to the north of the transition zone. For general purposes, however, and at regional or circumpolar scales, the treeline is a good practical boundary.

In terms of climate, a commonly used boundary is the line at which the average temperature in the warmest month is 10°C, known as the 10°C summer isotherm. This line corresponds roughly to the treeline, though they may diverge by 100 km or more in some areas. Both the treeline and the 10°C summer isotherm diverge from the Arctic Circle. In northern Norway, both lines extend near or beyond 70°N, while in the area around Hudson Bay, Canada, they extend as far south as 55°N. The area north of the treeline is also typically underlain by permafrost-ground that remains frozen all year.

> BOX 2. THE ARCTIC COUNCIL

▶ The Arctic Council was established in 1996 as a high level intergovernmental forum for addressing the common concerns and challenges faced by the Arctic governments and the people of the Arctic. Priorities of the Council are protection of the Arctic environment and the promotion of sustainable development as a means of improving the economic, social, and cultural well-being of Arctic residents.

The members of the Council are Canada, Denmark, Finland, Iceland, Norway, the Russian Federation, Sweden, and the United States of America. The Russian Association of Indigenous Peoples of the North (RAIPON), the Inuit Circumpolar Conference (ICC), the Saami Council, the Aleutian International Association (AIA), the Arctic Athabaskan Council, and the Gwich'in Council International are Permanent Participants in the Council.

Several non-Arctic states, inter-governmental and inter-parliamentary organizations, and non-governmental organizations are Observers to the Council.

The Council meets biennially at the ministerial level. The Chair and Secretariat of the Council rotate every two years among the eight Arctic States. Canada served first from 1996-1998, followed by the United States from 1998-2000. Finland has assumed the Chair for 2000-2002.

The environmental protection work of the Arctic Council builds on the Arctic Environmental Protection Strategy (AEPS) which was adopted by the Arctic States through a Ministerial Declaration at Rovaniemi, Finland in 1991.

The Council has currently five working groups to address its priorities. The first four mentioned below were established as a part of the AEPS and later incorporated into the Arctic Council:

– AMAP (Arctic Monitoring and Assessment Programme) provides reliable and sufficient information on the status of, and threats to, the Arctic environment from contaminants. It also provides scientific advice on actions to be taken in order to support Arctic governments in their efforts to take remedial and preventive actions relating to contaminants.
– CAFF (Conservation of Arctic Flora and Fauna) addresses and provides advice on the needs of Arctic species and their habitats. It is a forum for scientists, conservation managers and groups, and indigenous peoples of the North to tackle a wide range of Arctic conservation and sustainable use issues at the circumpolar level.
– EPPR (Emergency Prevention, Preparedness and Response) evaluates the adequacy of existing emergency and prevention arrangements in the Arctic, to improve cooperation for mutual aid in case of accidents, and to recommend necessary cooperative mechanisms.
– PAME (Protection of the Arctic Marine Environment) addresses policy and non-emergency pollution prevention and control measures related to the protection of the Arctic marine environment from both land- and sea-based activities. These include coordinated action programs and guidelines complementing existing legal arrangements.
– SDWG (Working Group on Sustainable Development) was established at the Second Arctic Council meeting in 1998 to oversee programs and projects aimed at protecting and enhancing the economies, culture, and health of the inhabitants of the Arctic, in an environmentally sustainable manner.

Other recently adopted major programs of the Arctic Council are:

– RPA (Regional Programme of Action for the Protection of the Arctic Marine Environment from Land-Based Activities), adopted in 1998 to strengthen regional cooperation and capacity building, particularly in relation to addressing regional priority pollution sources found in the Russian Federation.
– ACAP (Arctic Council Action Plan to Eliminate Pollution of the Arctic) adopted in 2000 to strengthen and support mechanisms that will encourage national actions to reduce emissions and other releases of pollutants.
– ACIA (Arctic Climate Impact Assessment), adopted in 2000 to evaluate and synthesize knowledge on climatic variability, climate change, and increased ultraviolet radiation and their consequences for Arctic ecosystems and societies.

In the marine environment, the Arctic boundary is also problematic. An oceanographic definition is the meeting point of the relatively warm, salty water from the Atlantic and Pacific Oceans and the colder, less salty waters of the Arctic Ocean. Like the treeline, the latitude of the ocean boundary varies greatly, from about 63°N in the Canadian Archipelago to 80°N near Svalbard. The subarctic marine environment can be defined as the area of mixed Arctic and Atlantic or Pacific water, extending as far south as 44°N off New-

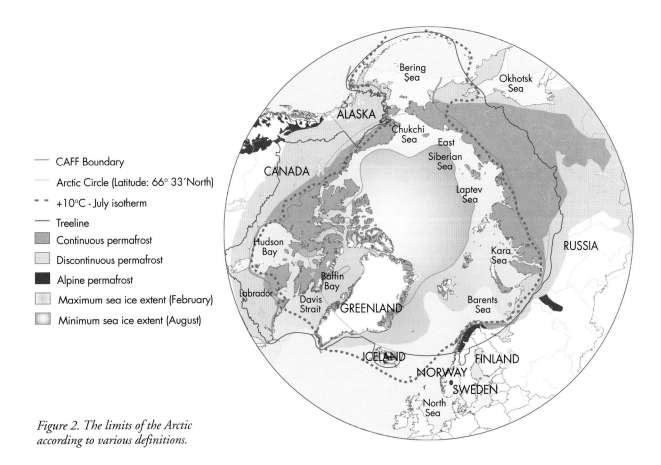

Figure 2. The limits of the Arctic according to various definitions.

foundland, Canada, and as far north as 68°N off Norway.

Finding or choosing one definition of the Arctic to satisfy all purposes is nearly impossible. Finland and Sweden both extend north of the Arctic Circle, but lie south of the treeline and the 10°C summer isotherm. Administrative districts relevant to conservation measures – provinces, territories, national parks, management units – rarely follow consistent biogeographic boundaries. Migratory species travel well outside the Arctic, and water and air currents carry substances from across the globe that affect plants and animals that reside exclusively in the Arctic. For the purpose of this report, we consider the species and issues found from the forest-tundra transition zone northwards on land, and in the marine area north of the CAFF boundary.

While a practical definition of the Arctic is necessary, if only to determine what physical and ecological processes should be covered by this report, the Arctic is intimately linked with the rest of the world. Ocean water circulates to and from the Arctic Ocean. Air masses move north and south. There is virtually no place on the earth that is not connected to the Arctic by migratory species. The Arctic is a distinctive region of the planet, and as such is worthy of a dedicated review of its environment, but its conservation and its future are tied to what happens throughout the world.

SOURCES/FURTHER READING

Murray, J.L. 1998. Physical/geographic characteristics of the Arctic. In: Arctic Monitoring and Assessment Program. *AMAP assessment report: Arctic pollution issues.* Oslo: AMAP. p. 9-24.

Stonehouse, B. 1989. *Polar ecology.* London: Blackie. viii + 222 p.

2. Ecology

> *Arctic flora and fauna are shaped by the patterns and variations of the Arctic climate. The characteristics of that climate, its ecological implications, and other aspects of Arctic biodiversity are the subject of this chapter.*

Climate

On a global basis, the Earth radiates into space as much energy as it receives from the sun. In tropical and temperate areas, where sunlight is most intense, more energy is received than is radiated. In the Arctic, sunlight comes from a low angle even in summer. Much of it is reflected by snow and ice, or is blocked by the frequent cloud cover. In winter, little sunlight reaches the Arctic at all. Annually, more energy is radiated from the polar regions than is received. The result is a flow of energy from the equatorial regions to the poles, carried by currents of air and water. This basic flow pattern causes global oceanic and atmospheric circulation, as heat flows from warmer to cooler areas. The flow of heat is inefficient, however, so that great differences in temperature remain between the tropics and the poles.

Shaping the climate

Air and water circulation patterns determine regional and local climates. The Gulf Stream warms Scandinavia, whereas at the same latitude the eastern Canadian Arctic is cooled by cold water and air flowing south through the Canadian Archipelago. The flow of relatively warm air and water along the eastern side of the Bering Sea keeps the Alaskan side of the Bering Strait warmer than the Russian side, which is influenced by cold air from northern continental Asia. In winter, the polar front contains a relatively stable, high-pressure air mass over the Arctic Basin and much of northern North America and Eurasia. Around the southern edge of this air mass, low-pressure systems produce winter storms in the North Pacific and North Atlantic.

Air and water currents interact and combine to affect climate, which in turn affects circulation patterns. These interactions produce regular climate patterns, trends, and extreme weather events. One pattern is the oscillation known as the North Atlantic Oscillation. In winters when low pressure systems dominate the area near Iceland and high pressure systems dominate the area near the Azores, strong westerly winds send warm air over northern Europe and into Asia, producing mild winters but with heavier snowfall. In winters when the air pressure near Iceland is higher and the westerly winds are weaker, northern Europe experiences cold, dry winters. In the 1990s, there was an increasing awareness that the oscillation tends to persist in one state or the other over decades, with marked biological consequences. In the North Pacific, the influence of the El Niño-Southern Oscillation cycle reaches northwards to Alaska and beyond. Patterns of storms and the quantity of moisture delivered to the Arctic are thus influenced by ocean circulation patterns in the tropical Pacific.

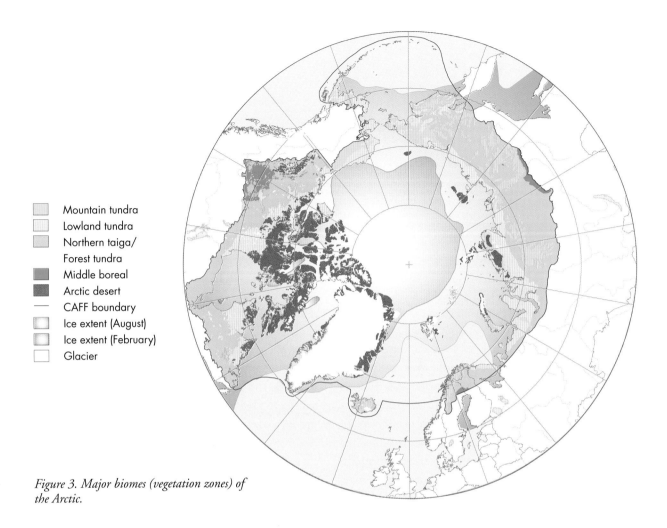

Figure 3. Major biomes (vegetation zones) of the Arctic.

Air and water currents transport other environmentally significant substances, especially contaminants. Industrial and agricultural pollution is transported by the air and rivers. Much of it reaches the Arctic, where it tends to stay. Ocean currents can also carry contaminants from areas of intense industrial and agricultural activity, particularly eastern Asia and western Europe, towards the Arctic. The impacts of these substances on Arctic flora and fauna are not fully understood, but chemicals such as persistent organic pollutants and heavy metals have reached concentrations in wildlife and people that raise concern.

Just as the Arctic climate is influenced by global climatic patterns, the Arctic influences the rest of the world. Global ocean circulation is driven by phenomena such as the cooling and downward flow of water in the North Atlantic, initiating a global flow of deep ocean water that ends with upwelling in the North Pacific. Flows of deep layers of water from and to the Arctic Ocean influence global climate, as does the transport of sea ice across the Arctic Basin. While these effects are beyond the scope of this report, it is important to note that the Arctic has a significant influence on global climate patterns.

Characteristics of the Arctic climate

The striking characteristics of the Arctic climate are severity, seasonality, and unpredictable variability.

- *Severity* refers to the persistent conditions that limit growth and survival, including not only cold winters but cool summers and low humidity. Low temperatures limit the presence and growth of individual species. Summer temperatures affect moisture, soil stability, nutrient availability, and habitat diversity.

- *Seasonality* is the contrast between the long, dark, cold winter and the brief, light, relatively warm summer. It can be indexed by the temperature difference between the coldest and warmest month.
- *Variability* can be measured by differences between extreme temperatures or snowfall and the long-term average. Average Arctic conditions are already close to the limits for growth or even survival of many species, and variability is typically high. Plants and animals must be able to adapt to survive and thrive.

Overall, the Arctic receives relatively little precipitation, most of which falls as snow. Some areas receive less than 50 millimeters per year. Because summer temperatures are relatively low, however, the rain does not evaporate quickly, and permafrost prevents water from draining into the ground. Thus, many areas are very wet in summer. Additionally, the winter accumulation of snow leads to a brief but intensive run-off in spring when the snow melts. The run-off takes organic matter from the land and injects it into the freshwater systems, producing a bloom of life in lakes and rivers. The run-off also carries sediments and pollutants. Lat-

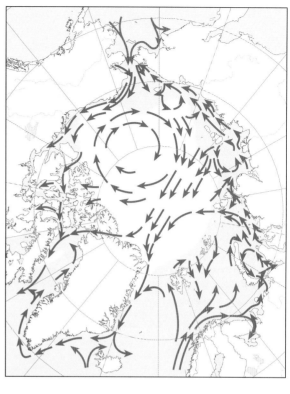

Figure 4. Surface ocean currents in the Arctic Ocean.

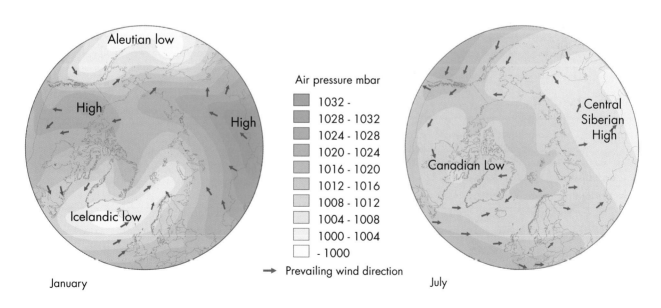

Figure 5. Average atmospheric sea-level air pressure in winter and summer.

BOX 3. SNOW: PHYSICS, ECOLOGY, AND CLIMATE CHANGE

▶ The snow pack has layers that can be distinguished, for example, by crystal size, crystal form, and bonding between individual snow crystals. Metamorphoses (reshaping the snow crystals and the bonding between them) are continuous thermo-mechanical processes, which begin soon after the snowfall and continue until the snow melts. These metamorphoses can change the layering, hardness, density, porosity, albedo, depth, compactivity, thermal conductivity, and thermal resistance of the snow. Snow pack qualities of ecological importance for mammals are depth, stratification, hardness, density, thermal resistance, and duration. Thermal resistance is the snow depth divided by its thermal conductivity, which depends on density and the metamorphic state of the individual snow crystals.

The presence or absence of trees has a large influence on snow pack, and the treeline marks the distinction between two broad classes of snowpack: *Taiga zone snow* is found in cold climates in forests where wind, initial snow density, and average winter air temperatures are all low. The snow pack is of thin to moderate depth (30 - 120 cm), cold, and porous. By late winter it consists of 50-80% depth hoar (loosely bonded, sugar-like snow) covered by new snow.

Tundra zone snow is found above or north of the tree line. The snow pack is thin (10-75 cm), cold, and windblown. It consists of a base layer of depth hoar overlain by multiple wind slabs. Melt features are rare.

Global climate change will affect both the frequency of thaws and the amount of snowfall. The combination of changes produces three general scenarios:

More thaws and increased snowfall
No significant changes in snow depth or snow cover duration are probable, but increases in hardness and density of the layers are likely. More ice layers but less depth hoar will develop in the snow pack. Thermal resistance will decrease due to increased snow density and changes in stratigraphy. These will cause colder conditions in the snow pack, ease movement on top, but make it difficult to live inside the snow pack and to find food under it.

Increased snowfall without thaws
Snow cover duration will increase. No ice lenses and less depth hoar will be formed. No significant changes in density and hardness will occur. Thermal resistance will increase slightly due to increased snow depth. These phenomena will cause warmer conditions and ease moving inside the snow pack, but make it more difficult to move on top of the snow and find food.

More frequent thaws with current snowfall
Snow depth and snow cover duration will decrease. There will be more melt-freeze layers and ice lenses but less depth hoar. Hardness and density will increase, and increased density and changes in stratigraphy together with thinning of the snow pack will cause thermal resistance to decrease greatly. These conditions will ease movements for some wildlife, but hinder others. For example, caribou that break through the crust will be easier prey for wolves that stay on top. It will be easier to find food for most of the year, but very difficult to dig under the snow. Hibernating animals and some mammals with protective coloring may have problems.

It is notable that a given change in one snow pack characteristic may have a positive effect on some mammal species and a negative effect on others. Ice lenses, for example, will likely become common in the taiga and tundra zones. They make it difficult for reindeer to forage under the snow and may also hurt the feet of digging animals, whereas for some mammals, for example arctic fox or ermine, hard icy layers make it much easier to move on top of the snow. For small mammals such as lemmings that winter under the snow, ice layers forming at the bottom of the snow pack make moving, nesting, and finding food extremely difficult. Ice layers above their habitat decrease the thermal resistance of the snow pack, making conditions colder. Ice layers may also inhibit ventilation and lead to dangerously high levels of carbon dioxide. Thus, climate-driven changes in snow pack characteristics may have a major impact on Arctic animals.

Tari Oksanen and Sirpa Rasmus, Department of Geophysics, University of Helsinki, Finland

SOURCE/FURTHER READING
Sturm, M., J. Holmgren, and G.E. Liston. 1995. A seasonal snow cover classification scheme for local to global applications. *J. Climate* 8(5): 1261-1283.

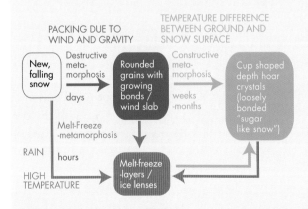

Figure 6. Metamorphoses of snow.

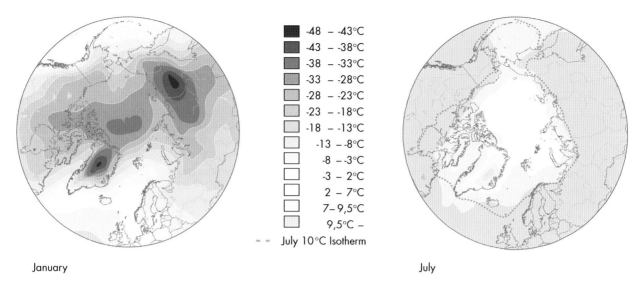

Figure 7. Average air temperatures in January and July.

er in summer, some areas become arid. The growth and survival of many plants are limited by the lack of moisture available in this, the driest part of the year.

Wind, too, plays a major role in climate. In winter, wind can blow snow off ridges, leaving the ground bare, while snowdrifts accumulate on downwind slopes and in depressions. In open areas, wind-packed snow is dense and hard. Wind makes low temperatures seem colder. In summer, wind can make plants and soils dry more rapidly. It also poses a physical challenge to plants. On the tundra, shrubs and dwarf trees grow low to the ground, often in sheltered areas where they will be buried by snow and thus protected from winter winds.

Spatial climatic variation

The Arctic climate varies strongly between regions. Water bodies, even when ice-covered, moderate temperatures in summer and winter. In the middle of continents, by contrast, temperatures are more extreme. Mountain ranges make climates more severe at higher elevations and also affect weather patterns downwind. Snow and ice reflect light, keeping the surface relatively cool, while bare ground and open water absorb light, increasing the local temperature. These processes have a significant effect both locally during the spring melt and globally for the net heat balance of the earth.

At scales of meters and even centimeters, microclimates vary sharply. Raised-center tundra polygons range from cool, saturated ground at their edges to warm, dry ground in their interiors. Air temperatures affect vegetation, but direct sunlight can warm plants and bare ground well above the temperature of the air. The plant layer also insulates the ground, keeping soils cooler in summer and warmer in winter than the air and the top layer of vegetation. Wet ground stays cooler than dry ground in summer, allowing a much thinner layer to thaw. The insulative properties of vegetation delay the peaks of warm and cold, so that deeper soil layers are typically warmest in late summer or early fall, after air temperatures have begun to drop.

Climate variation with time

Climate varies over time. The fossil record from the Canadian High Arctic includes a species of redwoods that grew thousands of kilometers north of the current tree line. Arctic fauna of 50-55 million years ago included lemurs and tortoises. Within the last two million years, four cycles of glaciation have shaped the landscape and, to a large extent, today's fauna and flora. Since the retreat of the last continental glaciers some 10,000 years ago, the world's climate has continued to change, if not in so dramatic a fashion.

BOX 4. THE EFFECTS OF PAST CLIMATES IN THE NORTH AMERICAN ARCTIC

▶ The Arctic is shaped not only by today's climate but also by the climate and its changes over the past 20,000 years. During the last Ice Age, vast ice sheets covered much of northern North America. Ice free areas north of the ice sheets-such as Beringia, which stretched from the present-day Yukon Territory through much of Alaska and across the Bering Land Bridge to what is now the Russian Far East-served as refugia for many tundra plants and mammals, which were crucial to recolonization after the ice retreated. The Arctic Ocean was a refugium for some marine organisms, including at least two mollusks, but most of today's high Arctic mollusks and marine mammals advanced northward from the Atlantic and Pacific Oceans. A particularly significant diversification of Arctic marine animals occurred when the Bering Strait submerged some 11,000 years ago, reconnecting the North Pacific and Arctic Oceans.

The Arctic has changed physically, as well, as the ice sheets melted. Without the weight of ice pressing it down, the land in much of the Canadian Arctic rebounded, in some cases up to 150 meters. As a result, whole islands emerged from the sea, such as King William Island, while others such as Victoria Island expanded significantly. This process continues in the areas that were beneath the thickest ice, which may rise another ten meters in the next few thousand years. Areas peripheral to the ice sheets, such as the Beaufort Sea coast and eastern Devon Island, have experienced submergence due to global sea level rise as the ice sheets melted.

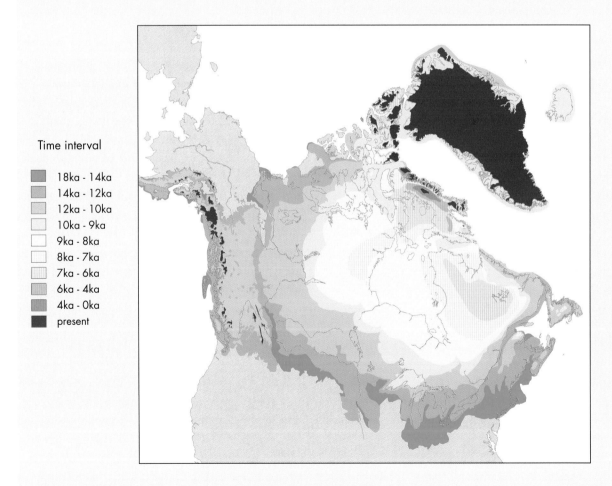

Figure 8. Deglaciation of North America in thousand year intervals before present (ka BP).

Ice core records and fossil pollen studies show that the Arctic climate has not changed smoothly since the last Ice Age, but has gone through several pronounced shifts as well as regional variations. Around 12,000 years ago, the Laurentide Ice Sheet produced a sudden flood of meltwater that changed ocean water circulation in the North Atlantic, resulting in a long cold period known as the Younger Dryas. After this period, temperatures for the next thousand years climbed to their highest levels since the ice age, driven by changes in solar radiation. Poplar and spruce trees reached the Arctic coast in western Canada, farther north than they are found today. Around 8,200 years ago, another sudden influx of meltwater from the Laurentide Ice Sheet caused changes in the circulation of the Labrador Sea, bringing another cold period to the eastern Canadian Arctic and North Atlantic. Treelines in the eastern Canadian Arctic have never been farther north than they are today, although the northern forest was denser some 6,000-3,000 years ago, before the climate entered another cooling period.

Fossil records from coastal waters show the expansion and contraction of ranges of whales, walrus, and warmth-demanding clams. The bowhead whale and the walrus expanded their ranges in the period 10,500-8,000 years ago and again 6,000-3,000 years ago. The first interval corresponds to the warmest period in the ice core records, and the second to the northernmost expansion of the treeline in the central Canadian Arctic. The fossil clam record shows an interval of generally extended ranges of warmth-demanding species such as the edible mussel, *Mytilus edulis*, along the Canadian side of Baffin Bay during the period 8,500-3,000 years ago. During that period, some species were found as far as 1,000 kilometers farther north than they are today.

As these records show, the Arctic climate is constant neither across time nor across space. Recent records also show variable patterns: the western Canadian Arctic and Alaska are warming, while the eastern Canadian Arctic is cooling. The distribution and abundance of Arctic biota are shaped by these forces, and on a millennial scale must be viewed as variable rather than fixed characteristics.

Arthur S. Dyke, Geological Survey of Canada, Ottawa, Canada

SOURCES/FURTHER READING

Barber, D., A. Dyke, C. Hillaire-Marcel, A. Jennings, J. Andrews, M. Kerwin, G. Bilodeau, R. McNeely, J. Southon, M. Morehead, and J. Gagnon. 1999. Forcing of the cold event 8,200 years ago by catastrophic drainage of Laurentide lakes. *Nature* 400: 344-348.

Broecker, W.S., J.P. Kennett, B.P. Flower, J.T. Teller, S. Trumbore, G. Bonani, and W. Wallfli. 1989. Routing of meltwater from the Laurentide Ice Sheet during Younger Dryas cold episode. *Nature* 341: 318-321.

Dyke, A.S., and V.K. Prest, 1987. The Late Wisconsinan and Holocene history of the Laurentide Ice Sheet. *Géographie physique et Quaternaire* 41: 237-263.

Dyke, A.S., 1998. Holocene delevelling of Devon Island, Arctic Canada: Implications for ice sheet geometry and crustal response. *Canadian Journal of Earth Sciences* 35: 885-904.

Dyke, A.S., J.E. Dale, and R.N. McNeely. 1996a. Marine molluscs as indicators of environmental change in glaciated North America and Greenland during the last 18,000 years. *Géographie physique et Quaternaire* 50: 125-184.

Dyke, A.S., J. Hooper, and J.M. Savelle. 1996b. A history of sea ice in the Canadian Arctic Archipelago based on the postglacial remains of the bowhead whale *(Balaena mysticetus)*. *Arctic* 49: 235-255.

Dyke, A.S., J. Hooper, C.R. Harington, and J.M. Savelle. 1999. The Late Wisconsinan and Holocene record of walrus (*Odobenus rosmarus*) from North America: A review with new data from Arctic and Atlantic Canada. *Arctic* 52: 160-181.

Gajewski, K., S. Payette, and J.C. Ritchie. 1993. Holocene vegetation history at the boreal-forest - shrub-tundra transition in north-western Québec. *Journal of Ecology* 81: 433-443.

Hill, P.R., P.J. Mudie, K. Moran, and S.M. Blasco. 1985. A sea level curve for the Canadian Beaufort Sea. *Canadian Journal of Earth Sciences* 22: 1383-1393.

Koerner, R.M. 1989. Queen Elizabeth Islands glaciers. In: R.J. Fulton, ed. *Quaternary Geology of Canada and Greenland*. Geological Survey of Canada, Geology of Canada, No. 1. p. 464-473.

Morlan, R., A.S. Dyke, R. McNeely, and M. Ballard. 1999. Mapping ancient history. http://wwwims1.gsc.nrcan.gc.ca/projects/mabasyquad/radiocarbon/indexC14.htm

Moser, K.A. and G.M. MacDonald. 1990. Holocene vegetation change at treeline north of Yellowknife, Northwest Territories. *Quaternary Research* 34: 227-239.

Richard, P.J.H., A. Larouche, and M.A. Bouchard. 1982. Age de la déglaciation finale et histoire postglaciaire de la végétation dans la partie centrale du Nouveau-Québec. *Géographie physique et Quaternaire* 36: 63-90.

Ritchie, J.C. 1984. *Past and present vegetation of the far northwest of Canada*. Toronto: University of Toronto Press. 251p.

Veillette, J.J. 1994. Evolution and paleohydrology of glacial lakes Barlow and Ojibway. *Quaternary Science Reviews* 13: 945-971.

In the past millennium, the Arctic has undergone a warm period (known as the Little Climatic Optimum), a cold period (sometimes referred to as the Little Ice Age), and now another warm period. These temperature changes were probably mirrored by changes in moisture, as warm periods tend to be wetter and cool periods drier. During the first warm period, Icelandic seafarers reached Greenland and established farms there. They disappeared after the 14th Century, perhaps victims of a colder climate that made livestock herding impossible. By the nineteenth century, a warming trend had brought relatively lush vegetation back to southern Greenland and a return to sheep farming in the fjords. This time it was by the Inuit, whose marine mammal hunting had allowed them to survive the intervening centuries of colder weather.

In the 21st Century, climate change is expected to lead to higher average temperatures in much of the Arctic. These increases could be as high as several degrees in the far North, with much warmer winters. Some areas, such as western Greenland and eastern Canada, are expected to become cooler. In warmer areas, precipitation is expected to increase, as are the frequency and intensity of extreme weather events, including storms and drought. Some large processes such as the Gulf Stream are likely to be affected, with profound impacts for much of the Arctic and, indeed, much of the world. There is evidence that such changes have already begun. Sea ice extent and thickness across the Arctic Ocean have decreased in the past four decades. Permafrost in Alaska has warmed and, in some areas, thawed.

On shorter time scales, climate can vary from decade to decade. Long-term conditions determine plant distribution and abundance. Shorter-term variations in weather affect annual plant productivity and its timing. Some decadal patterns are the result of shifts in circulation patterns of air and water, such as the North Atlantic Oscillation. These often have considerable influence on fish, birds, and mammals. These changes vary regionally across the Arctic, so that a warm period may occur in one area while another area has a period of cold.

During the daily and annual cycles of the Arctic, there are extreme changes in air temperature and other aspects of the physical environment. In the high Arctic and in continental areas, air temperatures can vary from -60°C to +40°C in extreme cases during the course of the year. Annual variations of up to 80°C are relatively common. In maritime areas, temperatures tend to be more consistent, but are still capable of large, rapid swings. From day to day, weather patterns can shift suddenly as well, producing many freeze-thaw cycles in a short period.

Life in the Arctic

The physical setting

■ The Arctic's landscapes range from bare rock to swamp, glacier to meadow, mountain to lowland plain. Large areas are still recovering from the effects of glaciation during the last Ice Age. Mosses, lichens, and microbes begin soil formation by fixing nitrogen from the air, making it available to other plants. Higher plants and animals continue the process. While some areas are still barren or in the initial stages of colonization, other areas have extensive vegetation. Some parts of the Arctic were not covered by the icecap, and tend to have deeper organic soils, especially in wetter areas. However, soil depths rarely exceed 75 centimeters and most soils are shallow.

Within the different landscapes, the terrain is often highly variable. In the shallow layer of ground that thaws in summer, the cycle of freezing and thawing disturbs the ground. This process, known as cryoturbation, causes most of the surface features characteristic of the Arctic. Surface stones are sorted, forming circles, nets, and polygons. Non-sorted patterns and features include mud boils, tundra polygons, and pingos (small hills with an ice core). The small-scale variation in surface features creates considerable variability in the underlying soil types and the vegetation that they support.

BOX 5. THE NORTH ATLANTIC OSCILLATION AND REINDEER HUSBANDRY

▶ The Fennoscandian climate is regulated by the Gulf Stream and by the global air pressure system. Low pressure near Iceland and high pressure near the Azores produce warm westerly winds in Fennoscandia in winter. High pressure near Iceland is connected with easterly winds and more extreme climate in northern Europe. Typically, one condition prevails for a decade or more, and then the system switches to the other condition for a similar length of time, a pattern known as the North Atlantic Oscillation (NAO). The NAO index has a high value when low pressure prevails around Iceland and vice versa. High NAO values mean increased precipitation in northern Fennoscandia with rainy summers and autumns, deep snow in winter, and more consistent and moderate temperature. Low NAO values indicate less precipitation, colder winters, deeper frost in the ground, warmer summers, and sudden shifts from one temperature extreme to the other.

The NAO phenomenon affects nature in many ways. There is evidence, for example, that both extremely high and low NAO indices have an adverse effect on reindeer. The calving success (calf/female ratio) of reindeer in western Finnish Lapland between 1961 and 1998 was related to the snow depth from the previous winter. If the snow was deep, the poor condition of the females resulted in an increase in calf mortality. Another important factor affecting availability of winter food was the formation of an ice layer on the ground in early winter. Such conditions are associated with thawing and subsequent frost, and they seem to be common while the NAO index is high.

Interestingly, low NAO values were also associated with poor calving success. The phase is characterized by low temperatures and high winds hardening the snow in mid- and late winter, reducing food availability and hence, survival of reindeer.

The recent high NAO phase, beginning in the late 1980s, has produced the highest index values since the

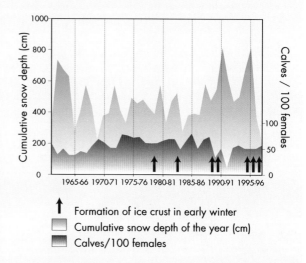

Figure 9. Reindeer calf/female ratio in relation to annual snow accumulation.

record began in 1870. The resulting increase in snow depth has caused great difficulties in reindeer husbandry because herders have had to rely on expensive supplemental feeding in winter to maintain the herds. Most herders in western Lapland, who had earlier relied on natural winter pastures, have now adopted this practice. At the moment, however, it is the only way to maintain the herds and protect the livelihood of the herders.

Timo Helle and Mauri Timonen, Finnish Forest Research Institute, Rovaniemi Research Station, Finland

SOURCE/FURTHER READING
Helle, T., A. Niva, and I. Kojola. 2000. Reindeer husbandry and changing climate. *Poromies* 66 (3):15-17. (in Finnish).

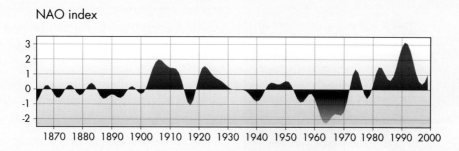

Figure 10. The NAO index indicates relative air pressure in Iceland and the Azores. A high NAO index (low pressure in Iceland and high in the Azores) means more westerly winds, resulting in warmer, moister weather in northwestern Europe.

Effects of permafrost: a pingo (top) and an ice lens in the soil (bottom).

A key characteristic of most land areas in the Arctic is permafrost. This permanently frozen ground reaches depths of 1500 meters in Siberia, and extends under most of the tundra and polar desert and well into the boreal forest. Its significance is most apparent when it melts. If the insulating layer of vegetation and soil is disturbed – by fires, weather, or human activity – permafrost can melt, deforming the ground. These changes, called thermokarst, are a common consequence of vehicle tracks or deforestation. Sudden melting can cause dramatic changes, including the draining of lakes or ponds. They are almost impossible to reverse over the short term.

Glacier, Baffin Island, Canada.

Polar desert, Cornwallis Island, Canada.

Polar semi-desert, Northeast Greenland.

Cottongrass wetland, Northeast Greenland.

Mountain treeline, northwestern Russia.

Taiga forest and mires near the treeline, Finnish Lapland.

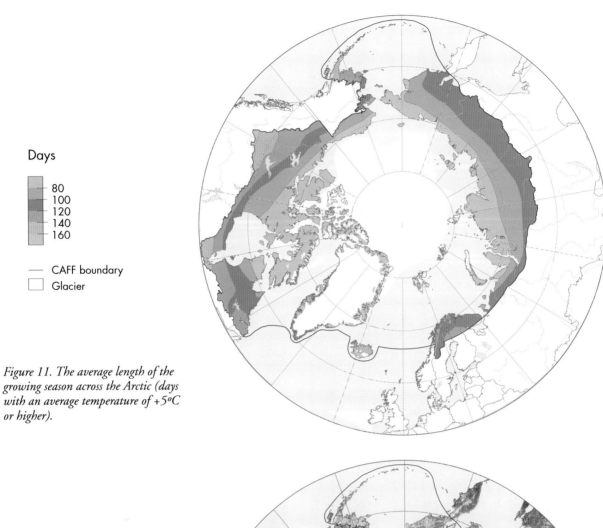

Figure 11. The average length of the growing season across the Arctic (days with an average temperature of +5ºC or higher).

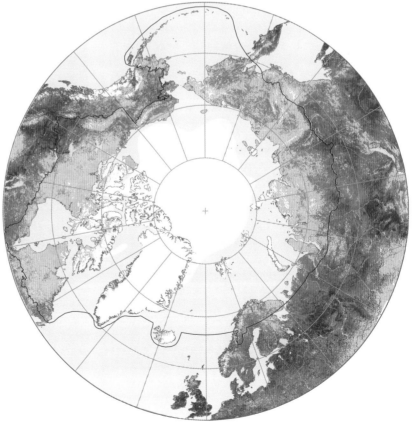

Figure 12. The Normalized Difference Vegetation Index (NDVI), or Greenness Index, indicates vegetation productivity. It is measured from satellite images on the basis of chlorophyll content (greenness) of vegetation.

The presence and quality of ice and snow on land and on lakes and rivers help to govern the productivity of those areas. Where snow has accumulated to great depth, such as in a gully or the lee side of a ridge, it may remain for most of the summer. Lakes with thick ice cover may not thaw until later in summer, limiting the amount of sunlight that reaches the plants and animals inhabiting them. Where snow and ice melt early, insects and grazing animals may be attracted to relatively small areas, using them intensively until larger areas are available. In fall, hard snow and layers of ice created by late rains can make it difficult or impossible for grazing animals to get to plants. Deep snow and hard, broken crusts make travel difficult for many large animals, reducing their winter survival. But, snow also insulates vegetation and can protect it from strong winds. Winters with little snow can be particularly hard on many plants, as well as on small animals such as lemmings that normally live under the snow.

In the ocean, the dominant feature of the Arctic is sea ice. In the central Arctic Basin, the pack ice remains throughout the year. Around the periphery of the Arctic Ocean and in the marginal seas, the ice forms and melts annually. The ice edge is a critical habitat for many species, and the annual melt of sea ice and glacial icebergs releases fresh water and nutrients to make a productive feeding zone. For many marine mammals, sea ice offers protection from predators and a surface for resting or bearing young. Even in the areas of permanent ice cover, a variety of plants and animals survive, including seals, polar bears, and beluga whales. Nearer the shore, sea ice can scour the ocean bottom, digging up nutrients and disturbing the benthic or sea-bed communities.

Responding to climate

Arctic ecology is shaped by the severity of the climate and its variability in time and space. Not only must Arctic species be able to survive long periods when food is limited or unavailable, they must be able to respond quickly when conditions are right.

The growing season, though brief, is intense. In the ocean, huge phytoplankton blooms are grazed by zooplankton, which become food for crustaceans, fish, and marine mammals. The plankton blooms are often the result of intense competition for available nutrients when sunlight reaches the water at the end of winter. By the end of spring, the primary producers have used most of the nutrients. On land, the growth of plants spurs a similar summer feast for animals as they breed, feed their young, and store energy for the coming winter. Insects and some small mammals have seasonal patterns of reproduction, producing several generations during summer when food is plentiful, and then slowing down or becoming dormant in winter. Voles on Svalbard, for example, are capable of producing four large litters in a season, and they breed as early as one month of age. Their populations are thus able to expand rapidly when conditions are favorable, making up for poor years.

Winters bring a different set of conditions, the key stresses being reduced food and cold. Plants store nutrients and energy over the winter so that they can grow rapidly when spring arrives, making full use of early spring sunlight to extend the growing season. Some sedges and grasses even start growing under the snow, where the springtime freeze-thaw cycles create tiny ice greenhouses around individual plants. Freeze-tolerant arthropods have biochemical mechanisms to allow them to freeze without causing damage to their cells. Freeze-avoiding arthropods make anti-freeze compounds to prevent ice formation in their tissues. Birds and mammals have the ability to store energy in the form of fat or blubber, to change their metabolic rates to minimize energy use during winter months, or to grow fur and feathers for insulation.

Finding favorable habitats

In space as well as time, wide variability in conditions can create highly patchy patterns of growth and productivity. Where moisture is a limiting factor, the presence of a lingering snowfield can allow lush plant growth on an otherwise dry hillside. Low areas and gullies that are protected from winds can be relatively benign habitats, while neighboring ridges and exposed areas sustain only low-lying and hardy plants. For instance,

BOX 6. VARIATIONS IN SNOW COVER AFFECT PLANTS

▶ Snow cover is distributed unevenly on a scale of meters or tens of meters, and its variations are one of the greatest influences on the life of plants in the Arctic. Snow depth depends largely on the topography of the landscape, and specific sites fall into one of three broad categories:

- Wind-exposed or elevated sites supporting only a thin layer of snow
- Sheltered slopes with moderate snow cover that melts in spring
- Snowbeds in depressions and on slopes where deep snow melts only in summer

At each site, snow depth modifies the ecological conditions, and plant communities are distinctly different:

Risto Virtanen, Dept. of Biology, University of Oulu, Finland

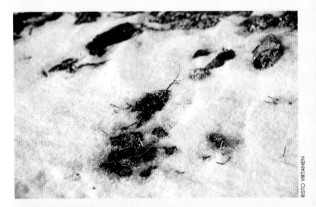

A wind-swept ridge in winter.

*Lapland diapensia **Diapensia lapponica**, a circumpolar cushion plant found at wind-exposed sites.*

Snowbed in July. Grasses and mosses dominate the vegetation.

Table 1. Effects of variation in snow cover.

TYPICAL ECOLOGICAL FACTORS				
		Wind-exposed ridge	Sheltered Slope	Depression
Snow cover	Depth (cm)	0-20	20-80	80+
	Length of snow free period (typical values in months)	4-5	3-4	<3
Soil	Moisture	Dry	Dryish-mesic	Mesic-moist
	Freeze-thaw processes	Intense	Low-Moderate	Low-Moderate
	Depth of soil active layer	Deep	Moderate	Moderate-thin
	Nutrient status	Low	Moderate	Low-moderate
Temperature	Variability	High	Moderate	Moderate
	Winter minima near soil surface	Very low	Moderate	Moderate - near zero
Desiccative stress in spring		High	Moderate	Low
Herbivore activity in winter	Reindeer	Moderate–high	Moderate	Low
	Microtines	Low	Moderate – high	Moderate – high
	Rock ptarmigan, mountain hare	High	Low	Low

TYPICAL PLANT FORMS				
		Wind-exposed ridge	Sheltered Slope	Depression
Plant group	Vascular plants	Trailing dwarf shrubs, cushion plants	Dwarf shrubs	Low herbs, dwarf willows
	Bryophytes	Small, compact, drought-tolerant species	Mesophytic mosses bryophytes	Meso-hygrophytic
	Lichens	Wind-hardy fruticose species	Fruticose reindeer lichens	

TYPICAL SPECIES				
		Wind-exposed ridge	Sheltered Slope	Depression
Plant group	Vascular plants	*Arctostaphylos alpina, Loiseleuria procumbens*	*Betula nana, Vaccinium uliginosum*	*Salix polaris*
	Bryophytes	*Racomitrium lanuginosum*	*Hylocomium splendens*	*Kiaeria starkei*
	Lichens	*Alectoria nigricans*	*Cladina mitis*	

SOURCES/FURTHER READING

Miller, P.C. 1982. Environmental and ecological variation across a snow accumulation area in montane tundra in central Alaska. *Holarctic Ecology* 5: 85-98.
Schaefer, J.A., and F. Messier. 1994. Scale-dependent correlations of arctic vegetation and snow cover. *Arctic and Alpine Research* 27: 38-43.

BOX 7. HOW LAND ANIMALS ADAPT TO COLD AND SNOW

▶ The Arctic winter is typically cold and snowy. Staying warm requires good insulation or shelter, or expending a great deal of energy. The snow cover offers shelter but may limit movement and access to food. To survive the harsh conditions of the far north, Arctic animals have developed many behavioral and morphological adaptations.

For large animals, thicker and denser fur or plumage may be adequate to hold an insulating layer of still air against their bodies. If so, they need to produce little or no extra heat to stay warm in cold weather. Small animals, however, cannot carry thick enough fur, and thus depend on other strategies. For instance, they use snow as shelter throughout the winter. Snow is an excellent insulator, and the temperature deep in the snow can be much higher than that in open air. Many northern grouse species stay in snow burrows during cold nights and poor weather. Even when the temperature above the snow is as low as -17°C, the grouse's burrow may be above freezing. By creating a warm burrow, the grouse reduces its energy demands and thus its need for food. Even large mammals use snow for shelter. Brown bears hibernate under the snow, and pregnant polar bears make snow

The female rock ptarmigan **Lagopus mutus** *(above) molts earlier and blends in with the tundra on her nest, while the male (right) is still partly white.*

*Mountain hares **Lepus timidus**: sometimes the color is just wrong!*

dens for the winter. Male polar bears make temporary dens during storms.

Another strategy for coping with high energy demands when food is scarce is to accumulate large fat deposits when food is available. Two extreme cases are the Svalbard ptarmigan and the Svalbard reindeer. In autumn, up to 30% of their body weight is fat. Moreover, many passerine birds, mammalian predators, and beavers store food to eat during winter.

While snow provides shelter for some animals, it causes problems for others. Deep and soft snow make it difficult for many animals to move, and may prevent them reaching the plants on which they graze. Some species have adapted by developing large feet for their body weight. The mountain hare, for example, has a foot that is twice the relative size of the foot of the European hare, helping it move over the snow without sinking. Lynx and Arctic wolves have large feet and long legs for the same reason.

By providing a white background, snow makes many animals highly visible in winter, even if they are well camouflaged in summer. Many northern species, such as ptarmigan and hares, have evolved to change the color of their fur or plumage to white in winter so that they retain their cryptic coloration year round. Some predators, like weasels and the arctic fox, also change their fur color between summer and winter. Whether this is to hide them from their prey or from potential predators is not clear.

Osmo Rätti, Arctic Centre, University of Lapland, Rovaniemi, Finland

SOURCES/FURTHER READING

Davenport, J. 1992. *Animal life at low temperature.* London: Chapman & Hall.
Johnsgard, P.A. 1983. *The grouse of the world.* Lincoln, Nebraska: University of Nebraska Press.
Marchand, P.J. 1991. *Life in the cold: An introduction to winter ecology.* Hanover, New Hampshire: University Press of New England.
Marjakangas, A., H. Rintamäki, and R. Hissa. 1983. Thermal responses in the capercaillie (*Tetrao urogallus*) and the black grouse (*Tetrao tetrix*) roosting in snow burrows. *Suomen Riista* 30:64-70.
Ridley, M. 1993. *Evolution.* Boston: Blackwell Scientific Publications.
Skelton, P. 1992 *Evolution: a biological and palaeontological approach (Book 1).* Kingston-upon-Hull, U.K.: The Open University.

on the Killik River in northern Alaska, 100 kilometers north of the treeline, stands a grove of poplar trees protected by a high bluff. In the ocean, the presence or absence of sea ice, which can vary greatly from year to year, helps to determine an area's productivity. Although sea ice limits the amount of sunlight reaching the water below, it also offers protection to fish and marine mammals from certain predators. Even while overall productivity is low, some species thrive under the ice.

Over larger distances, a combination of favorable conditions might occur in a relatively limited area.

Plant adaptations: the purple **Saxifraga oppositifolia** *showing a cushion-form growth;* **Senecio monstrosus** *with hairy leaves and stem; and the viviparous alpine meadow-grass* **Poa alpina** *var.* **vivipara**.

Arctic oasis: luxurious vegetation can be found under bird colonies.

Arctic oases – areas of rich vegetation in a sparse landscape – are one such example. Here, the availability of water allows a variety of plants to grow vigorously during the summer, and the richer soils and thicker vegetation in turn allow better retention of moisture. Often located in low-lying areas, these oases enjoy higher temperatures than their surroundings as the warmth of sunlight is retained in the plants and soil. In some cases, the presence of moisture creates wetlands and bogs, which produce more organic matter than can decompose in the short summer. There, a thick layer of plant matter builds up over time. While the lower layers of such areas are typically locked in permafrost today, a warming climate could release the nutrients and gases contained in the frozen ground.

In the ocean, winds and currents create and maintain recurring areas of leads (fractures through sea ice) and open water, called polynyas. Some polynyas, such as the North Water between Greenland and Ellesmere Island and the Great Siberian Polynya in the Laptev Sea, occur regularly, predictably supporting a great abundance of marine life. Others polynyas are less regular, but their presence stimulates plankton and fish, which attract many birds and mammals. Upwellings bring nutrients from the depths to levels where sunlight can stimulate the growth of phytoplankton and algae. Meltwater at the sea ice edge provides stable habitat for marine species, stimulating productivity in this area as well. Large rivers, which carry a great deal of warmth and nutrients, create productive zones in their estuaries and in the plumes of nutrient-rich water that extend into the ocean.

BOX 8. INSECT ADAPTATIONS IN THE HIGH ARCTIC

▶ Only a few hundred insect species are found in the high Arctic, where they must cope with a harsh and highly seasonal environment. Many adaptations are a response to the brevity of the summer season, allowing insects to make the most of the time when conditions are favorable. Other adaptations such as cold hardiness are essential for winter survival. In some respects, winter survival may be easier than ensuring enough development during the very short summer: some species overwinter in sites that are very exposed, but hence warm early in spring. The wide variety of seasonal adaptations in high Arctic insects is instructive in showing how the many different responses can be combined and coordinated to overcome the difficult conditions.

Hugh V. Danks, Biological Survey of Canada (Terrestrial Arthropods), Canadian Museum of Nature, Ottawa, Canada

*The woolly bear caterpillar of the moth **Gynaephora groenlandica**. The species is known for its long life cycle, which can vary from 7 to 14 years.*

Table 2. Insect adaptations in the High Arctic.

Adaptation	Supposed advantage	Examples
COLOR AND STRUCTURE		
Melanism	Solar heat gain in cool environments	Many insects
Hairiness	Conserves heat gained, in larger insects	Many insects
Robustness	Moisture conservation	Some ichneumonids
Reduced wings	Energy saved (flight limited by cool conditions)	Several flies, moths
Reduced eyes	Normally nocturnal forms are active in light in Arctic	Some noctuid moths
Reduced antennae of males	Detection of females during swarming flights is not necessary when flight is limited	Some chironomid midges
Small size	Species can use limited resources	Many species
ACTIVITY AND BEHAVIOR		
Selection of warm habitats for activity or progeny production	Activity allowed in cool environments in locally favorable sites	Most species
Selection of protected sites for overwintering (e.g. plant clumps)	Protection against cold or wind-driven snow	Many species (but some species overwinter fully exposed)
Low temperature thresholds for activity	Activity allowed in cool environments	Several species
Opportunistic activity	Activity whenever temperatures are permissive	Bumble bees, midges, etc.
Basking	Heat-gain in sunshine	Many butterflies, etc.

Adaptation	Supposed advantage	Examples
PHYSIOLOGY AND METABOLISM		
Adjustment of metabolic rate	Life-processes continue in cool environments	Some species of various groups
Resistance to cold	Survival during cold winters	All species
Resistance to starvation	Survival when food supplies are unpredictable	Some moth larvae, some springtails
LIFE-CYCLES AND PHENOLOGY		
Multi-year life-cycles	Life-cycle completed although resources are limited	Moth larvae, midge larvae, etc.
Rapid development	Life-cycle completed before winter supervenes	Mosquitoes, etc.
Abbreviation of normally complex life-cycles	Life-cycle completed before winter supervenes	Aphids, etc.
Earliest possible emergence in spring	Life-cycle completed before winter supervenes	Midges, etc.
Brief and synchronized reproductive activity	Life-cycle completed before winter supervenes	Many moths, etc.
Dormancy	Activity stops before damaging winter cold; individuals held dormant through summer to emerge as early as possible the next spring	Chironomid midges, etc.
Prolonged dormancy for more than one season	"Insurance" against a summer unsuitable for reproduction	Various species of flies, moths, sawflies, etc.
FOOD RANGE		
Autogeny (development of eggs without a blood-meal in biting flies)	Offsets shortage of hosts; or short season	Mosquitoes
Different food plants	Survive although a particular food plant is absent	Few species
Polyphagy	Survival despite low density of foods	Many species, especially saprophages/detritivores and predators
GENETIC ADAPTATION		
Parthenogenesis	Reduced mating activity in harsh conditions; well-adapted genotypes buffered against change	Midges, black flies, caddisflies, mayflies, scale insects, etc.

SOURCES/FURTHER READING

Danks, H.V. 1981. *Arctic insects. A review of systematics and ecology, with particular reference to the North American fauna.* Ottawa: Entomological Society of Canada. 608p.

Danks, H.V. 1990. Arctic insects: instructive diversity. In: C.R. Harrington, ed. *Canada's missing dimension: science and history in the Canadian arctic islands.* Vol. II. Ottawa: Canadian Museum of Nature. p. 444-470.

BOX 9. STAYING WARM IN A COLD CLIMATE

▶ Mammals and birds are homeothermic animals, maintaining a constant body temperature. In cold weather, they must either reduce heat loss or increase heat production. Reindeer, for example, have developed highly effective winter fur to insulate their bodies. Thick guard hairs with air-filled cavities protect woollen underfur that prevents the movement of air against their skin. Their legs are less insulated than their torso, and are maintained at a lower temperature. The nasal cavities are similarly cooled to reduce heat loss from breathing. As large animals, reindeer also have a smaller surface area to body mass ratio, which further reduces heat loss. By means of these various adaptations, reindeer can tolerate ambient temperatures at or below -30°C without needing to increase heat production.

Reindeer calves face a particular challenge at birth. They are typically born in early spring when snow still covers much of the landscape and cold rains are common. At birth, the calf may experience a drop in temperature of 50-60°C. Of necessity, newborn reindeer are well developed and can soon stand and walk. They have an effective capacity for regulating their body temperature, largely by activating the heat production of brown fat, a mechanism known as non-shivering thermogenesis. Brown fat is a unique mammalian tissue specially adapted to generate heat. Its color is caused by a rich network of capillaries and cell organelles, called mitochondria, where thermogenesis occurs.

Brown fat is the dominant adipose tissue type in the newborn reindeer, and occurs in the major body cavities and around vital organs such as the kidneys and heart. The heat produced by the various sites of brown fat is carried by the blood to the rest of the body. Brown fat plays an important role in the cold resistance in many newborn Arctic mammals, including muskox and harp seal pups. Brown fat is also present in adult cold-adapted small mammals and hibernators. In reindeer, the brown fat is most active shortly after birth and declines rapidly during the first month of life, gradually turning into a tissue that resembles white fat.

Several mammalian species in the Arctic evade the problems of cold and scarcity of food by spending the winter in torpor. They store large quantities of white fat during summer and autumn and use them as their energy source during winter. Species such as bear and badger spend their winter sleeping under the snow. Their body temperature is usually decreased by 4-5°C. Smaller mammals such as ground squirrels, bats and mice hibernate by decreasing their body temperatures much further, near to the temperature of their surroundings. They may periodically wake up from hibernation and excrete waste products from their body. These small hibernators have well-developed thoracic brown fat that is essential in warming the heart when awakening from hibernation.

Päivi Soppela, Arctic Centre, University of Lapland, Rovaniemi, Finland

SOURCES/FURTHER READING

Duchamp, C., F. Marmonier, F. Denjean, J. Lachuer, T.P.D. Eldershaw, J.-L. Rouanet, A. Morales, R. Meister, C. Bénistant, D. Roussel, and H. Barré. 1999. Regulatory, cellular and molecular aspects of avian muscle non-shivering thermogenesis. *Ornis Fennica* 76:151-165.

Irving, L., and J. Krog. 1955. Temperature of the skin in the arctic as a regulator of heat. *J. Appl. Physiol.* 7:355-364.

Johnsen, H.K., A.S. Blix, L. Jørgensen, and J.B. Mercer. 1985. Vascular basis for regulation of nasal heat exchange in reindeer. *Am. J. Physiol.* 249:617-623.

Nilssen, K.J., J.A. Sundsfjord, and A.S. Blix. 1984. Regulation of metabolic rate in Svalbard and Norwegian reindeer. *Am. J. Physiol.* 247:R837-841.

Soppela, P. 2000. Fats as indicators of physiological constraints in newborn and young reindeer, *Rangifer tarandus tarandus* L. *Acta Univ. Oul. A* 349:1-64.

Soppela, P., M. Nieminen, S. Saarela, and R. Hissa. 1986. The influence of ambient temperature on body temperature and metabolism of newborn and growing reindeer calves (*Rangifer tarandus tarandus* L.). *Comp. Biochem. Physiol.* 83A:371-386.

Timisjärvi, J., M. Nieminen, and A.-L. Sippola. 1984. The structure and insulation properties of the reindeer fur. *Comp. Biochem. Physiol.* 79A:601-609.

Brown fat provides the heat newborn reindeer calves need to stay warm.

The importance of movement

As conditions change seasonally and daily, many animals move to find better areas. Caribou, for example, travel great distances both in their migrations and over the course of their daily grazing. Some of their predators travel with them, including wolves and ravens that scavenge the carcasses of caribou that have died or been killed. On continental land masses that offer wide areas of similar habitat, there is usually enough space and enough variability in conditions to allow areas that have suffered from poor conditions or catastrophes to eventually be repopulated from other areas. On islands, or in similarly isolated areas, such re-population might be much more difficult if not impossible, and thus the number of species that can survive over decades is likely to be lower. This effect may explain, for example, the absence of lemmings in southwest Greenland.

The diversity of habitats across large and small distances both helps and hinders the survival of species and individuals. A range of habitat types can offer a range of options, if a species is capable of adapting. On the other hand, the isolation of small patches of a given habitat type, interspersed with other habitats, can also mean that plants and animals with limited abilities to move and spread are stuck with whatever is available in the local area. Much then depends on the ability of species to adapt quickly and to spread widely, buffering themselves against unfavorable conditions in a single area. As conditions change across an entire region, however, these adaptations may be insufficient.

Ecological relationships

The aggregation of species and their interactions create the overall ecosystem of a region. One way to analyze ecological relationships is by trophic levels – the relative positions of species in the food web. Plants convert the raw materials of energy and minerals into living matter, which is food for herbivores. When plants die and herbivores die or defecate, decomposers – microbes, fungi, and invertebrates – break the organic matter down into inorganic material, which is then available to more plants. Herbivores are eaten by carnivores, some of which also eat other carnivores. Omnivores, such as brown bears, feed at several trophic levels, eating plants, herbivores, and carnivores, hence increasing their flexibility to adapt as conditions require. Parasites live off animals throughout the food web.

At each level, energy is used to fuel the life cycle of the species in question, so that only a small portion is converted to the species in the next trophic level. As a result, far more total energy is stored in plants than in herbivores, and far more in herbivores than in carnivores. The short cool summer limits the time available for nutrient cycling. Most of the Arctic's nutrients are thus held in the soil, unlike many other ecosystems. Consequently, the Arctic food web is complex, with an abundance of decomposer organisms.

Another way to look at the structure of an ecosystem is through its functions. Nearly all the species in a system depend to some degree on the functions carried out by other species. Only certain plants can take nitrogen from the air and make it available in the soil for other plants to use. Microbes, fungi, and invertebrates process organic matter into inorganic forms that can be used by other plants as they grow. Herbivores depend on the growth of plants, and carnivores depend on the animals that convert plants into meat and fat.

Ecological function is more than the production of food. It includes competition and predation between the various species that exist in that system. Some predators perform crucial functions that affect the overall structure of a system. Sea otters in the Aleutian Islands, for example, eat sea urchins, which in turn prey on kelp. The sea otters control the populations of sea urchins. When sea otters disappear, the urchins flourish and severely reduce growth in the kelp beds. This undermines the species that depend on the kelp beds for protection and food.

The success of most species depends largely on the food web of which they are part. Some species are able to thrive under many conditions by adapting through significant differences in body size and behavior. Other species occupy different niches within the food web at different stages of their life cycle. The resilience of a

BOX 10. THE VOLCANIC ISLAND SURTSEY: A NATURAL EXPERIMENT IN COLONIZATION

▶ On November 14, 1963, fishermen on a boat off the south coast of Iceland witnessed a rare event. A volcanic eruption had given birth to an island, later name Surtsey, after Surtur, the giant of fire in Nordic mythology. Episodic eruptions continued on Surtsey until June 1967, by which time the newest of the Westman Islands had reached a size of 2.7 square kilometers and projected 170 meters above sea level. Since 1965, Surtsey has been protected under the Icelandic Nature Conservancy Act to avoid human disturbance and allow plants and animals to colonize the island as naturally as possible. The Surtsey Research Society is the caretaker of the island and organizes scientific expeditions every year.

After the island was formed, microbial moulds, bacteria, and fungi soon became established in the fresh volcanic rock and ash. In the summer of 1965, the first vascular plant was found. Mosses became visible in 1968, and lichens were first found on the lava in 1970. Vascular plants have been of greatest significance in the development of vegetation on Surtsey. The first twenty years were characterized by the invasion and spread of the coastal species *Honckenya peploides*, *Leymus arenarius* and *Mertensia maritima*, which formed a simple community on the island's unfertile sands. Seventeen other species were discovered on the island during that period. Of these, however, only seven became established, and they remained insignificant in the vegetation cover.

Seabirds started breeding on Surtsey in 1970 but their numbers and influence were limited until 1986 when a colony of lesser-black backed gulls and herring gulls was established on the southern part of the island. The colony has since grown to several hundred breeding pairs. Seven species of seabirds now breed on Surtsey. The vegetation on Surtsey changed considerably after the formation of the gull colony. The nutrient status of the soil has improved, enabling nutrient-demanding plants to become established and improving conditions for species already on the island. The colony area has particularly benefited the grasses *Poa annua* and *Puccinellia distans* and the forbs *Cochlearia officinalis*, *Stellaria media*, and *Cerastium fontanum*.

By the summer of 2000, 55 species of vascular plants had been discovered on Surtsey, 47 of which still lived on the island. The number of species on Surtsey has surpassed that of the other outer Westman Islands, which are considerably smaller. Since its birth, however, coastal erosion has taken a toll on Surtsey. By 2000, the island was only half its maximum size. In a few hundred years, Surtsey will probably be much like its small neighbours: a high, steep stack, with a grassy top, home to five to ten plant species and sea-birds. Surtsey will always be distinguished, however, for the remarkable story it has made.

— Total number of species recorded
— Species present each year

Figure 13. Number of vascular plant species found on Surtsey between 1965 and 2000. In the first decade, the island was colonized mainly by coastal plants dispersed by the sea. A new wave of plant invasion followed the formation of a gull colony in 1986. The invasion rate slowed down again after 1997.

Borgthór Magnússon, Agricultural Research Institute, Reykjavik, Iceland

SOURCES/FURTHER READING

Fridriksson, S. 1975. *Surtsey. Evolution of Life on a Volcanic Island.* London: Butterworths. 198p.

Fridriksson, S. 1994. *Surtsey. Lífríki í mótun* (in Icelandic). Reykjavík: Hið íslenska náttúrufræðifélag og Surtseyjarfélagið. 112p.

Magnusson, B., and S.H. Magnusson. 2000. Vegetation succession on Surtsey during 1990-1998 under the influence of breeding gulls. *Surtsey Research* 11:9-20.

BOX 11. THE ROLE OF FUNGI

▶ Fungi play an essential role in enhancing nutrient availability for plants and other living organisms in the Arctic. Parasitic and saprophytic fungi – those that use living and dead plants and animals for food – are essential for decay. They reduce dead organic material to mineral forms that plants can live on, while at the same time improving soil structure and removing materials that hamper renewed plant life. Lichens, the symbiotic association of fungi and algae, are regularly among the pioneer organisms colonizing bare rock and ground. Some lichens have secondary associations with cyanobacteria, and thus are important for nitrogen fixing. Various types of mycorrhiza – fungi in symbiotic relationships with plant roots – are especially important in stress situations, such as those caused across much of the Arctic by cold and nutrient limitations. In mycorrhiza, the fungi supply mineral nutrition and other essentials to vascular plants in exchange for carbohydrates.

The impact of the mycorrhizal fungi can extend from the individual plant to the whole plant community. Sugars from the plants and various mineral nutrients captured by the fungi are distributed through mycorrhizal links between individual vascular plants of the same and different species, and from healthy and well-established individuals to delicate seedlings. More than 80% of vascular plants depend on mycorrhizae. Such mycorrhizal plants rapidly populate virgin Arctic soils, indicating that the mycorrhizae can propagate efficiently.

Trees and higher fungi associate through the special kind of mycorrhiza called ectomycorrhizae, in which the fungus does not penetrate into the root cells. One of the more astonishing facts about fungal biodiversity in Arctic regions is the high diversity of ectomycorrhizal fungi, despite the low number of woody plants. On a small Arctic island, Edgeøya in the archipelago of Svalbard, one single dwarf willow species, *Salix polaris*, is known to be associated with more than one hundred different ectomycorrhizal fungal species. The biological importance of this diversity is unknown.

The biomass of fungi in Arctic tundra soils, based on estimates from Alaska, is similar to that found in temperate and tropical biomes such as grasslands and deciduous and coniferous forests. Most of the fungi forming ectomycorrhiza produce mushrooms. These visible fruitbodies, abundant as they are in some locations, constitute only a minor part of the biomass of the fungi: 50-90% of the fungal standing crop biomass is generally made up of the mycorrhizal roots. While lichens and fungal fruitbodies are choice food for wildlife from reindeer to ground squirrels, fungi play an even greater role as food for numerous soil arthropods and bacteria.

For many groups of fungi, taxonomic studies have been neglected in Arctic regions. One of the better known regions is Svalbard, for which a species catalogue of all known plants, fungi, and cyanobacteria has recently been published. Although most research to date has been on vascular plants, more than 1,200 fungi, including lichens, are known, compared with 173 vascular plant species that have been documented in the area.

Gro Gulden, Natural History Museums and Botanical Garden, University of Oslo, Norway
Esteri Ohenoja, Botanical Museum, Department of Biology, University of Oulu, Finland

*All tree species at the Arctic treeline and also many tundra plants have symbiotic mycorrhiza, which helps in nutrient and water uptake. A mycorrhiza formed by a **Cortinarius** fungus and Scots pine **Pinus silvestris** root.*

SOURCES/FURTHER READING

Allen, M.F. 1991. *The ecology of mycorrhizae.* Cambridge: Cambridge University Press. 184p.

Dahlberg, A. 1991. *Ectomycorrhiza in coniferous forest: structure and dynamics of populations and communities.* Uppsala: Swedish University of Agricultural Sciences. 38p.

Elvebakk, A., and P. Prestrud. 1996. *A catalogue of Svalbard plants, fungi, algae and cyanobacteria.* Norsk Polarinstitutt Skrifter nr. 198. 395 pp.

Jalink, L.M., and M.M. Nauta. 1989. Paddestoelen op Spitsbergen. *Circumpolar Journal* 4:1-17.

Laursen, G.A., and M.A. Chmielewski. 1982. The ecological significance of soil fungi in Arctic tundra. In: G.A. Laursen, and J.F. Ammirati, eds. *Arctic and alpine mycology.* Seattle: University of Washington Press. p. 432-488.

Read, D.J. 1983. The biology of mycorrhiza in the Ericales. *Canadian Journal of Botany* 61: 985-1004.

Trappe, J.M., and D.L. Luoma. 1992. The ties that bind: fungi in ecosystems. In: G.C. Carroll, and D.T. Wicklow, eds. *The fungal community: its organization and role in the ecosystem.* New York: Marcel Dekker. p. 17-27.

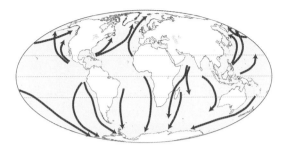

Figure 14. Migration routes of the humpback whale **Megaptera novaeangliae.**

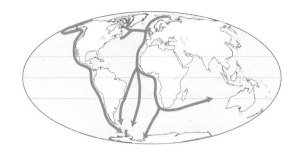

Figure 15. Migration routes of the arctic tern **Sterna paradisaea**.

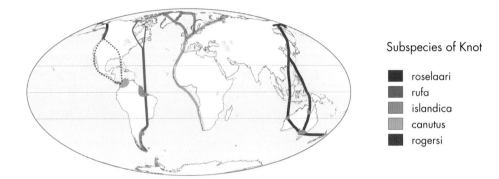

Figure 16. Breeding areas, migration routes (full color), and staging/wintering areas of different subspecies of knot **Calidris canutus** *(Data source: Piersma and Davidson 1992).*

system as a whole is often a matter of its complexity. If it depends heavily on a few key species, changes to them will affect all the plants and animals in that ecosystem. Arctic systems typically have few species and are susceptible to this type of disruption. For example, the explosion of lesser snow geese in Canada has caused radical changes to the functioning of ecosystems where the geese graze in summer. Before, the geese fertilized the grazing areas, enhancing plant growth. Now, the vast numbers of geese have overgrazed large areas of salt marsh, increasing the salinity of the soils and destroying the vegetation that was once excellent goose food.

Ecosystem relationships

Physical boundaries, functions, and food webs tie the plants and animals of certain regions together. Ecosystems can be considered in relatively small terms, such as watersheds, or in large terms, such as an entire sea. Thinking in terms of ecosystems is to approach conservation from the perspective of ecological relationships, through considering the many factors that contribute to the health of the species that comprise the system. The boundaries of ecological processes, however, are rarely definitive and consistent.

Ecosystems interact in many ways. Birds nest on land but may feed in the ocean or in lakes and rivers. Salmon and certain other fish are anadromous – they live some of their life in the ocean, but reproduce in freshwater. Reindeer and caribou may migrate between the tundra and the taiga. Nutrients are carried by runoff from land into lakes and rivers, and then into the ocean. Marine plants put iodine into the atmosphere, making it available for life on land, which could not survive without it. In short, the distinctions between systems are often a matter of convenience for the observer, while in reality their functions cannot be separated.

The complex relationships between the different systems within the Arctic are matched by the complex relationships of Arctic ecosystems with other

*The Arctic tern **Sterna paradisaea** and the knot **Calidris canutus** make the longest known seasonal migrations.*

BOX 12. THE ARCTIC AS A THEATER OF EVOLUTION

▶ Species-poor Arctic systems can rightly be described as "evolutionary theaters," where early acts of speciation are expected. One important feature of species-poor environments is the fact that the few species found there are presented with a diversity of habitats and food resources that otherwise would be occupied by competing species. The existence of such uncontested resources promotes the evolution of diversity within species as individuals adapt to different conditions. This process can promote the formation of discrete *morphs*: groups of individuals within the same population that differ markedly in behavior and ecological role. Recent studies have shown that such resource polymorphism is particularly widespread among northern freshwater fishes and can lead to the evolution of new species.

The arctic char has been examined specifically in this respect, as have several other fish species such as the three-spined stickleback and various whitefish. There are indications that we can also expect numerous examples of this phenomenon in birds, mammals and freshwater and terrestrial invertebrates. For instance in Iceland, stone flies and trichoptera have been shown to occupy unusually diverse habitats.

These observations clearly suggest that species-poor Arctic territories and islands are important theaters for the evolution of biological diversification. The perception, conservation, and management of biodiversity in these areas must take into account the dynamic, process-oriented nature of these systems, rather than following the typical static, pattern-oriented approach. An understanding of this process of generating biodiversity is of global importance, considering the rapid extinctions of species and populations in numerous ecosystems worldwide. Clearly, our understanding of these processes needs to be improved. More research must be conducted across species and geographic areas, in particular towards polymorphism and speciation in the Arctic.

*Skúli Skúlason, Hólar Agricultural College,
Skagafjordur, Iceland
Sigurdur S. Snorrason, University of Iceland,
Reykjavik, Iceland*

SOURCES/FURTHER READING

Bernatchez, L., A. Chouinard, and G. Lu. 1999. Integrating molecular genetics and ecology in studies of adaptive radiation: whitefish *Coregonus* sp., as a case study. *Biological Journal of the Linnean Society* 68:173-194

Gíslason, G.M. 1981. Distribution and habitat preferences of Icelandic Trichoptera. In: G. Moretti, ed. *Proceedings of the 3rd international symposium on trichoptera*. Amsterdam: Dr. W. Junk.

Gíslason, D., M.M. Ferguson, S. Skulason, and S.S. Snorrason. 1999. Rapid and Coupled phenotypic and genetic divergence in Icelandic Arctic char (*Salvelinus alpinus*). *Canadian Journal of Fisheries and Aquatic Sciences* 56, 2229-2234.

Lillehammer, A., M. Johannsson, and G.M. Gislason. 1986. Studies on Capnia vidua Klapalek (Capniidae, Plecoptera) populations in Iceland. *Fauna Norv. Ser. B.* 33, 93-97.

Robinson, B.W., and D. Schluter. 2000. Natural selection and the evolution of adaptive genetic variation in northern freshwater fishes. In: T.A. Mousseau, B. Sinervo, and J.A. Endler, eds. *Adaptive Genetic Variation in the Wild*. Oxford University Press.

Schluter, D. 1996 Ecological speciation in postglacial fishes. *Phil. Trans. R. Soc. Lond.* 351:807-814.

Skúlason, S., and T.B. Smith. 1995. Resource polymophisms in vertebrates. *Trends Ecol. Evol.* 10:366-370.

Skúlason, S., S.S. Snorrason, and B. Jónsson. 1999. Sympatric morphs, populations and speciation in freshwater fish with emphasis on arctic char. In: A.E. Magurran and R. May, eds. *Evolution of Biological Diversity*. Oxford; Oxford University Press. p. 71-92.

Slatkin, M., K. Hindar, and Y. Michalakis. 1995. Processes of genetic diversification. In: V.H. Heywood, ed. *Global Biodiversity Assessment*. Nairobi: United Nations Environmental Programme (UNEP). p. 213-225.

Smith, T.B., M.W. Bruford, and R.K. Wayne 1993. The preservation of process: the missing element of conservation programs. *Biodiversity letters* 1:164-167.

Snorrason, S.S., S. Skúlason, B. Jonsson, H.J. Malmquist, and P.M. Jonasson. 1994. Trophic specialization in Arctic char *Salvelinus alpinus* (Pisces: Salmonidae): morphological divergence and ontogenetic niche shifts. *Biol. J. Linn. Soc.* 52: 1-18.

World Conservation Monitoring Center. 1996 IUCN list of threatened animals. http:\\www.unep-wcmc.org.

---

regions of the world. In addition to the physical links of air and water currents, many species of birds and mammals migrate from the Arctic to temperate and tropical regions. The Arctic tern even migrates to the Antarctic. In short, the Arctic cannot be considered in isolation.

The concept of biodiversity in the Arctic

■ Attempts to describe and define the importance of various regions, areas, and sites often refer to bio-

BOX 13. MICROBES EVERYWHERE

▶ Microbes, including bacteria, algae, fungi, and protozoans, are responsible for a considerable part of the primary production and decomposition that occurs in Arctic ecosystems. Microbes also constitute the majority of Arctic biodiversity, encoded in the genetic material of a vast number of species. Nonetheless, this major strand in the ecological web has not even been assessed, let alone understood or put to use.

Conditions in the Arctic are often at the extremes of what life can endure, for microbes as well as larger plants and animals, and the growing season tends to be short. But microbes are extremely diverse and adaptable, so much so that they are the dominant life form in large areas of the Arctic. While most microbes are too small to be seen with the naked eye, some grow in masses that become large. The most conspicuous of these are lichens, the symbiosis of fungi and algae, a characteristic life form of the terrestrial Arctic.

Less conspicuous microbes abound both on land and in water, where algae are the main engine of primary production. Aquatic bacteria, fungi, and protists recirculate nutrients to the photosynthesizing primary producers. Cold ocean waters harbor a widespread but enigmatic group of microbes, the archaea, which were discovered a few years ago and whose ecology is still unknown.

Submarine hot springs create habitats of extreme heat and chemistry. Great diversity can be found at hot springs in coastal areas, where tides and currents create highly variable conditions, and in deep water, where hydrostatic pressure raises the boiling point of sea water to 250°C and above. Such hot springs are found near Iceland, and have proven to be a rich and novel source of biodiversity. A red pigmented bacterium of the genus *Rhodothermus*, for example, was discovered in a coastal hot spring at Isafjardardjup. A new order of hyperthermophilic bacteria, the Aquificiales, was discovered at Kolbeinsey, at depths of 100 meters in waters north of Iceland. The members of Aquificiales can grow in water up to 95°C, and gain energy by the oxidation of hydrogen or sulfur under microaerobic conditions. They are the oldest phylogenetic branch in the bacterial domain.

Thermophilic, or heat-tolerant, microbes are the subject of intense scientific and commercial interest. They promise new discoveries in evolution and development, and new tools for biotechnology, industrial processes, improved health care, and food production. In this sense, the diversity of the biosphere is an invaluable source of material, especially in the all-too-often overlooked world of microbes.

Ólafur S. Andrésson, University of Iceland, Reykjavík, Iceland
Viggó Þór Marteinsson, Prokaria Inc., Reykjavík, Iceland

---

diversity. The idea of diversity is important at many scales, from genetic variation within a species to the number of species present, from the differences in functions among the assembled species to the variety and range of habitats in a given area. Biodiversity is often oversimplified to refer solely to the number of species present, but such a simplification considers neither distinctiveness nor function. Areas with only a few species may be distinctive in terms of distribution, genetic diversity, rarity, or other characteristics.

The Arctic contains many species not found elsewhere, and many habitats and ecological processes and adaptations that are unique. These include the seasonal bursts of life on land and in the ocean, the ability of some plants to survive extreme cold and dryness, the physiological features that allow mammals to maintain body heat through an Arctic winter, and the presence of life within sea ice. Furthermore, some groups such as willows, sawflies, and sandpipers are found in greater diversity in the Arctic than anywhere else. In a global context, the Arctic is a significant component of the diversity of life on Earth.

In conserving Arctic flora and fauna, several aspects of biodiversity offer useful concepts:

- *Diversity within species.* Such diversity can be great, reflecting adaptations to specific local conditions, and can help protect a species from the impacts of environmental changes. The greater the range of conditions under which a species can survive, the more likely it is to do so in a changing environment. In the Arctic, with relatively few species

BOX 14. THE DIVERSITY OF ARCTIC CHAR

▶ The name "arctic char," used broadly, encompasses several closely related species and many more distinct life history types. The high diversity in the group shows a wide variety of adaptations to various environmental conditions and illustrates the need for specific research to catalogue and understand the meaning of this diversity and develop appropriate conservation measures.

Strictly speaking, the arctic char is the species *Salvelinus alpinus*, found in many forms around the Arctic and south to the northeastern United States and the Alps in Europe, and generally adapted to lake habitats, as are several related species in Russia. The Dolly Varden char, *Salvelinus malma*, is found in the North Pacific basin and, along with several closely related species, is generally adapted to river habitats. There remains considerable disagreement about the number of species within this group, between extremes of regarding all char as members of one species to defining up to 19 different species.

Below the level of species distinctions, there is considerable diversity in the life history, ecology, size, and stock structure of arctic char. Among the life history types are anadromous and freshwater populations. Anadromous char migrate from rivers and lakes to the ocean to feed and back to freshwater to overwinter and reproduce. Freshwater char may be blocked from the ocean, or may find the freshwater environment favorable in terms of productivity, temperature, or reproductive strategy.

In some areas, multiple forms of char exist in the same place but remain distinct morphologically, genetically, or ecologically. These variants are known as ecophenotypes, and typically occur in large, deep lakes containing a diversity of habitats or multiple basins. Perhaps the greatest ecophenotypic variety for chars occurs in Thingvallavatn, Iceland. Here, four distinct forms are known: small benthic, large benthic, pelagic piscivore, and pelagic planktivore. The four forms likely originated from anadromous char after the lake was isolated from the sea by lava flows 9,600 years ago. While they remain closely related genetically after such a relatively short period, they are quite different with respect to morphology, habitat association, and diet.

In many populations of char in lakes, there are distinctive size variants among mature fish. Most fish are small, but a few individuals become very large, with few fish found at the intervening lengths. An explanation for this divergence is that the growth of the small fish is limited by the availability of food, and so despite differences in age, the fish are all roughly the same size. A few fish, however, switch from a planktivorous diet and become cannibals, growing rapidly as a result. This situation may be the beginning of diversification into distinct ecophenotypes, which

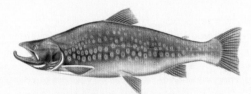
Dolly Varden char, anadromous male

Dolly Varden char, anadromous female

Dolly Varden char, residual male

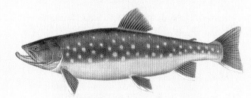
Arctic char, anadromous male

Arctic char, anadromous female

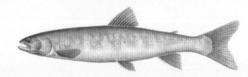
Arctic char, relictual

has not yet occurred because of lack of time, lack of ecological diversity in the environment, or other factors.

Char stocks that are isolated from one another may develop distinct biological characteristics such as abundance, size structure, age at maturity, habitat use, and so on. These characteristics are important for management and conservation, but more research is needed to understand the basis for such distinctions and to identify the population varieties that are found. At least five distinct genetic stocks of anadromous Dolly Varden char have been found in four river systems on the north slope of the Yukon Territory in Canada, and more stocks are likely to be present.

The diversity exhibited by the Arctic char group is theoretically important because it provides a range of evolutionary options in the face of the highly variable and changing Arctic environment. Arctic char have colonized the postglacial Arctic in a relatively short period, and thus have probably not had time to evolve fully. Nonetheless, the varieties of char have diversified to provide stability to Arctic aquatic ecosystems. As a species complex, they fill roles that in southern latitudes would be occupied by several species. From this perspective, as well as their inherent importance, char in all their diverse forms deserve high levels of protection and preservation, as well as being worthy of substantive research.

James D. Reist, Fisheries and Oceans Canada, Winnipeg, Canada

SOURCES/FURTHER READING

Babaluk, J.A., J.D. Reist, V.A. Sahanatien, N.M. Halden, J.L. Campbell, and W.J. Teesdale. 1998. Preliminary results of stock discrimination of chars in Ivvavik National Park, Yukon Territory, Canada using microchemistry of otolith strontium. *In*: N.W.P. Munro and J.H. Martin Willison, eds. *Linking Protected Areas with Working Landscapes Conserving Biodiversity.* Proc. Third Int. Conf. Science and Management of Protected Areas, 12-16 May 1997. Wolfville, Canada: SAMPAA. p. 991-998.

Babaluk, J.A., N.M. Halden, J.D. Reist, A.H. Kristofferson, J.L. Campbell and W.J. Teesdale. 1997. Evidence for non-anadromous behaviour of Arctic charr (*Salvelinus aplinus*) from Lake Hazen, Ellesmere Island, Northwest Territories, Canada, based on scanning proton microprobe analysis of otolith strontium distribution. *Arctic* 50: 224-233.

Griffiths, D. 1994. The size structure of lacustrine Arctic charr (Pisces: Salmonidae) populations. *Biol. J. Linn. Soc.* 51: 337-357.

Hammar, J. 1998. Evolutionary ecology of Arctic char (*Salvelinus alpinus* L.) – intra- and interspecific interactions in circumpolar populations. Unpublished D. Phil. dissertation, University of Uppsala, Sweden.

Jensen, J.W. 1981. Anadromous Arctic char, *Salvelinus alpinus*, penetrating southward on the Norwegian Coast. *Can. J. Fish. Aquat. Sci.* 38: 247-249.

Johnson, L. 1980. The Arctic charr, *Salvelinus alpinus. In,:* Balon, E.K., ed. *Charrs, Salmonid fishes of the Genus Salvelinus.* The Hague: Dr. W. Junk. p. 15-98.

Johnson, L. 1995. Systems for survival: of ecology, morals and Arctic charr *(Salvelinus alpinus). Nordic J. Freshw. Res.* 71: 9-22.

Jonsson, B., and K. Hindar. 1982. Reproductive strategy of dwarf and normal Arctic char (*Salvelinus alpinus*) from Vangsuatnet Lake, western Norway. *Can. J. Fish. Aquat. Sci.* 39: 1404-1413.

Parker, H.H., and L. Johnson. 1991. Population structure, ecological segregation and reproduction in non-anadromous Arctic charr, *Salvelinus alpinus* (L.), in four unexploited lakes in the Canadian High Arctic. *J. Fish Biol.* 38: 123-147.

Reist, J.D. 1989. Genetic structuring of allopatric populations and sympatric life history types of charr, *Salvelinus alpinus/malma*, in the western Arctic, Canada. *Physiol. Ecol. Japan*, Spec. Vol. 1: 405-420.

Reist, J.D., J.D. Johnson, and T.J. Carmichael. 1997. Variation and specific identity of char from Northwestern Arctic Canada and Alaska. *Amer. Fish. Soc. Symp.* 19: 250-261.

Sandlund, O.T., K. Gunnarsson, P.M. Jonasson, B. Jonsson, T. Lindem, K.P. Magnusson, H.J. Malmquist, H. Sigurjonsdottir, S. Skulason, and S.S. Snorrason. 1992. The arctic charr *Salvelinus alpinus* in Thingvallavatn. *Oikos.* 64: 305-351.

Savvaitova, K.A. 1989. *Arctic chars: the structure of population systems and prospects for economic use.* Moscow: Agrompromizdat. (Original in Russian, Translated by Canada Institute for Scientific and Technical Information, National Research Council, Ottawa, Ontario, Canada, 246 pp.).

Skulason, S.S., S.S. Snorrason, D.L.G. Noakes, and M.M. Ferguson. 1996. Genetic basis of life history variations among sympatric morphs of Arctic char, *Salvelinus alpinus. Can. J. Fish. Aquatic Sci.* 53: 1807-1813.

Svenning, M.A., and R. Borgstrom. 1995. Population structure in landlocked Spitsbergen Arctic charr. Sustained by Cannibalism? *Nordic J. Freshw. Res.* 71: 424-431.

BOX 15. NORTH-SOUTH TRENDS IN TERRESTRIAL SPECIES DIVERSITY

▶ As a general trend in northern regions, the number of terrestrial species declines as latitude increases, although the magnitude of the decline varies. The standard explanation is that fewer and fewer plants or animals can survive as the environment becomes increasingly severe. While this pattern is broadly true, the details can become more complex depending on what is examined.

The Taimyr Peninsula in Russia offers a classic example of the typical downward trend. As the level landscape shifts from forest-tundra in the south to polar desert in the north, the number of bird species decreases from 80 to 12, the number of vascular plants from 250 to 50, and the number of lichens from 230 to 160. Such changes are driven primarily by summer temperatures.

By contrast, a study of benthic macro-faunal communities along a transect from the North Pole to the Yermak Plateau in the Arctic Ocean found a total of 108 species at depths between 560 and 4,411 meters, with the highest number of species observed at intermediate depths. The species richness found in the Arctic at these depths was comparable to observations in tropical regions and in the Antarctic, in contrast to the terrestrial trend of decreasing diversity with increasing latitude.

The general rule of declining terrestrial diversity at higher latitudes has some other significant exceptions. Among certain types of shorebirds, for example, diversity increases as one moves northward. Of the 206 shorebird species known worldwide, 69 species occur in the Arctic region. In the family of *Scolopacidae* (snipes, sandpipers, and phalaropes), 55 of 87 species are found in the Arctic. For sandpipers, all 24 species occur in the Arctic, 17 of which breed exclusively in the Arctic region.

Changes in overall numbers may mask different trends among groups of species, as is shown in a comparison of major groups of terrestrial arthropods in Canada. Of more than 33,000 named species nationwide, 90% are insects, 6% are mites, and 1% are springtails. Among 381 named species in the Queen Elizabeth Islands in the high Arctic, the proportions are very different: 64% are insects, 20% are mites, and 12% are springtails. Overall, the number of insect species has declined by 99%, whereas mites have declined 96%, and springtails only 85%.

Such differences are also seen among flora, where the Arctic species diversity of non-vascular plants (fungi, mosses and lichens) is generally much higher than the diversity of vascular plants (ferns, conifers, flowering plants). Classical explanations for this include release from competition and differences in adaptation and dispersal mechanisms (e.g., microscopic, light-weight spores versus seeds).

Genetic and behavioral diversity can be high within arctic species. A single species, for instance the arctic char, may adapt to fill a variety of ecological niches and functions that in other ecosystems would be occupied by several different species. In short, while most Arctic ecosystems have fewer total species than comparable ecosystems in temperate or tropical latitudes, the measure of overall species diversity conceals considerable intra-species variation and complexity.

Christoph Zöckler, UNEP-World Conservation Monitoring Centre, Cambridge, UK
Henry P. Huntington, Huntington Consulting, Eagle River, Alaska, USA
Yuri Chernov, Russian Academy of Sciences, Moscow, Russia

SOURCES/FURTHER READING
Danks, H.V. 1990. Arctic insects: instructive diversity. In: C.R. Harrington, ed. *Canada's missing dimension: science and history in the Canadian arctic islands.* Vol. II. Ottawa: Canadian Museum of Nature. p. 444-470.
Groombridge, B., and Jenkins, M.D. 2000. *Global Biodiversity: Earth's Living Resources in the 21st Century.* Cambridge: World Conservation Monitoring Centre. 46p.
Matveyeva, N., and Y. Chernov. 2000. Biodiversity of terrestrial ecosystems. In: M. Nuttall and T.V. Callaghan, eds. *The Arctic: environment, peoples, policy.* Amsterdam: Harwood Academic Publishers. p. 233-273.
Kröncke, I. 1998. Macrofauna communities in the Amundsen Basin, at the Morris Jesup Rise and at the Yermak Plateau (Eurasian Arctic Ocean). *Polar Biol.* 19:383-392
Walker, M.D. 1995. Patterns and causes of Arctic plant community diversity. In: F.S. Chapin III and C. Körner, eds. *Arctic and alpine biodiversity.* Ecological Studies 113. Berlin: Springer-Verlag. p. 3-20.
Wilson, D. E., and D. M. Reeder (eds). 1993. *Mammal Species of the World.* Washington. Smithsonian Institution Press. 1206 pp.

Table 3. Estimated numbers of species in selected groups worldwide and in the Arctic.

Group	Total	Arctic	Arctic proportion %
Fungi	65,000	5,000	7.6
Lichens	16,000	2,000	12.5
Mosses	10,000	1,100	11.0
Liverworts	6,000	180	3.0
Ferns	12,000	60	0.5
Conifers	550	8	1.2
Flowering plants	270,000	3,000	1.2
Spiders	75,000	1,000	1.2
Insects	950,000	3,000	0.3
Vertebrates	52,000	860	1.6
Fishes	25,000	450	1.8
Reptiles	7,400	4	>0.1
Mammals	4,630	130	2.8
Birds	9,950	280	2.8

Data based on expert evaluation and literature: Groombridge and Jenkins 2000, Wilson and Reeder 1993, Matveyeva and Chernov 2000.

compared with other parts of the world, genetic, morphological, and behavioral diversity may be especially significant components of biodiversity.

- *Species population ranges.* Plants and animals living at the extremes of their geographic and climatic ranges may be essential for the long-term survival of their species. In the marginal conditions of the Arctic, such outlying groups may die out when conditions change, or they may provide a critical refuge when something affects the main population of the species. In either case, extreme groups are sensitive indicators of environmental change.

- *Species flexibility and vulnerability.* The presence of relatively few species may mean that certain key ecological functions depend on only one or two species, rather than several species with overlapping roles, as might be found in lower latitudes. While individual species may be very resilient to changing conditions, the system as a whole may be vulnerable if key functions are disrupted by great changes to certain species, as shown in the snow goose example above.

- *Habitat diversity.* The diversity of habitats across small and large distances offers a crucial buffer against change, and allows species to use alternative habitats for different purposes or under different conditions. This applies to insects seeking relief from cold and wind from day to day as well as to caribou and geese seeking grazing areas from year to year. Habitat diversity is essential in the Arctic, with its extremes of seasonal and annual variation.

While population trends and other such data are valuable, they do not describe what species do and how they interact. Much also remains to be learned about the functions and processes that occur in the Arctic. Later chapters attempt to describe the ecological context in which the available data should be used and understood.

SOURCES/FURTHER READING

Abbott, R.J., L.C. Smith, R.I. Milne, R.M.M. Crawford, K. Wolff, and J. Balfour. 2000. Molecular analysis of plant migration and refugia in the Arctic. *Science* 289:1343-1346.

Adamczewski, J. 1995. *Digestion and body composition in the muskox.* Unpublished Ph.D. thesis. University of Saskatchewan, Saskatoon, SK.

AMAP (Arctic Monitoring and Assessment Program). 1998. *AMAP assessment report: Arctic pollution issues.* Oslo: AMAP. p. 25-116.

Bernes, C. 1996. *The Nordic Arctic environment: unspoilt, exploited, polluted?* Copenhagen: Nordic Council of Ministers. 240p.

Estes, J.A., and D.O. Duggins. 1995. Sea otters and kelp forests in Alaska: generality and variation in a community ecological paradigm. *Ecological Monographs* 65(1):75-100.

Fitzpatrick, E.A. 1997. Arctic soils and permafrost. *British Ecological Society Special Publication Series* 13:1-39.

French, H. 1976. *The periglacial environment.* London: Longman. 309p.

Kouki, J. 1999. Latitudinal gradients in species richness in northern areas: some exceptional patterns. *Ecological Bulletin* 47:30-37.

Larsen, J.A. 1980. *The boreal ecosystem.* New York: Academic Press. xvi + 500p.

Parker, K., R.G. White, M.P. Gillingham, and D.F. Holleman. 1990. Comparison of energy metabolism in relation to daily activity and milk consumption by caribou and muskox neonates. *Canadian Journal of Zoology* 68: 104-114.

Piersma, T. and N. Davidson, eds. 1992. *The migration of knots.* WSG Bull. 64 (suppl.) 209 p.

Post, E., and N.C. Stenseth. 1999. Climatic variability, plant phenology, and northern ungulates. *Ecology* 80: 1322-1339.

Saari, L. 1995. Population trends of the Dotterel *Charadrius morinellus* in Finland during the past 150 years. *Ornis Fennica* 72:29-36.

Stenseth, N.C., O.N. Bjønstad, and T. Saitoh. 1998. Seasonal forcing on the dynamics of *Clethrionomys rufocanus:* modeling geographic gradients in population dynamics. *Researches on Population Ecology* 40: 85-95.

Stöcklin, J., and C. Körner. 1999. Recruitment and mortality of *Pinus sylvestris* near the nordic treeline: the role of climate change and herbivory. *Ecological Bulletins* 47: 168-177.

Summers, R.W., and L.G. Underhill. 1987. Factors relating to breeding populations of Brent Geese (*Branta b. bernicla*) and waders (Charadrii) on the Taimyr Peninsula. *Bird Study* 34: 161-171.

Underhill, L.G. 1987. Changes in the age structure of Curlew Sandpiper populations at Langebaan Lagoon, South Africa, in relation to lemming cycles in Siberia. *Transactions of the Royal Society of South Africa.* 46: 209-214.

Underhill, L.G. 1989. Three-year cycles in breeding productivity of Knots *Calidris canutus* wintering in South Africa suggest Taimyr Peninsula provenance. *Bird Study* 36: 83-87.

Väisänen, R.A. 1998. Current research trends in mountain biodiversity in NW Europe. *Pirineos* 151-152:131-156.

Weatherhead, E.C., and C.M. Morseth, eds. 1998. Climate change, ozone, and ultraviolet radiation. In: Arctic Monitoring and Assessment Program. *AMAP assessment report: Arctic pollution issues.* Oslo: AMAP. p. 717-774.

Weller, G. 2000. The weather and climate of the Arctic. In: M. Nutall and T.V. Callaghan, eds. *The Arctic: environment, people, policy.* Amsterdam: Harwood Academic Publishers. p. 143-160.

Zöckler, C. 1998. *Patterns in Biodiversity in Arctic Birds.* WCMC Bulletin No. 3. Cambridge: World Conservation Monitoring Centre. 15p.

3. Humans

Humans have long been a part of the Arctic environment. This chapter provides a brief history of human settlement and recent developments, and highlights the importance of living resources for today's Arctic peoples.

The First People

For thousands of years, the Arctic has been home to humans. After the last Ice Age, groups of people in Eurasia moved northward following the retreating glaciers, reaching the shores of the Arctic Ocean. The early stone-age culture in northern Scandinavia, known as the Komsa Culture, existed from 9,000 to 4,000 B.C. In northeast Asia and North America, people lived in unglaciated areas, and moved east and south across North America. Archeological sites in the Yukon Territory are 15,000 years old, and in Arctic Alaska date back 11,000 years. Some current village sites have been inhabited for several millennia. Several waves of people and cultures spread across the Arctic coast, from paleo-Eskimos thousands of years ago to the Thule Culture several centuries ago. Some reached the High Arctic as early as 2,000 B.C., relying primarily on muskoxen.

Patterns of human settlement and resource use reflect environmental and ecological factors. Protection from severe weather, storms, surges of sea ice, floods, and other destructive forces are clearly critical considerations in selecting a place to live. Access to fresh water, to building materials, and to food are also essential. Settlements are often found along the migration corridors of fish and mammals or in areas where firewood is readily available. The great seasonal variations in the

An Evenk fisherman, Sakha Republic (Yakutia), Russia.

Ruins of an Inuit stone dwelling, Northeast Greenland.

Arctic environment meant that many groups would move among different locations during the year as the resources of each area became abundant. These patterns continue today as families move to fish camps in summer, marine mammal hunting parties live on the sea ice in spring, and herders move their animals between summer and winter grazing areas.

The locations of settlements in times past can often tell us a great deal about the environmental conditions that prevailed at the time they were inhabited. Changes in the courses of rivers may turn a boun-

BOX 16. DRIFTWOOD IN THE ARCTIC

▶ Driftwood is common in many parts of the Arctic. It has long been used for building material, for making boats, sleds, and weapons, and for fuel in places far from the nearest tree. The journey of the wood – from inland forests, down rivers, across the ocean, and to beaches where the wood is used by people or rots to bring forth new life – is a fascinating demonstration of connections across the Arctic.

The earliest systematic investigation of driftwood was make by Olav Olavius, who in 1780 traveled around Iceland making an economic survey of the amount, type, and usefulness of the driftwood he found. Olavius concluded that the wood had come from Siberia, brought by the currents that come to Iceland from the north. Subsequent studies and evidence from ships caught in the ice confirmed what is now known as the Transpolar Drift, currents that flow from Alaska and eastern Russia across the Arctic to Fram Strait in the North Atlantic.

Arctic driftwood originates in the boreal forest, and is carried northwards to the ocean by large rivers. The Yukon and Mackenzie Rivers in North America bring primarily spruce and poplar, most of which comes from trees along naturally eroding riverbanks. In Russia, however, extensive forestry operations lose at least four percent of their annual volume of wood during river transport, thus contributing far more wood to the Arctic Ocean than comes via erosion. Driftwood from Russia includes pine, larch, spruce, birch, and poplar.

Sea ice is essential to the widespread distribution of driftwood. Logs in the open ocean will sink after a period of roughly 10-20 months, depending on the type of tree. Logs trapped by sea ice, on the other hand, remain afloat until after the ice melts, allowing them to make a multi-year journey across the Arctic Basin. Thus, beaches near the zone of melting ice in the North Atlantic tend to accumulate far more wood than coasts farther north that rarely see open water.

Recent studies of the origin of driftwood found around the Arctic indicate that its journey, not surprisingly, mirrors that of the ice that carries it. Although the Transpolar Drift carries ice from both the Beaufort and Kara Seas, the two branches appear to remain distinct, so that driftwood from North America reaches Greenland but not Iceland, whereas wood from Russia remains on the eastern side of the Transpolar Drift, reaching Svalbard and Iceland.

Ólafur Eggertsson, Iceland Forest Research Station, Iceland

Driftwood in the Lena Delta, Russia.

tiful fishing spot into a poor one. A colder climate may make a previously suitable site uninhabitable. Shifts in resource patterns may force changes in hunting practices. Villages that were productive for centuries may have been unable to adapt to sudden or drastic changes. Hence, the record of human habitation in the Arctic is one of recurring change, involving a constant search to find better conditions in a marginal environment. Greenland, for example, was first inhabited some 4,000 years ago, during a warm climatic phase. By 3,000 years ago, however, the climate had cooled and people disappeared from the island for a thousand years.

Before modern transportation, people had to depend on local materials. Driftwood and whale bones served as structural supports for houses and boats along the coast. Animal skins provided clothing, boat covers, and rope. Teeth, bones, ivory, and stone made hunting tools. Survival meant having access to all these materials, either locally or through trade.

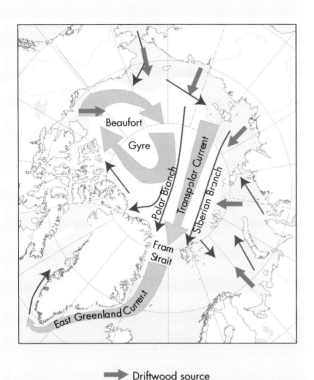

Figure 17. Places of origin and routes of transport for Arctic driftwood.

SOURCES/FURTHER READING

Eggertsson, Ó. 1994a. Origin of the Arctic driftwood, a dendrochronological study. *Lundqua Thesis* 32:13 + 4 appendices.

Eggertsson, Ó. 1994b. Mackenzie River driftwood, a dendrochronological study. *Arctic* 47:128-136.

Eggertsson, Ó. 1994c. Driftwood as an indicator of relative changes in the influx of Arctic and Atlantic water into the coastal areas of Svalbard. *Polar Research* 13:209-218.

Eggertsson, Ó., and D. Laeyendecker. 1995. Origin of driftwood in Frobisher Bay, Baffin Island, a dendrochronological study. *Arctic and Alpine Research* 27:180-186.

Gordienko, P.A. and L. Laktionov. 1969. Circulation and physics of the Arctic Basin waters. *Annals of the International Geophysical Year, Oceanography* 46: 94-122.

Nilsson, S., O. Sallnäs, M. Hugosson, and A. Shvidenko. 1992. *The Forest resources of the former European USSR.* New York: IIASA, Parthenon Publishing Group. 407p.

Olavius, O. 1780. *Oeconomisk reise igiennem de nordvestlige, nordlige og nordöstlige Kanter af Island.* Köbenhavn: Gyldendals Forlag. 614 p.

Figure 18. The seasonal cycle of hunting, fishing, and gathering for Polar Inuit in 1980. (Source: Arctic Centre, Rovaniemi, Finland).

BOX 17. REINDEER AND THE PEOPLING OF THE YAMAL

▶ The textbook view of indigenous communities in the Arctic describes stability, low subsistence pressure, and overall static relations between prehistoric human societies and their environment. This view has been challenged in recent decades by newly excavated archaeological sites, searches through historical documentary records, and the collection of oral tradition and historical memories of native elders. As our understanding of subsistence transitions in the Arctic grows, an entirely new history is emerging – one that abounds in migrations, rapid subsistence shifts, the development of new technologies and economic practices, and other social and cultural transitions.

Recent discoveries by Russian archaeologists in the Central Siberian Arctic prove that such transitions took place across the entire Eurasian Arctic. Unfortunately, these data, published and discussed primarily in the Russian sources, are far less known within the wider Arctic science community. Recent excavations on the Yamal Peninsula by a team from the Institute of History and Archeology in Ekaterinburg produced pieces of the earliest reindeer harnesses and transportation sleds, dated to the Sixth Century AD. By this time, groups of prehistoric caribou hunters and fishermen crossed the central Yamal river valleys and expanded into the northern portion of the peninsula. Eventually, they reached the Arctic coast along the northernmost tip of Yamal, at or above 74°N. Their few domesticated transport reindeer allowed them to explore the various resources of the Arctic coastland, including huge amounts of driftwood, beached walrus ivory, and the living stocks of walruses, seals, and polar bears.

While the northernmost portion of Yamal became the new resource frontier, life in the interior tundra zone was

A Dolgan man driving a sledge, Taimyr, Russia.

also changed substantially by domesticated reindeer. Inland summer settlements and camps expanded, their populations grew, and housing patterns and the annual subsistence cycle were modified. According to a reconstruction, people were then able to make annual migrations from the more hospitable and resource-stable areas along the northern treeline (roughly at 66°N in Southern Yamal) to the lakes and river valleys of central Yamal with their abundant summer fish resources. At first, people tried to build small dug-out houses at these summer camps, as they had done in the forest zone. Later, they switched to portable tents in which the hunters and their families would live for five or six months. Dozens of wood and bone toys confirm that, for the first time, children were able to follow adult hunters on long summer treks, thanks to transport reindeer and sleds.

As the original small herds of domesticated transport reindeer grew, the Yamal people – who were, most probably, ancestral to the present-day Nenets – gradually developed a more mobile lifestyle, with greater reliance upon long seasonal migrations to open tundra pastures in the north. The next subsistence revolution came up around 1650-1750 AD, as animal husbandry opened the way to a pastoral culture supported by a reliable source of food and clothing – the domesticated reindeer. The stocks of a few dozen to barely a hundred transport reindeer were replaced by family herds of up to several thousand animals each. This transition occurred under the influence of many factors, including climatic cooling during the Little Ice Age and new governmental policies introduced by the Russian and Swedish states in their colonial expansion to the North. Strict taxation, land appropriation, pressure for precious northern pelts, and enforced pacification of the native tribes – all played their role in pushing Arctic peoples farther northward into the tundra zone and into the full-scale nomadic economy of the reindeer herder.

Igor Krupnik, Arctic Studies Center, Smithsonian Institution, Washington, DC, USA
Natalya Fedorova, Institute of History and Archaeology, Ekaterinburg, Russia

Dolgans in reindeer-skin clothes, Taimyr, Russia.

SOURCES/FURTHER READING

Fedorova, N.V. 2000a. Reindeer, dog, the "Kulai Phenomenon," and the Legend of Sikhirtya. *Drevnosti Yamala* 1:54-66. (In Russian.)

Fedorova, N.V. 2000b. Seven years of the Yamal Archaeological Expedition: some results and prospects for the future. *Nauchnyi Vestnik* 3:4-12. (In Russian.)

Fedorova, N.V., and W. Fitzhugh. 1998. Ancient Legacy of Yamal. In: I. Krupnik and N. Narinskaya, eds. *The Living Yamal*. Moscow: Sovetskii Sport. p. 50-55.

Fedorova, N.V., P.A. Kosintsev, and W.W. Fitzhugh. 1998. *"Gone to the hills": ancient culture of the Yamal Coastal Dwellers during the Iron Age*. Ekaterinburg: Ekaterinburg Publishers. (In Russian.)

Krupnik, I.I. 1993. *Arctic adaptations: Native whalers and reindeer herders of northern Eurasia*. Hanover, New Hampshire: University Press of New England.

Krupnik, I.I. 1998. Understanding reindeer pastoralism in modern Siberia: ecological continuity versus state engineering. In: J. Ginat and A. Khazanov, eds. *Changing nomads in a changing world*. Brighton: Sussex Academic Press. p. 223-242.

Krupnik, I.I. 2000. Reindeer pastoralism in modern Siberia: research and survival during the time of crash. *Polar Research* 19(1):49-56.

Reindeer herd in Chukotka, Russia.

Gathering reindeer in Finnish Lapland.

56 HUMANS

Black smoke from oil drilling operations, North Slope, Alaska.

A nickel smelter in Norilsk, Taimyr, Russia.

Current Settlement Patterns

The economic and political developments of the 20th Century caused major changes in the settlement patterns of Arctic residents. Schools, wage-earning jobs, and other institutions of modern life have made communities more sedentary because most people no longer have the time to spend much of the year traveling. The introduction of government land management systems has often undermined older systems of communal access and use. Settlements have been moved or combined as a result of economic or political forces. In the eastern Canadian Arctic, for example, nearly all the Inuit settlements were established after World War II as a result of government policies for the provision of social services. Although they were usually located at the sites of existing trading posts, moving permanently to the settlements required a major shift in Inuit seasonal patterns.

Today, the human population of the Arctic regions is probably declining overall due to deteriorating economic conditions in Russia, which have compelled many people to move away from remote northern areas. In most other regions, the human population is increasing as a result of large family sizes, improved health and health care, and the continued migration

Fuel barrels left from World War II, Northeast Greenland National Park.

BOX 19. CO-MANAGEMENT IN THE INUVIALUIT SETTLEMENT REGION

▶ The Inuvialuit inhabit the northwestern part of the Northwest Territories and the North Slope of the Yukon Territory in Canada. In 1984, they reached a land-claims agreement with the Government of Canada, establishing the Inuvialuit Settlement Region. As part of the agreement, a co-management system was established for all matters relating to the management of living resources in the region, including habitat.

The Inuvialuit Game Council (IGC), established under the land-claims agreement, has six members, six alternates, and a chairman, all of whom are Inuvialuit. The six communities in the region each have a Hunters and Trappers Committee (HTC). Each committee appoints a member and an alternate to comprise the IGC, which represents the collective Inuvialuit interest in wildlife. The IGC also appoints the Inuvialuit members to the co-management bodies established pursuant to their land-claims agreement. On the government side, various federal and territorial agencies participate in the co-management groups. The five co-management groups are:

– Fisheries Joint Management Committee (FJMC)
– Wildlife Management Advisory Council (Northwest Territories) (WMAC(NWT))
– Wildlife Management Advisory Council (North Slope) (WMAC(NS))
– Environmental Impact Screening Committee (EISC)
– Environmental Impact Review Board (EIRB)

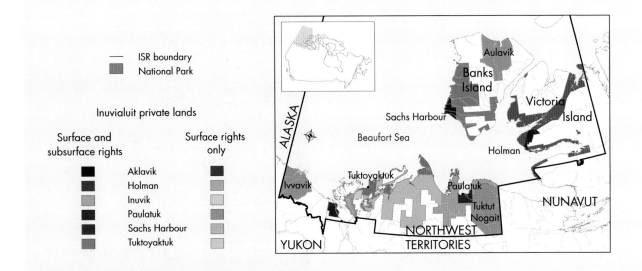

Figure 22. The Inuvialuit Settlement Region (ISR). The six communities own surface rights to some areas and surface and sub-surface (mineral) rights to other areas.

of non-indigenous people to the Arctic. In these areas, standards of living are improving, narrowing the gap between Arctic areas and the more populous regions of Arctic countries. These changes often alter patterns of local resource use and result in increased use of imported goods.

Despite the relatively small and sparse population, contemporary land use in the Arctic occurs over enormous areas. The North Slope Borough of Alaska, for example, encompasses more than 225,000 square kilometers and has approximately 7,000 residents among its eight communities. Nearly all of this area is or has

The EISC and the EIRB collectively make up the screening and review process for the Settlement Region. The FJMC deals primarily with fish and marine mammals. The two WMACs deal, in their respective territories, primarily with birds and terrestrial wildlife. Each group is comprised of 50% Inuvialuit (appointed by the IGC) and 50% government members (appointed by the agencies), with a mutually agreed upon chairman. This composition guarantees the inclusion of indigenous knowledge and cultural sensitivity in the process. Furthermore, all the groups include extensive community consultation elements in their operating procedures.

In order to promote the efficient and informed operation of these co-management groups, the Inuvialuit approach to sustainable development and environmental management includes support for a full range of programs, including wildlife research, traditional knowledge studies, harvest monitoring, the establishment of protected areas, and the maintenance of biodiversity. The IGC also participates in a variety of international fora, and in 1988 signed an agreement with the Iñupiat of the North Slope of Alaska concerning management of the shared population of polar bears in the Beaufort Sea. The cohesiveness of the Inuvialuit co-management system is enhanced by the establishment of a Joint Secretariat to provide technical and administrative support to the co-management groups and to the IGC.

Duane Smith, Chairman, Inuvialuit Game Council, Inuvik, Northwest Territories, Canada

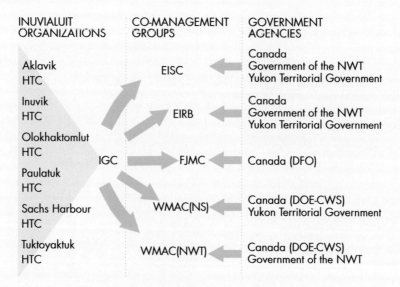

Figure 23. Composition and Relationship of Renewable Resource Management organizations in the Inuvaluit Settlement Region. DFO refers to Department of Fisheries and Oceans Canada, DOE-CWS refers to Environment Canada-Canadian Wildlife Service, other abbreviations are explained in the text.

been used in recent years or decades, the patterns of use in a given year depending on the availability of fish and wildlife and other factors. The situation is similar across most of the rest of the Arctic. Mapping studies in the Canadian Arctic in the 1970s showed that Inuit were familiar with the entire region. The extent of their land use formed the basis for extensive land claims settlements.

In Eurasia, land access for reindeer herding is a crucial factor for traditional communities. The introduction of national boundaries in Scandinavia in the eighteenth century began the fragmentation of herd-

BOX 20. ENVIRONMENTAL IMPACTS OF DIAMOND MINING IN THE CANADIAN ARCTIC

▶ As late as 1990, no one foresaw diamond mining in the Arctic. But in 1991, diamond-bearing volcanic shafts known as kimberlite pipes were discovered about 300 kilometers north of Yellowknife, in Canada's Northwest Territories. Within a year or two, over 20 million hectares were claimed by miners using aircraft and helicopters for access. By 1998, at least 200 new kimberlite pipes had been discovered in an area that had no all-weather roads or permanent human settlement and only one gold mine, served by airplane. By 2000, one mine was producing diamonds (BHP's Ekati) and a second (Diavik) was under construction – the first diamond mines in North America. The environmental assessments for the mines involved many public meetings – 250 meetings between 1994 and 1998 in the case of Diavik – to involve stakeholders, especially aboriginal communities. Dene, Metis, and Inuit have long traditions of using the area for subsistence hunting and trapping. The primarily aboriginal communities close to the mines have unemployment rates of 50 to 89%, with few opportunities for economic growth and strong dependence on caribou and fish for subsistence.

The Ekati and Diavik mines are on the northern shore of the 570-square-kilometer Lac de Gras, which discharges its water into the Coppermine River and thence to Coronation Gulf. The kimberlite pipes are mined as open pits, and the processed kimberlite is held in tailings ponds created by emptying small lakes. The mines are supplied by air or seasonally by truck on an ice road from Yellowknife. The road is open between 50 and 78 days per year in late winter, and traffic varies from 100 to 150 transport trucks/day. Local hunters also use the ice road heavily, because it crosses the taiga winter range for the Bathurst Herd of barren-ground caribou. The herd migrates through the Lac de Gras area during spring and fall. As many as 100,000 caribou are in the area of the mines during spring, summer, and fall. With 350,000 animals, the Bathurst Herd is one of the largest in North America. Its present range covers 250,000 square kilometers and extends from below the treeline as far south as Saskatchewan across the barrens to the Arctic coast.

The two mines are expected to cause the loss or change of vegetation over about 100 square kilometers, although reclamation after mine closure may reduce that figure. The effects on wildlife are largely unknown at this time. The greatest uncertainties are the cumulative effects, especially as the number of active mines in the area is expected to increase. During the environmental assessment hearings, aboriginal elders expressed concerns for cumulative effects on caribou and the quality of freshwater. The hearings concluded that, with mitigation efforts, there would be no major significant effects, although a high degree of monitoring would be necessary. The existing two mines are working with aboriginal organizations and governments to mitigate and monitor any effects.

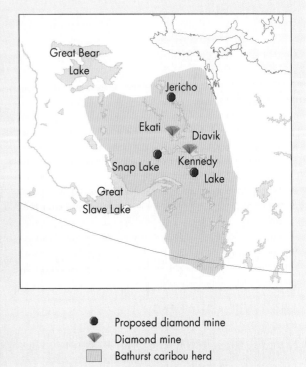

Figure 24. Diamond mines on the range of the Bathurst caribou herd in the Northwest Territories, Canada.

Anne Gunn, Northwest Territories Department of Resources, Wildlife and Economic Development, Government of the Northwest Territories, Yellowknife, Northwest Territories, Canada

---

ing territories and patterns. In Russia today, herders are often restricted by industrial development, by official regulations, or by the difficulty of accompanying herds to remote areas for extended periods. In Fennoscandia, development of hydroelectric dams, competition for land use from forestry and tourism, and questions over land-use rights and reindeer ownership have led to recent conflicts. The need to be able to move to new areas as conditions demand often conflicts with the modern practice of codifying land uses.

A fishing settlement in Sakha Republik (Yakutia), Russia (top). Kangiqtugaapik (Clyde River), Baffin Island, Canada (middle). A small fishing village in the Maniitsop municipality north of Nuuk, Greenland (bottom).

Use of Living Resources

Traditional hunting, fishing, and gathering are a mainstay of indigenous communities and many other local settlements throughout the Arctic. Today's activities continue patterns that developed long ago, though often in ways that have been modified by modern technology and lifestyles. In most areas, snowmachines are preferred to dog teams and reindeer for winter travel. Rifles and shotguns have replaced the bow and arrow, and outboard motors propel boats once driven by paddles and wind. This equipment has increased hunting efficiency, in some cases placing a greater strain on resources. But despite changes in harvest methods, the importance of food from the local land and waters remains strong. Production in

Inuit hunters skinning a seal in Grise Fiord, Ellesmere Island, Canada.

In spring, Inuit hunters live on the sea ice for several weeks. Women and children travel in the small cabin on the sledge. Pond Inlet, Nunavut, Canada.

BOX 21. THE TŁĮCHǪ AND THE DETSỊ̀ÌLÀÀʔEKWǪ̀: SHOWING RESPECT FOR CARIBOU

A long time ago when the caribou did not travel the trails to this area, the Tłįchǫ were starving. A man had a dream, and the next day he walked straight to the barrenlands, and invited the detsı̀ı̀làà?ekwǫ̀ to follow him to this land...

Romie Wetrade, age 78 (as quoted by Legat et al. 1995)

▶ The Tłįchǫ, known in English as the Dogrib, traditionally occupy the area between Tıdeè (Great Slave Lake) and Sahtì (Great Bear Lake), extending well past Kòk'èetì (Contwoyto Lake), Ts'eèhgooı (Alymer Lake), and ʔedaàtsotì (Artillery Lake) in the barrenlands to the Dehtsotì (Mackenzie River) in the west. The Tłįchǫ state that they and the detsı̀ı̀làà?ekwǫ̀ (caribou that travel between the barrenlands and the boreal forest) have a very close and respectful relationship. The detsı̀ı̀làà?ekwǫ̀ did not always migrate to Tłįchǫ territory and the Tłįchǫ want to ensure they will continue to migrate on the land they now share.

Respect is shown by only taking what is needed, using all parts of the harvested animals, and discarding any unused parts in respectful ways. Respect is also shown by having and sharing knowledge of the detsı̀ı̀làà?ekwǫ̀. A lack of knowledge, and therefore respect, will result in the detsı̀ı̀làà?ekwǫ̀ migrating elsewhere and a population decline. The elders' knowledge of the detsı̀ı̀làà?ekwǫ̀ is collected through harvesting activities, verified through discussions with other harvesters, and shared through oral narratives in association with the general truths. Examples include:

- detsı̀ı̀làà?ekwǫ̀ have unpredictable migration patterns.
- detsı̀ı̀làà?ekwǫ̀ return to the same birthing grounds,
- ʔedzıe (middle-aged cows) take turns leading during the migration,
- detsı̀ı̀làà?ekwǫ̀ have extremely good memories,
- detsı̀ı̀làà?ekwǫ̀ have a strong sense of smell, and
- ʔadzìı (lichen) is the most important food for detsı̀ı̀làà?ekwǫ̀, kwetsi (rock tripe) puts on fat, and the variety of vegetation on the barrenlands in spring and summer is important to the health of the detsı̀ı̀làà?ekwǫ̀.

The Tłįchǫ elders are concerned there is a lack of knowledge and therefore a lack of respect among young people and among decision-makers associated with governments and industry. The elders believe that if respect through knowledge is shown, the detsı̀ı̀làà?ekwǫ̀ habitat would be better cared for. For example:

- Smells from exhaust suggest to the detsı̀ı̀làà?ekwǫ̀ that there is a fire. This confuses them and does not allow them to smell locations where there is lush vegetation.
- Dust and ash-like particulates are found on vegetation near industrial areas and associated infrastructure. The elders know that these particles affect the vegetation and therefore the detsı̀ı̀làà?ekwǫ̀.
- Constructing roads and bridges where important detsı̀ı̀làà?ekwǫ̀ water crossings are located shows lack of knowledge and lack of respect for these places and for the detsı̀ı̀làà?ekwǫ̀ who learn, remember, and depend on these places for ease of travel.

The elders are aware that small changes are taking place and consider these indicators of potential long-term and major problems in the future. They believe that contaminants, of which dust is one, are already affecting the vegetation and animals, and predict that the population of the detsı̀ı̀làà?ekwǫ̀ will decline or is already declining.

Allice Legat, Whaèhdǫǫ̀ Nàowo Kǫ̀ (Traditional Knowledge and Heritage Program), Dogrib Treaty 11 Council, Rae, Northwest Territories, Canada

SOURCES/FURTHER READING

Legat, A., and G. Chocolate. 2000. Caribou migration and the state of their habitat. Yellowknife: Final Report for West Kitikmeot Slave Study Society. (www.wkss.nt.ca).

Legat, A., S.A. Zoe, C. Fooball, and M. Chocolate. 1995. Traditional methods of directing [for harvesting] caribou. Rae, Northwest Territories: Whaèhdǫǫ̀ Nàowo Kǫ̀, Dogrib Treaty 11 Council.

Parlee, B. 1998. Community based monitoring: traditional knowledge study on community health, community based monitoring (cycle one). Yellowknife: Final Report for West Kitikmeot Slave Study Society. (www.wkss.nt.ca).

---

some villages averages over a kilogram of meat per person per day. Other villages may use up to 80 species among the various households in the community.

Arctic cultures reflect the significance of such foods, especially in the value they place on sharing with others. Successful hunters may provide for several families, with some households producing thousands of kilograms of food per year, most of which is then distributed throughout the village and beyond. In Greenland, this pattern now includes a cash com-

An Evenk hunter skinning a fox, Sakha Republik (Yakutia), Russia.

ponent as professional hunters sell some of their catch at small, local markets. The net result is the same: people who for one reason or another cannot get their own meat still have access to local foods, provided by those who are particularly skillful at obtaining it.

Reindeer herding societies show a similar blend of the traditional and the modern. Herding in Eurasia today is controlled by a number of government regulations concerning access to land and the right to herd. In Russia, economic changes have reduced the viability of meat production in remote areas. Support systems in the form of living quarters near grazing ranges, schools for the children of herders, and transportation of reindeer meat to markets have disappeared. Herders have had to fall back on their own resources or watch the herds decline. In Fennoscandia, some 600,000 rein-

A wooden fox trap used by Evenks, Sakha Republik (Yakutia), Russia.

70 HUMANS

deer are owned by Saami, although traditional herd management practices have been displaced by government systems. Herding in some regions has suffered from contamination by radionuclides from Chernobyl and then from overpopulation and overgrazing. Nonetheless, reindeer herding remains an important component of the identity of many indigenous groups.

The ability of reindeer herding to reliably provide food led to the introduction of herding in Alaska and Canada in the early 20th Century and in Greenland in 1952 as a remedy for food shortages created by declines in caribou. Saami herders were brought from Fennoscandia to teach the techniques of herding and husbandry. Today, only a few herds remain in North America, but they are important for several villages. Herd management is complicated by the presence of

Dried arctic char **Salvelinus alpinus**, *Canada.*

Ice cellar in the permafrost, Sakha Republik (Yakutia), Russia.

BOX 22. WHY PINE TREES HAVE DIFFERENT NAMES: A KHANTY FOLK TALE

▶ At one time all the trees were people. Between the steppe and the tundra a powerful Pine-tree tribe was travelling. Its chief had twelve daughters and one son. As a result of long reflection, this father-chief came to the decision to marry off seven daughters all at once. He announced this to the cranes and to other different birds, and said that they should spread this news to other lands to the south.

Suitors came from almost all the parts of the world, not in the same fashion as the northern people, fishermen and hunters, but along with their retinues and servants. And as usual, among the retinues there was gossip and slander: this daughter was more beautiful, and that one was brighter, and so on. At the wedding feast they were drinking wine, *kumys, araka, pangk* [1] – and also having arguments: whose drink was better? They understood each other, as brother Cedar said, "only every fifth word."

Brother Cedar and sister Pinia.

Though their languages were similar, still they were different. Long ago, they had gone off in different directions like branches from the same tree.

So everybody began to talk in his own way, and the longer they talked, the louder they became.

And that's how they came to quarrel.

And, though the chief had promised to give his daughters half a kingdom as a dowry, all his sons-in-law went back with their wives to their own lands. And four more daughters were persuaded to run away or were kidnapped by the members of the retinue, and they ran away with them, but nobody knew where. Only the youngest sister stayed in the North with the brother Cedar.

For a long time the sisters lived apart, and they ceased to understand each other completely. And then their husbands started to become hostile to the Northerners. And when the people of the North united and elected young Cedar in place of the aged chief, they turned against each other. Impostors appeared.[2]

"I'm Cedar!" declared the husband of the older sister in far-away Lebanon.

"And I'm Cedar! The Tsar," declared another one in the Himalayas.

Numi-Torum (Supreme God) looked once, looked twice at this mess, and turned the first people on the whole Earth into the trees.

In their childhood the sisters had spoiled their only brother, and when he became a tree, his branches were decorated with the big cones with oily, tasty nuts. And people began to call descendents of the brother "cedars," and the sisters' offspring, "pine trees". In order not to confuse the new people – the descendents of Adam and Eve – they gave to the southern females different names: Pitsundskaya, Veimutova, Pinia[3]. The sister who had stayed in her harsh native land with her brother Cedar was dryly named the Common Pine tree, although she herself is the most uncommon one, a faithful companion to Cedar, who is generous and kind to people, birds, and animals.

Recorded in 1963 by E.Osokin[4] from V.P.Mytnichenko in the village of Negodka, Kargasokskii Raion, Tomsk Oblast. English translation by Olga Balalaeva.

[1] Various intoxicating beverages made out of fermented mare's milk, a distillation from sour milk, and the fly agaric mushroom. All are not customary among Ob'-Ugrians; the geographic range of the drinks, as the question at the heart of the suitors' discussion implies, is probably meant to suggest the geographic distribution and different nationalities of the suitors.

[2] Scientific name of the Cedar – Siberian stone pine (*Pinus sibirica*). "True" cedars – Atlas-type, Lebanese, Himalayan – are not relatives of our cedar, just the namesakes.

[3] Pitsundskaya, Veimutova, Pinia. The narrator attempts to make distinctions: Pitsundskaya, refers to a place in Crimea, so this is probably a popular name for a local variety of pine; Veimutova is unknown, perhaps another varietal name; Pinia, of course, is the Latin family name.

[4] Severnaya Kniga: Krayevedcheskii i literaturno-publitsisticheskii sbornik. Assotsiatsiya korennykh malochislennykh narodov Severa. 1993, - 300 s.,ill. (*Northern Book: A Collection of Local History, Literature and Journalism.* Moscow: Association of the Indigenous Peoples of the North, 1993. 300 pp. Illus.)

wild caribou herds. Individual reindeer or entire herds may be absorbed into the caribou herds, to the detriment of the herders. In Canada, there are also conflicts with muskoxen, which are perceived to disturb both caribou and reindeer. Reindeer were also introduced in Iceland, where a feral herd remains in the southeastern central highlands.

Commercial hunting, trapping, and fishing continue in some areas. Norway conducts commercial sealing and whaling for domestic markets, and seal products may be exported. In Greenland, whale and seal meat is sold primarily within the country, though some seal skins are exported. In Canada, local meat plants process caribou and muskox meat for both domestic sale and export. In several countries, guided sport hunts attract a number of clients to the Arctic in pursuit of game from ducks to caribou to polar bears. These hunts are significant contributions to local economies. The harvest of some 700 caribou from the Bathurst Caribou Herd in Canada's Northwest Territories, for example, annually generates about 3 million Canadian dollars. Some countries impose conditions on such hunting. In North Greenland, hunting must be done using dog teams or small boats. Sport hunters after polar bears in Canada must use local guides and dog teams rather than snowmachines, and community quotas limit the total harvest.

In the early 20th Century, trapping of arctic foxes was a source of trade and income in northern North America. In more recent decades, fur markets have gone up and down, faced with opposition from some animal rights activists and with competition from fur farms. In Russia, fur farming along the Arctic coast was a large enterprise meeting a high demand for furs in central Russia. Brigades of marine mammal hunters provided food for the foxes, and in the mid-20th Century many communities were organized around the collectives that ran the farms. In Fennoscandia, trapping of willow and rock ptarmigan was an important source of income as late as the 1950s, when most rural families engaged in this activity in the winter. A few trappers continue the practice. Today, trapping and fur farming across the Arctic are smaller in scale but still important in many areas.

Fishing is currently the largest and most profitable use of Arctic living resources. It employs many people directly and indirectly and provides the main source of income for many remote communities. Northern waters, including the Bering and Barents Seas and the waters around Greenland and Iceland, produce a significant portion of the world's marine catch. Fish and shellfish taken in the Bering Sea alone account for 2-5% of global fisheries, and 56% of U.S. fish production. In most areas today, there are conflicts over the allocation of fishing catches between small, coastal fishers and large, ocean-going vessels, in addition to disputes over the way fisheries should be managed. A number of approaches have been tried in various countries and areas, from quotas set aside for community development to the allocation of fishing catches by quotas given to individual fishers. Some people advocate the use of traditional practices. Saami fishers in northern Norway, for example, seek to establish a local management system for fjord and coastal fishing, built on traditional patterns of resource use.

Other conflicts over fisheries include the allocation of catches and fishing areas. On the Kola Peninsula, sport fishing for Atlantic salmon attracts fishers willing to pay high prices to fish in "virgin" Arctic rivers such as the Ponoi. The Saami who have fished these rivers for millennia, however, are now excluded from their traditional fishing sites and are greatly limited in their catches because local regulations favoring the sport fishing operations. Balancing the rights and interests of indigenous and other local residents against the attraction of income from outsiders is proving difficult.

Cultural Significance of the Environment

■ In Arctic cultures, the connection between humans and the environment is deep and multi-faceted. The physical tie is powerful and important, but no more

BOX 23. THE USE OF WILD PLANTS IN NORTHERN RUSSIA

▶ In the Russian North are many communities that traditionally consume local biological resources, including both indigenous groups and ethnic Russians whose ancestors moved north centuries ago. Many of these traditional practices changed during the 20th Century as boarding schools introduced new foods and customary patterns of hunting, fishing, and gathering were replaced with wage-earning occupations.

Many plants are used, including edible berries and roots, a variety of medicinal herbs, grasses for making baskets and articles of clothing, and the mushroom *Amanita muscaria*, which has hallucinogenic properties and was used in rituals and as a unit of exchange. Much of the knowledge concerning the identification, gathering, and preservation of these plants has been lost, although it is clear that there were distinct patterns in the extent to which various indigenous groups used plants traditionally.

The consumption of berries was always more important in the settled forest areas than on the tundra. Tundra nomads such as the Nenets, Dolgan, and Yakut, as well as taiga nomads such as the Evenk and Even, were typically busy with other activities and could not gather and store berries, though they ate them raw when the berries were ripe. More sedentary dwellers of the taiga spent the time needed to gather and store berries, and used a greater variety of plants. The gathering of berries and grasses was done traditionally by women and children. The only people who never used berries were the Nganasan, the northernmost tundra dwellers in the Taimyr, who regarded berries as bird food.

Berries were used in many ways. They were eaten raw, dried, and brewed to make beverages. They were added

Willow **Salix** *spp. bark is used to tan reindeer skins and dye wool.*

to fish products, and some were even dried and powdered to make a kind of flour. Among the berries that were used were cowberry, bilberry, bog bilberry, cloudberry, cranberry and many others.

In addition to berries, many other plants were used. Coastal marine mammal hunters used sea plants, such as the sea kale. Reindeer herders in northeastern Russia ate the partially digested lichen from the stomach of the animals they had killed, often making a soup of this lichen with the blood, fat, and intestine of the reindeer. The women of Chukotka and Kamchatka used to take edible roots from the winter caches made by small mammals, replacing what they took with dried meat and fish and covering it carefully with sod. Pine nuts, the inner bark of the willow, pine, and larch, and several other plant products were used by various groups across the Russian North. Tea was made from a variety of plants, including a birch fungus, wild rose, and fireweed.

The degree of plant use is also associated with the length of time a particular group had occupied a region. Relative newcomers like the Yakut and Dolgan, who arrived in the North only two or three hundred years ago, have a far less complex understanding of local flora than, for example, the Chukchi, who have lived on the tundra for over a millennium.

Today, plant use in general has declined, as has the knowledge required to find, identify, and gather such plants. Some species, however, continue to be used extensively, and commercial markets generate income for gatherers of cranberries, cow berries, and pine nuts.

Arkady Tishkov, Institute of Geography, Russian Academy of Sciences, Moscow, Russia
Nikolai Pluzhnikov, Institute of Ethnography, Russian Academy of Sciences, Moscow, Russia.

Cloudberries **Rubus chamaemorus** *are an important dietary supplement for many Arctic residents. They also provide cash income in Fennoscandia and northwestern Russia.*

so than the spiritual link between hunter and animal, person and landscape. These intangible aspects of the place of humans in the Arctic are a vital element of the way the Arctic system affects those who live as part of it.

Consuming traditional foods is a central part of life for many Arctic residents. Such foods are not only healthy, they also impart a sense of overall well-being to many indigenous people, a feeling that cannot be obtained from imported foods. Access to traditional foods is important, even – or perhaps especially – to people who have moved away from the places where they grew up. Returning to take part in hunting or gathering, or to celebrate a festival or holiday, keeps many people close to the land of their birth.

The connection also manifests itself through a detailed understanding of the local environment, the species that live there, and the way they function. Such knowledge is essential for survival in unforgiving conditions, but it is also a way of getting ever closer to the land and sea, to the plants and animals. In recent years, the study of indigenous knowledge has developed out of an appreciation for the depth of understanding that exists and a recognition of the importance of that knowledge to the people who hold it. The information contained in such knowledge is gradually being recognized as a useful indicator of Arctic environmental change.

Changes in the modern era have eroded many facets of Arctic cultures. In some communities, traditional ways have been displaced by the modern wage economy. In others, the basic underpinnings of Arctic ways of life are still intact or are being recovered. Among Saami, for example, their language and the traditional organization of reindeer herders are among the cultural features that have survived. In Canada, there are recent government subsidy arrangements such as hunter-support programs that encourage Inuit to stay on the land. While the end result of these efforts cannot be known, Arctic residents are pressing for a greater ability to determine for themselves how and how fast their cultures will change.

SOURCES/FURTHER READING

Beach, H. 1997. Negotiating nature in Swedish Lapland: ecology and economics of Saami reindeer management. In: E.A. Smith, and J. McCarter, eds. *Contested Arctic: indigenous peoples, industrial states, and the circumpolar environment.* Seattle: University of Washington Press. p. 122-149.

Freeman, M.M.R., ed. 1976. *Report of the Inuit land use and occupancy project.* 3 vols. Ottawa: Indian and Northern Affairs Canada.

Haetta, O.M. 1994. *Samene: historie, kultur, samfunn* (in Norwegian). Oslo: Gröndahl & Dreyers Forlag.

Hall, C.M., and M.E. Johnson. 1995. *Polar tourism.* Chichester, U.K.: John Wiley.

Helander, E. 1999. Sami subsistence activities: spatial aspects and structuration. *Acta Borealia* 2.

Helander, E. 2000a. Samiska rättsuppfattningar i Tana (in Norwegian). *Norske Offentlige Utredninger.*

Helander, E. 2000b. Rätt i Deanodat (in Norwegian). *Statens Offentlige Utredninger.*

Huntington, H.P., J.H. Mosli, and V.B. Shustov. 1998. Peoples of the Arctic. In: AMAP. *AMAP Assessment Report: Arctic pollution issues.* Oslo: Arctic Monitoring and Assessment Program. pp. 141-182.

Jernsletten, N. 1997. Sami traditional terminology: professional terms concerning salmon, reindeer and snow. In: H. Gaski, ed. *Sami culture in a new era.* Karasjok, Norway: Davvi Girji. p. 86-108.

Krupnik, I.I. 1993. *Arctic adaptations.* Hanover, NH: University Press of New England.

Lundmark, L. 1989. The rise of reindeer pastoralism. In: N. Broadbent, ed. *Readings in Saami history, culture, and language.* Miscellaneous Publications 7. Umeå, Sweden: Umeå University. pp. 31-44.

McGee, R. 1996. Ancient people of the Arctic. UBC Press, University of British Columbia, Vancouver, B.C. 244p.

McDonald, M., L. Arragutainaq, and Z. Novalinga. 1997. *Voices from the Bay: traditional ecological knowledge of Inuit and Cree in the Hudson Bay bioregion.* Ottawa, Ontario: Canadian Arctic Resources Committee and Environmental Committee of Municipality of Sanikiluaq. 98 p.

Paine, R. 1994. *Herds of the tundra: a portrait of Saami pastoralism.* Washington: Smithsonian Institution.

Peters, E. J. 1999. Native people and the environmental regime in the James Bay and Northern Quebec Agreement. *Arctic* 52:395-410.

Slezkine, Y. 1994. *Arctic mirrors: Russia and the small peoples of the North.* Ithaca, New York: Cornell University Press. xiv + 456p.

Solbakk, A., ed. 1990. *The Sami People.* Karasjok, Norway: Sami Instituhtta & Davvi Girji.

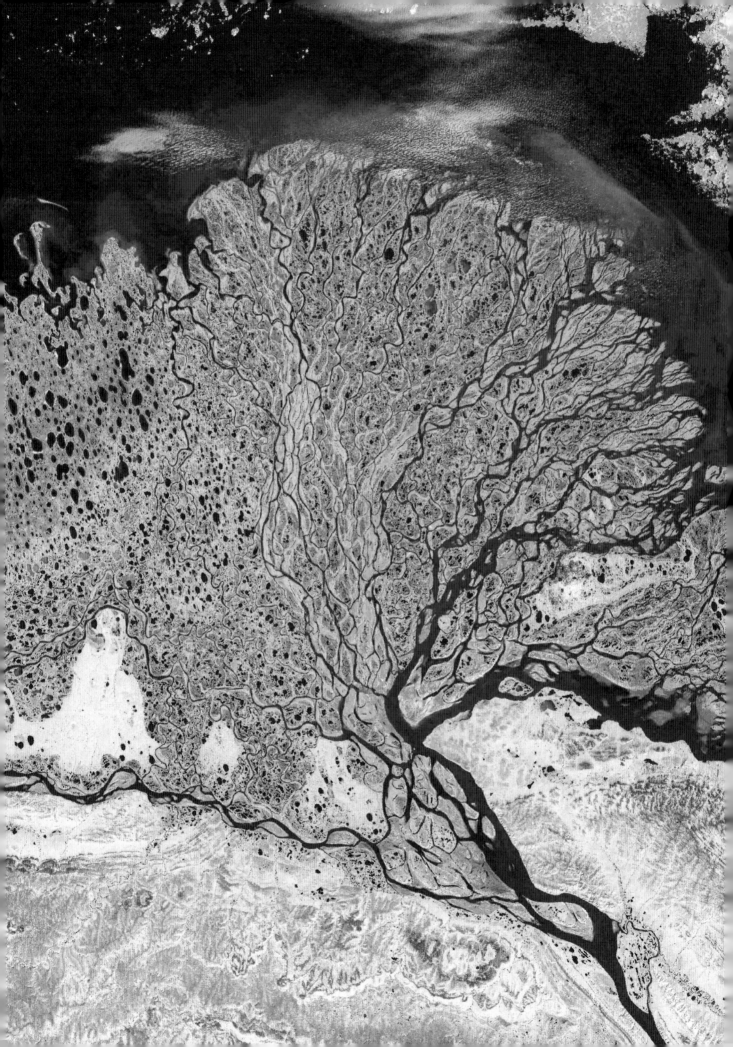

4. Conservation

Conservation relies on a variety of approaches designed to achieve specific and general goals. This chapter reviews some of the conservation issues, needs, and approaches in the Arctic.

Habitat

■ The creation of protected areas is one of the most common conservation approaches worldwide. Such areas include national parks, wildlife refuges, special habitat areas, sites of scientific interest, and sacred sites. The rationale behind protected areas is that they restrict human influence, allowing natural processes or areas of great beauty to remain undisturbed. Some areas protect specific things, such as nesting grounds. Others are intended to encompass a full range of ecological functions.

In the Arctic, there are currently 405 protected areas (including Ramsar sites), giving formal protection to approximately 2.5 million square kilometers, or 17% of the land area of the Arctic as defined by the CAFF Program. Roughly half of the total area, however, is the Northeast Greenland National Park, and the remainder is distributed unevenly across countries and biogeographic zones. Some 27% of unproductive zones such as glaciated areas and polar deserts is protected, compared with only 5.4% of the productive northern boreal forest. The percentage of land protected in the Arctic region of each country ranges from 9.5% in Canada to 50% in the United States. Within the CAFF Program, the Circumpolar Protected Areas Network (CPAN) has been established to support and promote protected areas.

Recently designated protected areas often include provisions for traditional hunting, fishing, trapping, and gathering. This trend recognizes the ways in which humans can, at times, function as part of the local ecosystem. A more flexible approach to the concept of protected areas has meant a greater willingness to accept them in many regions. In northern Canada national parks have been established in northern Yukon, and on Ellesmere Island and Baffin Island through land claim agreements between the federal government and Inuit organizations. Protected areas have also been sought for the benefit of local interests, such as the preservation of sacred sites, an issue of particular interest in the Russian Arctic.

The role of smaller areas

Some protected areas are large, spanning entire watersheds or ecosystems in order to protect whatever exists in those areas. Now, few large areas remain for which there is the political will to protect. Instead, the priority has largely shifted to protecting smaller but significant areas, including sensitive or rare habitats and areas threatened by human activity. Another area of emphasis has been to link protected areas to reduce the risk of isolating small areas.

The Lena Delta: a composite satellite image (Compiled by UNEP GRID-Arendal).

BOX 24. THE CIRCUMPOLAR PROTECTED AREAS NETWORK (CPAN)

▶ In 1996, CAFF completed a *Strategy and Action Plan* for a Circumpolar Protected Areas Network (CPAN). The primary functions of the network are to maintain in perpetuity the dynamic biodiversity of the Arctic region through habitat conservation in the form of protected areas, to represent as fully as possible the wide variety of Arctic ecosystems, and to improve physical, informational, and managerial ties among circumpolar protected areas. The *CPAN Strategy and Action Plan* specifies actions on both national and circumpolar levels, to fill gaps in habitat protection and to advance overall functioning of the network.

The rationale behind networking protected areas on a circumpolar basis includes:

- Many Arctic fauna species are migratory and no one country can ensure habitat protection for critical stages in the entire life cycle of all these species.
- Certain key areas ("hot spots") that are critical to maintaining the biodiversity and productivity of the entire Arctic ecosystem may be under multiple jurisdictions.
- Many Arctic countries have indigenous communities and local rural populations that depend completely or to a large extent upon the consumption of Arctic flora and fauna, and therefore upon maintaining the integrity of the local ecosystem.
- The countries have recognized and embraced the need to protect, as fully as possible, the wide variety of Arctic ecosystems across their natural range of variation, and to maintain viable populations of all Arctic species in natural patterns of abundance and distribution.
- In a global environmental context, the Arctic and much of its flora and fauna are still relatively intact, a natural heritage of increasing global significance as natural ecosystems and relatively undisturbed areas become ever more rare.

Since publication of the *CPAN Strategy and Action Plan*, many new protected areas have been established in the Arctic and others expanded or altered. Counting only protected areas larger than 10 square kilometers and corresponding to IUCN Management Categories I-V and including Ramsar sites, the number has increased to 405 over the last six years. Progress has been most striking in the Russian and Canadian Arctic.

Still to be addressed, however, is the imbalance in protection across ecosystems and habitats, ranging from over 27% for the Arctic desert and glaciers to 5.4% and 1.7% for the northernmost areas of the boreal forest and the Arctic marine environment. Enhancing protection of these underrepresented ecosystems will be a major challenge for CPAN and CAFF in the years to come.

Table 4. Protected areas in the Arctic classified in IUCN categories I-V, plus Ramsar international wetland sites, as of 2000. Areas smaller than 10 square kilometers are not included.

COUNTRY	NUMBER OF AREAS	TOTAL AREA (km²)	% OF ARCTIC LAND AREA OF THE COUNTRY
Canada	61	500,842	9.5
Finland	54	24,530	30.8
Greenland	15	993,070	45.6
Iceland*	24	12,397	12.0
Norway**	39	41,380	25.3
Russia*	110	625,518	9.9
Sweden	47	21,707	22.8
USA (Alaska)	55	296,499	50.2
Total	**405**	**2,515,943**	**17.0**

* Large marine components and several large protected areas in Russia included in this table have been designated on a regional level but not endorsed by federal authorities.
** Most of the area protected is located in Svalbard, only about 7% of the Arctic mainland is protected.

Snorri Baldursson, CAFF International Secretariat, Akureyri, Iceland
Christoph Zöckler, UNEP-World Conservation Monitoring Centre, Cambridge, UK

Please note: The table is not easily comparable with former figures (e.g., CAFF 1994, 1996) because some marine components have been withdrawn, some figures were revised according to new national calculations, and some areas were re-categorized according to the new IUCN system.

SOURCES/FURTHER READING

CAFF (Conservation of Arctic Flora and Fauna). 1994. *Habitat Conservation Report No. 1: The State of Protected Areas in the Arctic.* Trondheim: Directorate for Nature Management. 163p.

CAFF (Conservation of Arctic Flora and Fauna). 1996. *Habitat Conservation Report No. 2: Proposed Protected Areas in the Circumpolar Arctic.* Trondheim: Directorate for Nature Management. 196p.

CAFF (Conservation of Arctic Flora and Fauna). 1996. *Habitat Conservation Report No. 6: Circumpolar Protected Areas Network (CPAN) Strategy and Action Plan.* Trondheim: Directorate for Nature Management.

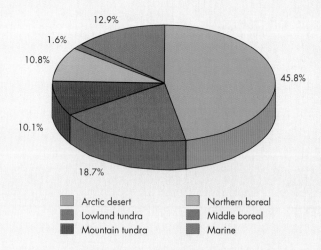

Figure 25. Percentage of the total protected area in the Arctic that is located in each biome.

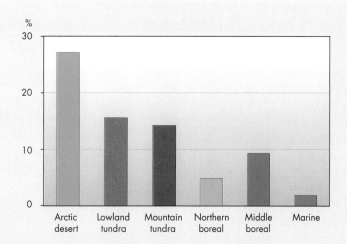

Figure 26. Percentage of the territory of each Arctic biome that is protected.

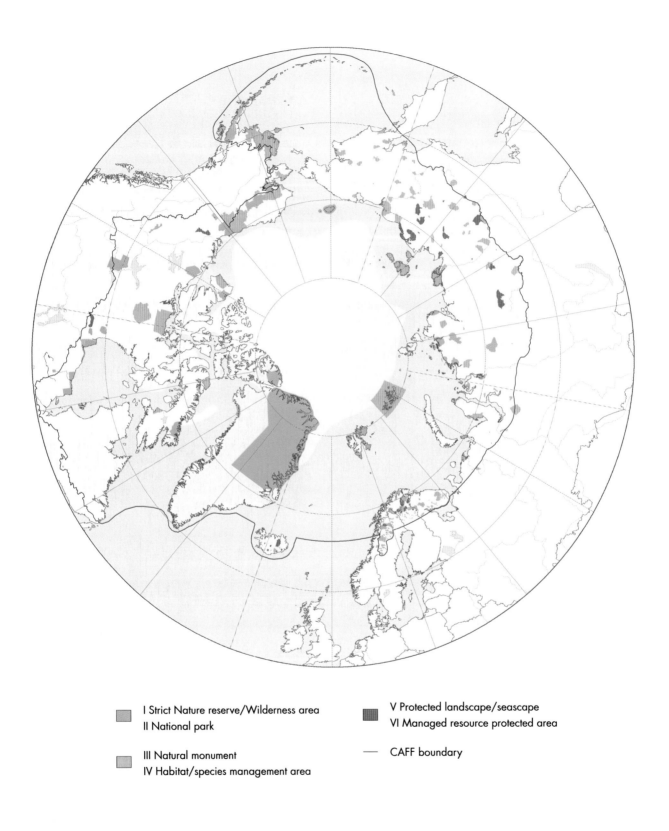

Figure 27. Protected areas (> 500 hectares) in the Arctic by IUCN Categories I-VI (Compiled by UNEP-WCMC).

80 CONSERVATION

Ramsar sites, for example, are designated to protect wetlands under the 1971 Convention on Wetlands of International Importance. Networks have been established along the flyways of migratory birds. Coordinating the networks and making sure that critical areas are protected can help migratory species throughout their range. The international Man and the Biosphere Program establishes biosphere reserves for research, monitoring, and training, as well as for conservation. The world heritage sites named under the 1972 Convention Concerning the Protection of World Cultural and Natural Heritage recognize areas of "outstanding universal value." In 2000, there were 44 Ramsar sites, six biosphere reserves, and three world heritage sites in the Arctic.

Protected areas can lay the foundation for conservation by identifying especially significant areas and by protecting large, unfragmented areas. They provide an ecological benchmark for monitoring human impacts, such as industrial development, and they serve as refugia for wild species that may be overharvested elsewhere. Nonetheless, protected areas cannot by themselves conserve the Arctic environment. Actions outside protected areas and outside the Arctic, particularly regarding migratory species, can undermine the significance of the protected area. Global problems such as pollution and climate change do not stop at the boundaries of a protected area. The displacement or overabundance of wildlife can cause local habitat changes. To be effective in the long run, conservation must use other approaches in addition to habitat protection.

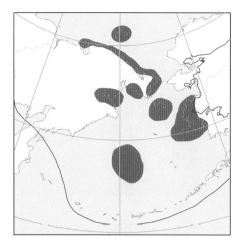
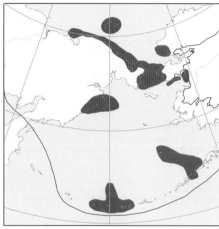
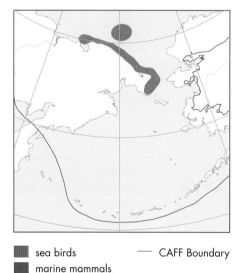

■ sea birds — CAFF Boundary
■ marine mammals
■ polar bear

Figure 28. Areas of high population densities of different groups of marine vertebrates in the Beringia Region. (Source: Banks et al. 1999).

Wildlife

■ Managing consumptive use – fishing, hunting, gathering – is the oldest form of conservation. In hunter-gatherer societies, where over-exploitation would lead to shortage and suffering, some limitations on use were explicit. Traditional rules on St. Lawrence Island, Alaska, and Iceland, for example, made sure

BOX 25. PROTECTED AREAS IN NORTHERN FENNOSCANDIA: AN IMPORTANT CORRIDOR FOR TAIGA SPECIES

▶ Finland is situated on the western margin of the large coniferous forest – the taiga – that extends across the whole of northern Russia. The western parts of the taiga belt are relatively intact on the Russian side of the border, with a diverse tree species composition and a high diversity of animals. On the Finnish side, by contrast, large areas especially in southern and central Finland are commercial forests, which are mostly monocultures of even-aged trees.

The effects of these differences are large. Species that prefer large, intact forest areas, including wolf, brown bear, wolverine, lynx, capercaillie, hazel grouse, and pine marten, are found in much higher densities on the Russian side. Species favored by early successional stages, such as moose and mountain hare, are more abundant on the Finnish side. The Russian forests also host a higher number of rare, wood-inhabiting invertebrates than do the forests in Finland. Birds such as the Siberian jay, the Siberian tit, and the three-toed woodpecker are specialized for old-growth coniferous forests. They have declined markedly in Finland in recent decades due to the impacts of large-scale forestry, and find refuge only in the remaining old-growth taiga forests.

In addition to commercial forests, Finland has a large, relatively continuous belt of protected areas that connects with protected areas in the mountains of Norway and Sweden. These areas are important habitat for taiga species in Fennoscandia. They serve as a corridor for gene exchange, and allow recolonization of habitats in the case of local extinctions. The wolf population of northern Finland, for example, depends largely on animals moving in from Russia. Lower plants and animals also exchange genes over large areas, and can suffer from genetic isolation. For instance, a study of the genetic structure of *Fomitopsis rosea*, a rare polyporous wood-inhabiting fungus, shows that isolated populations in southern Sweden have a narrower genetic structure than populations within continuous taiga forests of Russia.

The connection of Fennoscandian taiga to the larger taiga of Russia, and the protection of large intact forests in western Russia, are vital for the long-term survival of Fennoscandian taiga species. A joint Finnish-Russian effort strives to maintain this connection by establishing a network of protected areas that extends across the border.

Anna-Liisa Sippola, Arctic Centre, University of Lapland, Rovaniemi, Finland

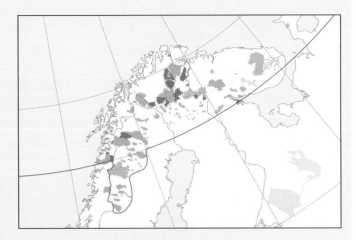

- I Strict Nature reserve/Wilderness area
- II National park
- III Natural monument
- IV Habitat/species management area
- V Protected landscape/seascape
- VI Managed resource protected area
- — CAFF boundary

Figure 29. Protected areas (> 500 hectares) by IUCN Categories I-IV in Fennoscandia.

SOURCES/FURTHER READING

Högberg, N. 1998. Population biology of common and rare wood-decay fungi. *Acta Universitatis Agriculturae Sueciae Silvestria* 53:1-38.

Lindén, H., P. Danilov, A.N. Gromtsev, P. Helle, E.V. Ivanter, and J. Kurhinen. 2000. Large-scale corridors to connect the taiga fauna to Fennoscandia. *Wildlife Biology* 6(3):179-188.

Punkari, M. 1985. Maankäytön ja luonnon tila. In: Punkari, M., ed. *Suomi avaruudesta*. Helsinki: Tähtitieteellinen yhdistys Ursa. p. 74-87. (In Finnish).

Rassi, P., H. Kaipiainen, I. Mannerkoski, and G. Ståhls. 1992. Report on the monitoring of threatened animals and plants in Finland. Ministry of Environment. *Komiteanmietintö* 199:30. 323p. (In Finnish with English summary).

Siitonen, J., and P. Martikainen. 1994. Occurrence of rare and threatened insects living on decaying *Populus tremula*: a comparison between Finnish and Russian Karelia. *Scandinavian Journal of Forest Research* 9:185-191.

Väisänen, R.A., E. Lammi, and P. Koskimies. 1998. Muuttuva pesimälinnusto. *Keuruu*. 567p. (In Finnish).

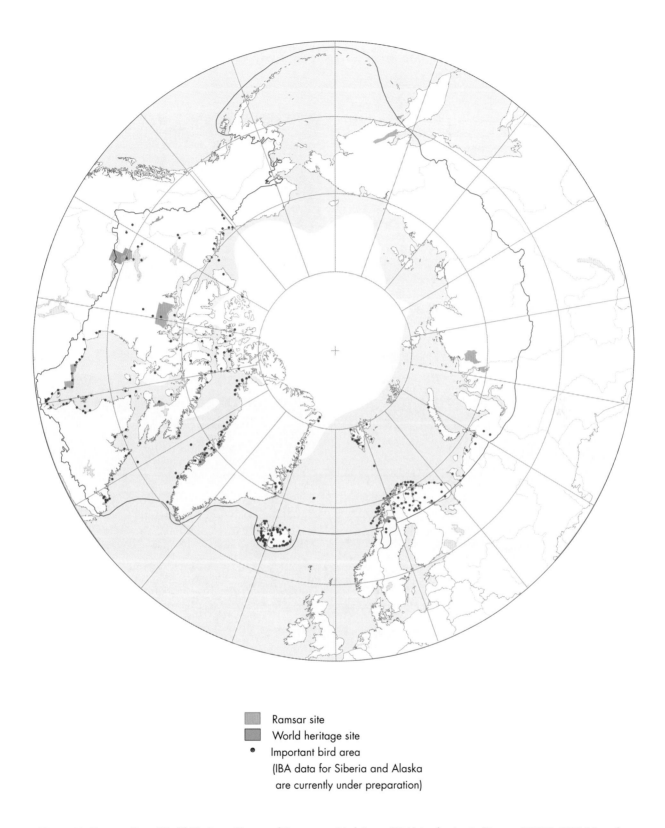

Figure 30. Ramsar Sites, World Heritage Sites, and Important Bird Areas (IBA) in the Arctic (Source: UNEP-WCMC and BirdLife International). Not shown are 11 Ramsar sites in Greenland, which have been recently approved but not formally implemented.

BOX 26. PROTECTING THE LENA DELTA – AN EXAMPLE OF THE RAPID GROWTH OF PROTECTED AREAS IN RUSSIA

▶ Since CAFF was established in 1991, Russia has added more protected areas than any other Arctic country. The total area included in the strictest category of nature reserve, the *zapovenik*, has doubled in ten years to 183,940 square kilometers. A similar area of tundra and coastal ecosystems came under other categories of protection, so that the total area protected is now larger than Norway. The Taimyr Autonomous Region started the main part of this effort. In 1993, it created the Great Arctic Reserve, one of the areas that has given this region Russia's densest network of *zapovedniks*. The neighboring Sakha Republic has made another great commitment to habitat protection. In 1995, it decided to designate at least 20% of its territory as nature reserves, totaling some 700,000 square kilometers. One beneficiary of this effort has been the Lena Delta.

Typical wet tundra of the Lena Delta. The bird density in this kind of habitat can be over 500 individuals per hectare.

A flock of black brent **Branta bernicla nigricans**.

Corydalis arctica and *Salix berberifolia* are Yakutian endemic plants, found in the Lena Delta Reserve.

The Lena River is the world's seventh largest, and its delta covers approximately 30,000 square kilometers. Interspersed with some 6,500 channels and 30,000 lakes, the land of the delta is largely wet tundra and is flooded each spring. Here, between 72° and 75°N, is some of the most productive territory in the far North. 368 vascular plant species have been identified in the delta, including 24 that are listed in the Red Data Book of Yakutia. Of the 96 species of rare vascular endemic plant in the Arctic, five are found in the delta, one of which – the poppy *Papaver leucotrichum* – is found nowhere else. Eight species of whitefish are found in the waters of the delta, leading scientists to believe that it is their evolutionary center. 122 bird species have been sighted in the delta, of which 67 have been found breeding at least once. The density of birds is unusually high for species that do not breed in colonies. Among mammals, the rare and endangered black-capped marmot lives in the mountains near the delta. This animal survives the region's long, harsh winters by hibernating nearly nine months of the year.

Recognizing the significance of the area, the Soviet Union established the Lena Delta Nature Reserve in 1986. In 1995, in cooperation with WWF, the Sakha Republic established the Lena-Nordenskjöld International Biological Station for research in the delta. In 1996, the Republic expanded the nature reserve by adding the New Siberian Islands and additional areas within the delta and in the adjacent Sokol area of the Kharaulakh mountains. This increased the size of the protected area to over 61,000 square kilometers, making it the largest in Russia and one of the largest in the Arctic.

The Lena Delta Nature Reserve is significant not only for its flora and fauna. An international workshop convened at the Lena-Nordenskjöld Station in July 2000 encouraged the development of the station and its research program, and recommended that the Lena Delta be nominated as a World Heritage Site. For research, the delta is an excellent site for monitoring changes in the presence and diversity of species, which may signal regional and global environmental changes. It can also be linked with other research stations and areas across the Arctic.

Altogether, the Russian Federal government, as well as regional governments should be congratulated for what they have achieved, in particular during the last decade. Increased awareness of these globally important steps and "gifts to the earth" should inspire future national and circumpolar decisions.

Peter Prokosch, WWF International Arctic Programme, Oslo, Norway

SOURCES/FURTHER READING

Gilg, O., R. Sané, D.V. Solovieva, V.I. Pozdnyakov, B. Sabard, D. Tsanos, C. Zöckler, E.G. Lappo, E.E. Syroechkovski Jr., and G. Eichhorn. 2000. Birds and mammals of the Lena Delta Nature Reserve, Siberia. *Arctic* 53(2):118-133.

Müller, H.H, P. Prokosch, and E. Syroechlovsky. 1993. *Nature Reserves on Taimyr*. Oslo: WWF Arctic Programme. 38p.

Prokosch, P. 1997. Nature reserve development in Russia. *Arctic Bulletin* 1.97:12-14.

Prokosch, P., and N.G. Solomonov.1998. *The Lena-Delta and New Siberian Island Nature Reserve*. Oslo: WWF Arctic Programme. 38p.

Talbot, S.S., B.A. Yurtsev, D.F. Murray, G.W. Argus, C. Bay, and A. Elvebakk. 1999. *Atlas of rare endemic vascular plants of the Arctic*. Conservation of Arctic Flora and Fauna (CAFF) Technical Report No. 3. Anchorage: U.S. Fish and Wildlife Service. iv + 73p.

BOX 27. THREATS TO ARCTIC BIODIVERSITY – OVERVIEW

▶ As illustrated in this report, human activities pose many threats to the Arctic environment. Global threats such as climatic change, ozone depletion, and long-range pollution may have widespread and severe effects. Physical and chemical disturbance by oil and gas exploitation, pollution from factories and smelters, and overexploitation of species are examples of local but severe threats. The list below is intended to provide the reader with a quick overview of major categories of threats to Arctic flora and fauna and a few examples to illustrate their effects. These threats are discussed in more detail elsewhere in the report.

THREAT EXAMPLES

Climate change
In a warmer climate, much of today's permafrost will thaw, and soil instability and erosion will increase. Water drainage and plant communities would be affected directly, in turn, causing changes to animal populations.

A northward shift of the treeline would lead to the loss of tundra habitats essential for many plants and animals.

In Canada, reduction in sea ice in Hudson Bay threatens the area's polar bears, which depend on a regular seasonal cycle of ice.

Climate change may disrupt air and water currents, thereby altering migration routes of marine mammals, fishes, and birds.

Ozone depletion
Increased ultraviolet radiation may cause photochemical damage in the DNA of living organisms.

Physical disturbance/Habitat fragmentation
Oil and gas infrastructure has altered tundra habitats and the distributions of large mammals in Alaska and Siberia.

Hydroelectric dams in northern Quebeck, northern Manitoba and northern Scandinavia and Iceland have fragmented habitats, flooded lands, and changed the fish communities of rivers and lakes.

Bottom trawling (e.g., for cod) and dredging (e.g., for scallops) can severely alter the structure and complexity of benthic communities by removing slow-moving or attached organisms, smoothing the sea floor, and increasing suspended sediment loads.

Chemical disturbance
Pollution from mining operations has killed trees and other vegetation around smelters on the Kola Peninsula and in Norilsk, Russia. The acidification of land and waters and the accumulation of heavy metals in food webs extend over wide areas.

PCBs are found in high levels in polar bears in many arctic regions, and are suspected to cause reproductive system abnormalities.

Species introductions
The Nootka lupine was introduced in Iceland, where it grows vigorously, pushing aside native vegetation and reducing the diversity of plant cover.

The opossum shrimp introduced to rivers and lakes in Norway and Sweden competes with native fish for zooplankton, causing declines in those fish populations.

Consumptive use/Overexploitation
Overfishing has led to declines of some fish species, changes in marine food webs and fisheries crises in many Arctic regions. For example, the starvation of many seals and the collapse of seabird colonies in the Barents Sea region in the 1980s have been associated with the overexploitation of capelin.

Overgrazing by sheep in Iceland has been a major contributor to extensive erosion and loss of soil.

Other uses
Tourism brings visitors to sensitive areas like seabird colonies and marine mammal rookeries.

The by-catch of seabirds from fishing operations is a significant cause of mortality in many areas.

---

that egg gathering did not harm bird colonies. Explicit rules were uncommon, however, and harvests were more likely to be limited by changes in the abundance and distribution of the species that were used. Traditional systems more typically addressed the spiritual relationship between humans and the environment, to make sure that animals remained available to hunters.

By the middle of the 20th Century, explicit systems of hunting regulations, including harvest limits and seasonal restrictions, had been introduced in much of the Arctic. These systems often conflicted with traditional hunting practices, such as the harvest of migratory birds in spring and summer. More recently, regulations have begun to reflect the distinct needs and values of indigenous and other users. The increasing political force and autonomy of indigenous peoples, particularly in North America and Greenland, have spurred this process.

As a consequence, wildlife management has become increasingly tailored to the needs of traditional users. Most notably, in northern Canada preference is given on the basis of need, with constitutionally guaranteed rights of harvest for indigenous peoples. In Alaska, despite considerable disagreement over defining subsistence users and allocating harvests, several small changes have made traditional practices legal. Hunting swimming caribou on the Kobuk River is now permitted, recognizing a practice that has existed for millennia. Transferable permits allow active hunters to take animals on behalf of relatives who cannot hunt, as is customary.

One of the most notable recent innovations is the involvement of hunters or fishers in wildlife management. In theory, hunters and fishers who help develop the regulations will better understand the rationale for them and be more willing to abide by them. In practice, this approach has enjoyed success in North America, where support for co-management has grown widely, although difficulties remain.

Economic importance of harvests

Harvesting for local use has economic as well as cultural significance. In Canada's Northwest and Nunavut Territories, replacing local foods with imported ones would cost an estimated 55-60 million Canadian dollars each year. In addition to individual and community uses of fish and animals, several species are harvested commercially in the Arctic, including caribou, muskoxen, furbearers, some marine mammals, and some birds.

Commercial fishing, however, is the most important both economically and biologically, and is the most elaborately managed harvest. Seasons, fish bycatch, size, methods and equipment, areas, and other aspects of fishing are all regulated to some degree, nationally or internationally. But, the existence of regulations is no guarantee of sound conservation. Many stocks and species crashed in the 20th Century. In addition, bycatch of birds and non-target fish along with habitat destruction by fishing gear remain serious problems, despite some recent improvements.

Profits are the main pressure for setting high catch limits, but over-capitalization and subsidies compound the issue. Open access fisheries on the high seas make enforcement of quotas impractical or impossible. Because the fish may be at different stages of their life cycle when they are available to different types of fishers, allocation disputes can complicate the setting of effective regulations. Given that many species experience natural and often unpredictable population cycles, fisheries regulations must be able to adapt and respond to new information. The recovery of herring populations in the northeast Atlantic, however, is evidence that fisheries management can be effective.

Not all wildlife management concerns harvests. Management is also directed toward the recovery of threatened and endangered species such as Peary caribou or peregrine falcon. Conversely, populations can also become too large. The creation of wildlife refuges, combined with changes in land use of wintering areas outside the Arctic, have triggered an explosion in most North American snow goose *Chen caerulescens* populations, leading to severe overgrazing impacts in their Arctic breeding areas.

Industrial and Commercial Development

■ Hydroelectric dams, mining, and oil and gas development and their associated roads and deep sea ports can cause intense local disturbances. They can also cause fragmentation of large intact areas and habitats into smaller units, isolating plant and animal pop-

BOX 28. UNSUSTAINABLE TAKE OF MURRES IN GREENLAND

▶ For as long as people have lived in the Arctic, they have depended on fish, mammals, and birds to provide food, clothing, and other materials. Greenland, for example, is rich in seabirds. In the past, bird harvests were kept sustainable in the long-term by the limitations imposed by hunting equipment and the short range of daily transport. Today, this balance no longer exists.

The highly productive seas along the west coast of Greenland and the polynyas of Thule and Scoresbysund support breeding colonies of millions of little auks, tens of thousands of thick-billed murres (also called Brünnich's guillemots), fulmars, and kittiwakes, and many other species. Even larger numbers of seabirds from other parts of the North Atlantic come to the open water areas of Southwest Greenland for the winter. As long as people have been living in Greenland, these rich living resources have been exploited by hunting and egg gathering. The sophisticated hunting technology of the Thule culture included special bird arrows, and seabirds even today constitute an important part of the Greenlandic menu. The take of thick-billed murres alone, for example, is 300,000-400,000 individuals per year.

Changes in hunting

Birds that breed in large concentrations in colonies on small islands and cliffs, as most seabirds do, will always be highly vulnerable to hunting and egg gathering. The short Arctic summer makes it problematic for birds to lay a second clutch of eggs if the first is taken. Seabirds are generally long-lived birds with a low reproductive rate, and thus their populations are highly sensitive to the take of breeding adults. This is well-illustrated by a theoretical example, where all young of a pair of murres and a pair of ptarmigans survive and reproduce for five years. During that time, the ptarmigans multiply to 3,000 individuals, while the murres on average multiply only to five and a half individuals.

Therefore, a human population that lives predominantly from hunting and gathering can easily reduce the populations of such slowly reproducing, colonial species. Hunting and egg gathering activities hundreds of years ago undoubtedly depleted many small seabird colonies near human settlements. At that time, however, means of transport and hunting weapons limited the ability of hunters to reach and take birds, and human impacts were probably confined to areas near settlements. But hunting in Greenland is no longer carried out using kayaks and bird arrows. Bird colonies that previously took many hours or days of hard physical effort to get to may nowadays be reached by speedboat after a day at work. The situation for the Greenlandic seabird populations has been turned upside down.

Murres and other seabirds are an important part of the Greenlandic menu. An estimated 300,000-400,000 birds are taken annually.

The take is no longer sustainable

Before the first reliable censuses of seabird colonies in Greenland in the 1960s, estimates of the sizes of many colonies had been made in the first half of the century. Based on these estimates, together with counts during the last 30-40 years, it can be stated that numbers of breeding thick-billed murres in central West Greenland have declined from around half a million to less than 12,000 individuals. A number of smaller colonies have disappeared altogether, as has one that used to have 100,000-150,000 birds. Another colony that had been home to between 25,000 and 50,000 birds has only 1,000 individuals today.

It is striking that the largest declines occur in the colonies that are closest to human settlements, while colonies that are situated in poorly populated areas are either intact or only a little reduced. In northernmost Northwest Greenland and in the Thule district, for example there are still colonies with more than 100,000 murres. Where murre populations have crashed, all evidence point to overhunting near the colonies during spring and summer as the prime cause for the declines.

And it is not only murres that have been overharvested. Large colonies of common eiders and Arctic terns have also disappeared or declined considerably during the same period.

No tradition of limitations

On the Faeroe Islands in the North Atlantic, similar large colonies of common murres exist. There have been problems in the Faeroes, too, but to a much lesser extent than in Greenland. The differences are remarkable. On the Faeroes, people have for centuries practiced strict rules for exploitation of the colonies to avoid overharvest. Netting and snaring schemes have been developed that target immature individuals instead of adult breeders. In the colonies, egg gathering and catching took place in different parts and in a rotation scheme to allow breeding success in different parts of the cliffs.

Such management has never existed in Greenland. The Arctic is so unpredictable that hunters need to take advantage of any chance to hunt birds and mammals to reduce the possibility of shortage when conditions change for the worse. Therefore, it is deeply ingrained in Inuit to take whatever possible – here and now.

In modern Greenland, many rules and regulations have been established to counteract the large declines in certain game populations. In Northwest Greenland, this includes a closed season for murre hunting from 1 June to 31 August. Much informational material has been produced, including posters, folders, brochures, television spots, and videos. But traditions for obeying such rules and regulations are only slowly established. "Hunting regulations only apply as long as you can be watched from town" is a popular interpretation of this phenomenon. An additional obstacle is that in Inuit culture, more so than in western culture, it is inappropriate to criticize the behavior of others or to tell them what to do and what not to do.

In Southwest Greenland the situation has begun to improve. Here, the murres are protected during half the year, a limitation that is respected to a high extent, and the colonies are stable, though relatively small. In Southwest Greenland, people have the advantage of being able to hunt the huge numbers of wintering birds outside of the breeding season. This is not as devastating to the populations because the take includes many immature birds and is distributed over far larger populations. In Northwest Greenland, on the other hand, people rely more on hunting for their food and livelihood. Here, murres are present only in summer, and people find it difficult to accept a ban on hunting during the months that the birds are present. Stopping the decline of murres and of other seabirds will be more difficult.

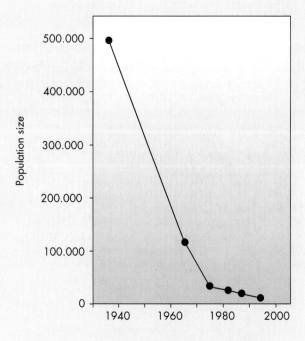

*Figure 31. Decline of the thick-billed murre **Uria lomvia** population in central West Greenland (Source: Knud Falk and Kaj Kamp).*

Hans Meltofte, National Environmental Research Institute, Department of Arctic Environment, Denmark

BOX 29. CAFF INTERNATIONAL MURRE CONSERVATION STRATEGY AND ACTION PLAN

▶ The two species of seabirds, known in North America as murres and in Europe and Asia as guillemots, inhabit coastal and offshore marine regions of all Arctic countries. The common murre frequents boreal and low Arctic areas, and the thick-billed murre inhabits more northerly areas. They are among the most numerous and widespread of Arctic seabirds, playing important roles in Arctic marine food webs and in the lives of people in coastal communities.

A number of human activities directly or indirectly reduce breeding success and survival in murre populations. By virtue of their biology and ecology, murres are particularly vulnerable to these impacts. Every year, hundreds of thousands of murres die in fishing gear, are hunted, or are fouled by oil pollution at sea. Breeding colonies are threatened by disturbance and coastal development in many countries. Because they do not breed until they are 4-5 years old and only lay one egg per year, murre populations are unable to rebound quickly from declines. Furthermore, their colonial and social habits make populations vulnerable to local impacts such as an oil spill. Population models have shown that the current level of mortality in certain murre populations is not sustainable.

As a result of their migratory patterns, most murre populations are shared by two or more Arctic countries. Effective conservation of murre populations thus requires a coordinated international effort. In 1996, CAFF produced an *International Murre Conservation Strategy and Action Plan*, providing for the first time a strong, circumpolar effort to conserve Arctic murre populations, and considerable progress has been made since its adoption. Most CAFF countries are now implementing national murre conservation plans. CAFF's Circumpolar Seabird Working Group (CSWG) has produced a circumpolar murre monitoring plan and is preparing a separate banding plan. In addition, a database of murre colonies in CAFF countries has been compiled and a murre colony map produced.

The real test of the effectiveness of the Strategy is the extent to which accidental and unsustainable mortality is prevented. Canada's restrictions on the large murre hunt in Newfoundland and Labrador, for example, have reduced the total kill by over 400,000 birds per year. Other important benefits include informing governments of international concerns for murres and sharing information and approaches to murre conservation among CAFF members. Finally, the Strategy has created momentum to take cooperative action on areas of common seabird concern, e.g., to reduce human disturbance at Arctic seabird colonies and to reduce impacts of incidental take (bycatch) of seabirds in fisheries operations.

Common murre **Uria aalge**, *Newfoundland, Canada.*

The process of producing and implementing the Murre Strategy and Action Plan has been used as a model for developing other species-level conservation plans under CAFF (e.g., the conservation strategy for the four species of eiders that occur in the Arctic.)

John W. Chardine, Canadian Wildlife Service and CAFF Circumpolar Seabird Working Group, Sackville, New Brunswick, Canada

SOURCES/FURTHER READING

CAFF (Conservation of Arctic Flora and Fauna). 1996. The International Murre Conservation Strategy and Action Plan. Ottawa: CAFF. 16p.

CAFF (Conservation of Arctic Flora and Fauna). 1998. *Technical Report No. 1: Incidental take of Seabirds in Commercial Fisheries in the Arctic Countries* (Vidar Bakken and Knud Falk, eds.) Akureyri: CAFF. 50p.

CAFF (Conservation of Arctic Flora and Fauna). 1998. *Technical Report No. 2: Human Disturbance at Arctic Seabird Colonies* (John Chardine and Vivian Mendenhall, eds.) Akureyri: CAFF. 18p.

CAFF (Conservation of Arctic Flora and Fauna). 2000 *Technical Report No. 7: Workshop on Seabird Incidental Catch in the Waters of Arctic Countries, 26-28 April 2000. Report and Recommendations* (John W. Chardine, Julie M. Porter, and Kenton D. Wohl, eds.) Akureyri: CAFF. 73p.

*Thick-billed murre **Uria lomvia**, Newfoundland, Canada.*

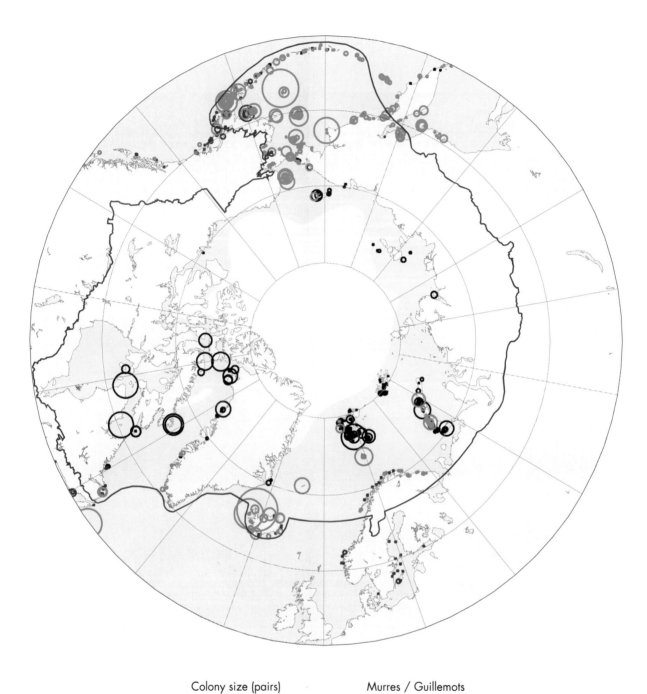

Figure 32. Murre colonies in the Arctic. (Source: CAFF-CSWG, Compiled: UNEP-WCMC)

BOX 30. SUCCESS WITH GEESE: PROOF THAT CONSERVATION IS WORKING?

▶ If the state of the Arctic environment can be measured in numbers of geese – an important group of its characteristic fauna – there are good reasons both for optimism and for targeted approaches to solve the remaining problems. Twelve of the world's 15 goose species breed in the Arctic; most show increasing numbers and have populations at the highest levels ever counted. On the species level, only the lesser white-fronted goose is sharply declining and globally threatened with extinction. From a circumpolar perspective, the sector of the Arctic deserving global concern is Eastern Siberia. There, in contrast to the European, Western Siberian, and North American regions, goose populations are generally small, declining, threatened, or of unclear status.

Of the more than 11 million geese world-wide (estimated in the mid-1990s), about 7.8 million (70%) breed in the Arctic region. The 12 Arctic goose species can be subdivided into 24 subspecies, and further into some 43 geographically distinct populations. The largest populations are found in North America. The huge snow goose populations in Northern Canada are the most numerous in the world, estimated at 2.6 million in the mid-1990s and now over 6 million. There is increasing concern that overgrazing of their summer range is causing habitat degradation. Intermediate-sized populations are found in Europe and Western Siberia, and the smallest are in Eastern Siberia. Of the 43 goose populations in the Arctic, 17 are increasing (40%), 12 are stable (28%), 8 are decreasing (18%), and for 6 the trend is unknown (14%) (see figure 84).

The highest research and conservation priorities should be given to safeguard the small and declining goose populations breeding and wintering in the Eastern Eurasia. Other large bird species in Eastern Siberia, for example various eiders, also show problems in that region. In North America and the Western Eurasia, national legal and regulatory instruments, as well as international agreements, either already exist or are in preparation. In the Eastern Palearctic no such international agreements yet exist.

Geese are a good example that nature conservation efforts can work. There are many correlations between strategies like limiting hunting pressure or the establishment of protected areas in habitat areas and the sudden recovery of formerly threatened populations. A better understanding of why many geese have significantly recovered could provide the knowledge, motivation, and political arguments needed to help the populations and regions that are still doing poorly.

Peter Prokosch, WWF Arctic Programme, Oslo, Norway

SOURCE/FURTHER READING
Madsen, J., A. Reed, and A. Andreev. 1996. *Status and trends of geese (Anser sp., Branta sp.) in the world: a review, updating and evaluation.* Gibier Faune Sauvage, Game Wildl. 13: 337-353.

*Bean goose **Anser fabalis**, Finland.*

*Barnacle goose **Branta leucopsis**, Greenland.*

*Snow goose **Anser caerulescens**, Bylot Island Bird Sanctuary, Canada.*

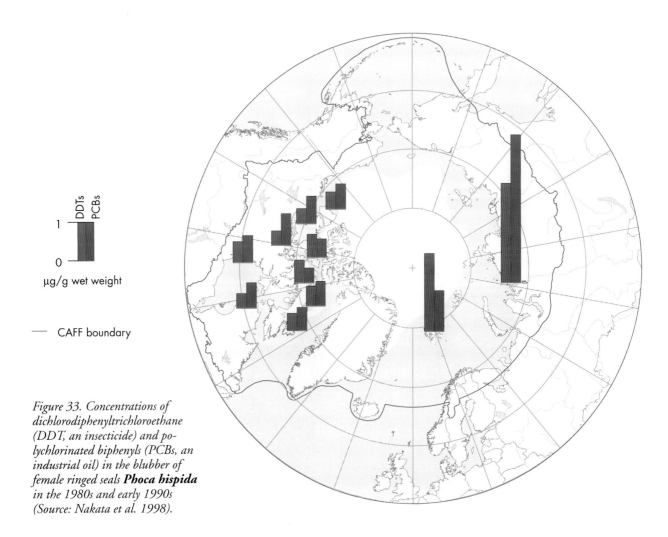

*Figure 33. Concentrations of dichlorodiphenyltrichloroethane (DDT, an insecticide) and polychlorinated biphenyls (PCBs, an industrial oil) in the blubber of female ringed seals **Phoca hispida** in the 1980s and early 1990s (Source: Nakata et al. 1998).*

ulations. Animals may avoid developed areas, changing their distribution patterns. Habitat quality can deteriorate because of pollution, changes in water flow, and the spread of dust. Hunting can increase due to improved access to remote areas. Species may be introduced to a region, intentionally or otherwise, causing major disruptions.

The development of mining operations in Svalbard in the 1920s, in northern Russia in the 1930s, and elsewhere in the Arctic in the following decades introduced large-scale industrial activity to the region. Beginning in the 1960s, both the Eurasian and the North American Arctic saw major oil and gas exploration and development, including building an extensive infrastructure of roads and pipelines. Hydro-electric dams are especially common in the Nordic region and in northern Quebec. In addition, transmission lines extend for thousands of kilometers. Offshore oil and gas development has another set of impacts including noise from seismic testing, ice breakers, and off-shore drilling.

Conservation responses to development include environmental impact assessments. Such assessments typically include analyses of direct and indirect impacts, although they are limited by the difficulty of predicting the complex ecological consequences of human activities over time. Cumulative impact assessments, analyzing the combined effects of several development activities, past and present, are a recent innovation to try to address this difficulty. One benefit of environmental assessments has been the creation of measures to minimize impacts. Directional drilling allows more oil to be reached from a single pad or platform. Caribou migrations can be monitored so that mining operations can be timed to minimize disturbances to the animals. Impact assessments and mitigation measures, however, cannot entirely prevent impacts or the risk of accidents. Industrial development remains a serious threat to the Arctic environment.

Tourism is another form of economic development that is increasing, with consequences for the environment. The transport of tourists to the Arctic, in itself, increases the volume of ship and airplane traffic. Many visitors want to see areas of great beauty or richness, such as bird colonies, marine mammal haul-outs, and caribou aggregations. Because there are relatively few places where such sights are accessible and reliable, tourist traffic is often concentrated. Arctic vegetation is typically unable to withstand repeated trampling, and paths of bare ground have appeared in some heavily visited spots. The animals that attract tourists may be disrupted, for instance when people get too close to nesting birds or resting walrus. Efforts are underway to develop effective management policies and guidelines for tourism in the Arctic, focusing on minimizing its impacts and respecting local cultures.

Diffuse Threats

■ Pollution is found throughout the Arctic. It comes both from local sources and from the long-range transport of industrial and agricultural pollutants from temperate and tropical areas. Acidification is a serious problem, particularly for freshwater systems. On the Kola Peninsula, the effects of pollution from nickel smelters have spread over 128,000 square kilometers, extending from Russia into Finland and Norway. Although harder to assess, long-range pollution can also have detrimental effects. The deposition of acidic nitrogen oxides and sulfur compounds is high at some locations in Greenland, Alaska, Norway, and Russia, and may affect plants. Significant levels of pollutants that arrive via long-range oceanic and atmospheric transport, such as persistent organic pollutants and some heavy metals, have been found in seals, whales, polar bears, migratory birds, and humans who eat significant quantitites of traditional, country food. Because these pollutants originate in many countries throughout the world, reducing emissions requires international action.

Ozone depletion occurs primarily in the polar regions, caused largely by human-made chemicals reaching the stratosphere and causing a chain of reactions that results in little or no ozone in spring. The loss of ozone allows more ultraviolet radiation to reach the earth's surface. Natural ultraviolet levels in the Arctic are low. The increased exposure to ultraviolet radiation that results from ozone depletion may harm the eyes and skin of animals, damage the surfaces of plants, and cause genetic damage. Experimental research on plants has shown different responses to increased exposure to ultraviolet radiation. Mosses grew more, but leaf and shoot growth decreased in evergreens. International conventions to eliminate ozone-depleting chemicals appear to be having some effect, but more monitoring is needed.

Climate change is a major concern. Models predict that the polar regions are likely to be affected soonest and to the greatest extent. Although the great year-to-year variability in the Arctic climate complicates detection, scientists and indigenous people are reporting signs of substantial change around the Arctic. As soils and permafrost warm and thaw, Arctic ecosystems that once absorbed carbon dioxide now release it. Long-term predictions include the loss of as much as half of the tundra as the treeline moves north, leaving tundra species with nowhere to go. Sea ice is expected to retreat, perhaps disappearing from the marginal seas of the Arctic. The ecological effects of climate change will lead to significant social, cultural, and economic changes. International negotiations are addressing emissions of carbon dioxide and other likely causes of climate change.

BOX 31. OIL DEVELOPMENT IMPACTS ON THE NORTH SLOPE OF ALASKA

▶ The exploration and development of oil on the North Slope of Alaska is one of the largest industrial undertakings in the Arctic. Since the discovery of the oil fields at Prudhoe Bay in 1968, development has spread east, west, and offshore, sending 12.8 billion barrels of oil south through the Trans-Alaska Pipeline. Such intensive and extensive activity has changed the environment on the North Slope in many ways. As North Slope oil development further expands, it is essential that we understand its potential impacts, particularly in sensitive areas.

The 19 producing oil fields and associated pipelines on the North Slope are spread over more than 2,500 square kilometers of tundra and wetland. Nearly 9,000 hectares have been excavated or filled during the construction of roads, pipelines (including the North Slope section of the Trans-Alaska Pipeline), drilling pads, airstrips, and other infrastructure. More than that amount has been disturbed by indirect and secondary effects, for example road dust and changes in water flow. Freshwater is used to build ice roads in winter, and seawater is pumped into the oil formations to force more of the lighter oil to the surface. Offshore, gravel docks and causeways extend out from the shore, and artificial gravel islands have been built as drilling platforms. Ship, vehicle, and air traffic bring supplies, conduct seismic exploration, and add to the noise and pollution of drilling.

The physical evidence of these activities may last for decades or longer, and is accompanied by biological and ecological changes. Causeways may impede fish passage along the coast, and fish overwintering areas in rivers and lakes may be harmed by gravel and freshwater extraction. Caribou from the Central Arctic Herd have increased in numbers over the period of oil development, but their distribution has changed and concentrated calving has shifted and been displaced away from development. Recent studies have

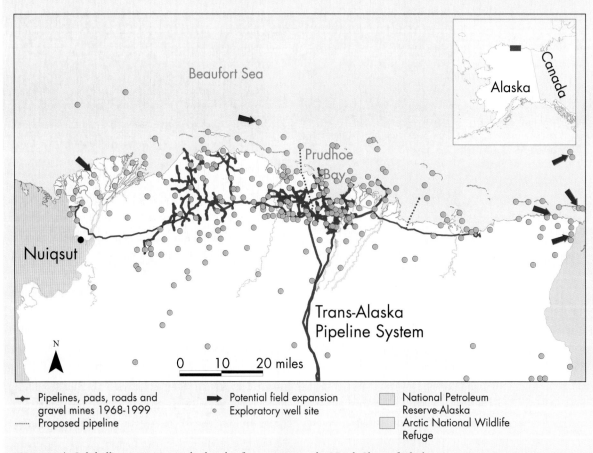

Figure 34. Oil drilling activities and related infrastructure on the North Slope of Alaska.

The oil development area, Prudhoe Bay, Alaska.

Oil drill in Prudhoe Bay, Alaska.

shown that caribou use of preferred habitat in a given area declines substantially as the density of roads increases, as is happening as development expands.

Offshore, there is reason for concern about development impacts, including the potential for oil spills, to fish and wildlife. Marine mammals inhabit nearshore waters and migrate along the coast. Hundreds of thousands of migratory birds, including threatened species, use the Beaufort Sea nearshore habitats during spring and fall migration, staging, brood-rearing, and molting. Because the locations of most offshore and onshore polar bear maternity dens are unknown, winter seismic surveys may cause disturbance and abandonment of dens. Early abandonment of dens would increase the risk of cub mortality. Other marine mammals may be affected by the noise associated with offshore exploration and development. In particular, the migration of bowhead whales may shift farther offshore, reducing opportunities for Iñupiat whalers from the village of Nuiqsut to carry out their traditional fall hunt.

Air pollution and wastewater from oil field activities plus frequent small spills of oil and other chemicals combine to spread heavy metals, nitrogen oxides, and petroleum hydrocarbons over the development area and beyond. Offshore development poses a great risk if an oil spill occurs. The consequences of a spill are far more serious if the oil contaminates waters with broken ice, where recovery techniques are unproven.

These points may not tell the whole story. There has never been a comprehensive environmental impact assessment for the development of onshore oil fields on the North Slope. An Environmental Impact Statement for the Trans-Alaska Pipeline conducted some analysis of North Slope development, but did not anticipate most of the nearby fields that are now producing oil. Offshore development has been subject to the Environmental Impact Statement process, but these have not adequately addressed cumulative impacts of onshore and offshore activities and infrastructure.

While some improvements in the methods used to extract oil have reduced environmental damage, many impacts are unavoidable. The U.S. National Research Council has begun a two-year review of information on the cumulative effects of oil development on the North Slope, which should be complete in 2002. Its results will better delineate the potential environmental impacts of oil development in Arctic Alaska.

Pamela A. Miller, Arctic Connections, Anchorage, Alaska, USA

REFERENCES

Amstrup, S.C., C. Gardner, K.C. Myers and F.W. Oehme. 1989. Ethylene glycol (antifreeze) poisoning in a free-ranging polar bear. *Veterinary and Human Toxicology* 31(4): 317-319.

Cameron, R.D., D.J. Reed, J.R. Dau, and W.T. Smith. 1992. Redistribution of calving caribou in response to oil field development on the arctic slope of Alaska. *Arctic* 45: 338-342.

Fechhelm, R.G. 1999. The effect of new breaching in a Prudhoe Bay causeway on the coastal distribution of humpback whitefish. *Arctic* 52(4): 386-394.

Jaffe, D.A., R.E. Honrath, D. Furness, T.J. Conway, E. Dlugokencky, and L.P. Steele. 1995. A determination of the CH_4, NO_x and CO_2 emissions from the Prudhoe Bay, Alaska oil development. *Journal of Atmospheric Chemistry* 20: 213-227.

Lawhead, B.E., and C.B. Johnson. 2000. *Surveys of caribou and muskoxen in the Kuparuk-Colville region, Alaska, 1999, with a summary of caribou calving distribution since 1993.* Final Report for Phillips, Alaska Inc. Fairbanks: ABR Inc.

Nellemann, C. and R.D. Cameron. 1996. Effects of petroleum development on terrain preferences of calving caribou. *Arctic* 49(1): 23-28.

Nellemann, C. and R.D. Cameron. 1998. Cumulative impacts of an evolving oil-field complex on the distribution of calving caribou. *Can. J. Zool* 76: 1425-1430.

U.S. Army Corps of Engineers. 1999. *Final environmental impact statement: Beaufort Sea Oil and Gas Development/ Northstar Project.* Anchorage.

Walker, D.A., P.J. Webber, E.F. Binnian, K.R. Everett, N.D. Lederer, E.A. Nordstrand, M.S. Walker. 1987. Cumulative impacts of oil fields on northern Alaskan landscapes. *Science* 238: 757-761.

Wolfe, S.A. 2000. Habitat selection by calving caribou of the Central Arctic Herd, 1980-1995. M.S. Thesis, University of Alaska Fairbanks.

BOX 32. HABITAT FRAGMENTATION: A THREAT TO ARCTIC BIODIVERSITY AND WILDERNESS

▶ Fragmentation is often defined as a decrease in some or all types of natural habitats in a landscape, and the dividing of the landscape into smaller and more isolated pieces. As the fragmentation process develops, the ecological effects will change. Fragmentation can be caused by natural processes such as fires, floods, and volcanic activity, but is more commonly caused by human impacts. It often starts with what are seen as small and harmless impacts. As human activity increases, however, the influence of fragmentation becomes greater. In the end, it leads to devastating effects on native species, a total change to the landscape, and the destruction of the area's wilderness heritage.

Today, in the vast circumpolar area, many ecosystems are still relatively undisturbed. They are large enough to allow ecological processes and wildlife populations to fluctuate naturally, and for biodiversity to evolve and adapt. These areas, which are among the last remaining wilderness areas of the world, therefore, constitute a global heritage.

However, Arctic ecosystems are under increased pressure from development activities. Road construction, pipelines, mining activities, and logging are among the significant causes of fragmentation. In the long run, the cumulative effects of such activities will cause a large loss of biodiversity and will reduce the Arctic's wilderness heritage considerably. For example, roads block the movement of small animals, expose large animals to heavy hunting pressure and poaching, cause sedimentation of rivers from erosion, and stimulate more development, thus creating further habitat fragmentation. Cumulative impacts may be more serious in the Arctic as the permafrost, underlying the thin biologically active layer, magnifies disturbances and makes restoration efforts difficult or impossible.

Norway as an example

In mainland Norway, much habitat has been lost through piecemeal development, particularly during the past 20 to 30 years. During the last century undisturbed or pristine wilderness areas (more than 5 km from roads and other infrastructures) have been reduced from 48% of the countryside nationwide in 1900 to 11.8% in 1998. In northern Norway only 24% of land was classified as wilderness in 1998. This trend is largely the result of agriculture, forestry, and hydroelectric development. In the Svalbard archipelago, the wilderness areas are still intact, but proposals for roads to connect coal mines are shadows on the horizon.

It is now increasingly realized that large pristine areas, in addition to being crucial to the conservation of biodiversity, are important to national identity and provide opportunities for outdoor recreation, ecological research, monitoring, and education. For the Saami, such areas are important for traditional reindeer herding. Recent political decisions in Norway reflect this changed attitude and place increased emphasis on preservation of wilderness areas. It is important for Arctic nations, which still are blessed with much undisturbed land, to control piecemeal development activities before it is too late.

Jan-Petter Huberth Hansen, Directorate for Nature Management, Trondheim, Norway

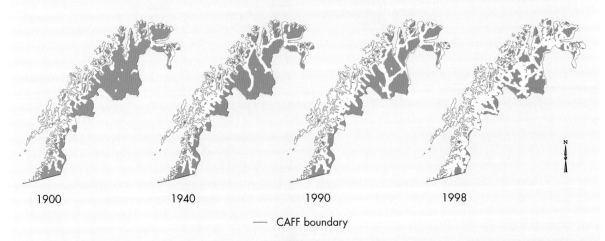

Figure 35. Decrease of wilderness areas in Norway north of the CAFF boundary. Wilderness is defined as an area lying more than five kilometers from roads, railways, and regulated water-courses (Source: Norwegian Mapping Authority).

BOX 33. IMPACTS OF INFRASTRUCTURE: THE ARCTIC 2050 SCENARIO

▶ Exploration for oil, gas, and minerals is the primary driving force of infrastructure development in the Arctic. Roads, pipelines, airstrips, ports, and other facilities affect wildlife by disrupting the physical environment, altering the chemical environment, helping the spread of exotic species, and, in particular, modifying animal behavior. The increase in human presence, especially by newcomers, is also a major factor, as development in infrastructure induces changes in the use of land, water, and resources over areas well beyond the physical footprint of development.

Between 1900 and 1950, less than 5% of the Arctic was affected by infrastructure. Between 1950 and 2000, the areas affected grew to 20-25% of the Arctic, mainly as a result of petroleum development in Alaska and Russia. In addition, most of Fennoscandia is affected by forestry, tourism, hydroelectric development and mining activities. A recent study by UNEP predicts that continued development of existing infrastructure at the current rate will leave few areas in the Arctic undisturbed in 50 years time. Depending on growth rates, 50-80% of the Arctic may reach high levels of anthropogenic disturbance by 2050. For Fennoscandia and parts of Russia, these levels may be reached within 20-30 years.

The spread of disturbance will lead to the loss of traditional lands for most of the indigenous people dependent upon reindeer husbandry and caribou hunting. Subsistence hunting, herding, fishing, and gathering remain a vital component of the socio-cultural systems of indigenous Arctic peoples. In northern Scandinavia and parts of Russia, the current growth of infrastructure is increasingly incompatible with the requirements of traditional reindeer husbandry. At the current rate of development, most traditional Saami reindeer husbandry will not be possible in 50 years, forcing major changes in lifestyle for today's herders. An influx of newcomers, attracted by work or by easier access, will create conflicts over the use of lands and resources for hunting, fishing, and other activities.

The extent of areas affected – even under conservative estimates that assume a 50% reduction in the rate of development – clearly illustrates the severity of the environmental impacts of infrastructure in the Arctic. Although the Arctic is among the least developed areas in the world, it is particularly sensitive to physical and other disturbances. If current policies and practices regarding development continue, the fragmentation and reduction of available habitats over the next half-century will seriously threaten biodiversity, ecosystem function, and the traditional lifestyles of many indigenous people.

Christian Nellemann, Norwegian Institute for Nature Research, Lillehammer, Norway

SOURCE/FURTHER READING
Nellemann, C., L. Kullerud, J. Vistnes, B.C. Forbes, G.P. Kofinas, B.P. Kaltenborn, O. Grøn, D. Henry, M. Magomedova, C. Lambrechts, R. Bobiwash, P.J. Schei, and T.S. Larsen. 2001. GLOBIO – Global methodology for mapping human impacts on the biosphere. *UNEP.*

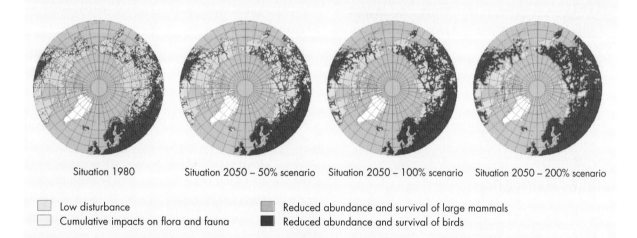

Situation 1980   Situation 2050 – 50% scenario   Situation 2050 – 100% scenario   Situation 2050 – 200% scenario

☐ Low disturbance
☐ Cumulative impacts on flora and fauna
■ Reduced abundance and survival of large mammals
■ Reduced abundance and survival of birds

Figure 36. Human impact on Arctic biodiversity and ecosystems in 1980 and projections to 2050 using three rates (50-100-200%) of increase in resource development and infrastructure growth. Note that the model overestimates impacts in the lower latitudes shown on the maps.

BOX 34. PRESERVING SACRED SITES IN THE RUSSIAN ARCTIC

▶ *In the tundra, sacred sites are scattered. These are most frequently hills or uplands. Often you will find there a heap of reindeer skulls, animal bones, sticks with multi-colored cloth belts, various rings attached or tied to the sticks by threads. In general it looks like trash or bright rubbish. Nenets people know that you cannot touch these things, and especially that you cannot take them away. The master of this dwelling (the site) may be cross or upset. Though spirits are invisible, they see everything. They are not kind to people. It is understandable – they are protecting their land, their tundra estates, lakes, rivers, and forests. Most often the Nenets try to mollify the spirits of the sacred place – the sanctuary – especially if they pass nearby. A Nenets will have a smoke, leave some tobacco on the ground, if there is a drink, pour some onto the ground, but will not touch the sacred sticks.*

These lines, written by Nadezhda Salinder in the local Yamal-Nenets newspaper Krasnyi Sever, can appear quite mystical to the outsider. The anti-religious policies of the Soviet Union before Perestroika resulted in the degradation and even destruction of the religions and the sacred sites of indigenous peoples. Only recently has the delicate issue of indigenous religion and sacred sites been seriously discussed.

In the Arctic, this task is especially urgent because large-scale industrial resource development threatens many important sites. In 1998, the Russian Association of Indigenous Peoples of the North (RAIPON) sent a preliminary inquiry on this subject to its regional members. The responses reflect the increasing concern of local peoples about the state of their historical, cultural, and spiritual heritage, which is primarily connected to sacred sites. In addition, the respondents expressed a great desire to protect this heritage. An analysis of the information that was gathered revealed that the characteristics of sacred sites and sanctuaries can vary greatly, depending on where they are, which indigenous group or groups use them, and whether they are primarily used by men or women or both.

In Arctic conservation efforts, there is an obvious gap with respect to the protection of sacred lands of indigenous peoples, a gap that needs to be filled. In this connection, it should be noted that:

- The list of sites of great historical, cultural, and spiritual value for the indigenous peoples of the Russian Arctic includes several hundred units and is growing. The majority are located outside protected areas, and many are found in places of permanent human settlement.
- An overwhelming majority of the sites are biologically diverse or are otherwise important for conservation. They frequently include scenic landscapes or other outstanding natural features.
- A significant number of sacred sites and sanctuaries are under pressure from human activity. Some have already been badly damaged or irrevocably destroyed.
- In the near future, human pressure on these sites will greatly increase, for example, from oil and gas exploration or from development of the Northern Sea Route.

The preservation of sacred lands of indigenous peoples can be advanced by establishing special protection regimes. "Ethno-ecological territories" or "traditional land-use territories," for example, could be established by law and incorporated into the national protected area network. A parallel effort can be undertaken at the international level, for example through CAFF's Circumpolar Protected Area Network (CPAN).

Russian Association of Indigenous Peoples of the North, Moscow, Russia

Understanding

■ Effective conservation requires accurate and timely information. Basic and applied research, to investigate fundamental processes and to address specific problems, are essential to understanding how the Arctic functions physically, ecologically, and socio-economically. Understanding how northern communities relate to their environment is vital to finding appropriate ways to manage that relationship in the long term. Monitoring complements research by periodically gathering data such as birth rates, survival, or other indicators over long periods in order to detect and interpret changes. It can help test predictions of how ecosystems, wild species, and human activities will respond to policy changes and management. Monitoring is an essential conservation tool to detect trends

> **BOX 35. CLIMATE AND CARIBOU:**
> **INUIT KNOWLEDGE OF THE IMPACTS OF CLIMATE CHANGE**
>
> ▶ Climate defines the Canadian Arctic and the people who live within it. For Inuit, frequent weather observations continue to be a daily part of hunting, traveling and other activities. As a result, Inuit have much knowledge to contribute to our understanding of climate change impacts. Recognizing this knowledge, four communities in the Kitikmeot region of Nunavut began to document and communicate Inuit knowledge of caribou and calving grounds. Thirty-seven interviews with elders and hunters were conducted between 1996 and 2000 as a part of the *Tuktu* (caribou) and *Nogak* (calves) Project.
>
> Inuit spoke of a changed climate in the 1990s compared with previous decades. Increasing temperatures with earlier spring-melts and later fall freeze-ups meant periods of longer summer-like conditions. Further, weather events have become variable and unpredictable, well beyond what people expect or consider "normal."
>
> This changed climate has had many important impacts on caribou. For example, migration routes and the locations of calving grounds have shifted. Caribou migrating across sea ice avoid swimming across leads because it is difficult for them to get out of the water and onto the ice. In the 1990s, sea ice melted earlier, forming especially wide leads, making it difficult for caribou to follow traditional ice crossings en route to their calving grounds.
>
> Food sources have sometimes been rendered inaccessible. Lately, Inuit have noted more frequent short-term changes in temperature, especially in freeze-thaw cycles. These cycles help form an icy layer on top of the snow or the tundra, preventing access to vegetation.
>
> "One spring, a lot of caribou died because of freezing rain and sleet. There were no areas for them to feed around . . . They had starved to death because of sleet. They had nowhere to eat. The ice was too thick . . . they could not dig through it." (Moses Koihok, Cambridge Bay, 1998.)
>
> Heat, too, can affect caribou. There were more days of record high temperatures in the 1990s than in any other period that Inuit remember. Sometimes temperatures were high enough to melt winter ice and snow in just a few days. Caribou are not well-adapted to these unusually warm conditions and can succumb to heat exhaustion.
>
> "Caribou would die from the heat of the sun. When the weather gets too hot, a lot of them would suffocate side by side. My wife and I have seen that at *Tahiluk*. They suffocated from the heat of the sun." (Mackie Kaosoni, Cambridge Bay, 1998.)
>
> These examples illustrate the way in which climate change has influenced the caribou of the Kitikmeot region. Through the Tuktu and Nogak Project, the tremendous knowledge base of those who live on the land has contributed to our understanding of the impacts of climate change in the Arctic.
>
> *Natasha Thorpe, Tuktu and Nogak Project, Victoria, British Columbia, Canada*
> *Qitirmiut Elders and Hunters, Tuktu and Nogak Project, Cambridge Bay, Nunavut, Canada*
>
> SOURCE/FURTHER READING
> Thorpe, N.L. 2000. *Contributions of Inuit ecological knowledge to understanding the impacts of climate change on the Bathurst Caribou Herd in the Kitikmeot Region, Nunavut*. M.R.M. Research Project No. 268, School of Resource and Environmental Management. Burnaby, BC: Simon Fraser University. www.rem.sfu.ca/pdf/nlthorpe.pdf

early enough to be able to make a timely response.

In the end, however, conservation depends not only on understanding the environment, but on the ability and will to make appropriate decisions. The best information is of little use if it is ignored. Too often, communication and understanding between researchers and decision-makers is poor. Adaptive management and co-management require and also foster better coordination between researchers, managers, and users. Improving such ties is a major challenge in Arctic conservation.

Decisions about the use of natural resources are complex, often involving several competing interests. Education can help build public support for conservation by creating awareness of what needs to be conserved and why. Such efforts can demonstrate the values, tangible and intangible, of the natural world. In short, Arctic conservation is not merely a matter of hunting regulations or national parks, but an issue of human attitudes towards the environment from the local to global scales.

BOX 36. TOURISM – THREAT OR BENEFIT TO CONSERVATION?

▶ Tourism is the world's largest and fastest-growing industry. The Arctic is an increasingly popular destination, with few limits to its growth. In 2000, more than 1.5 million tourists visited the Arctic, as defined by CAFF. If one includes the entire state of Alaska, the figure is twice as high, exceeding the number of Arctic residents. More than a million tourists crowd into the Arctic Fennoscandia, making use of the Arctic's densest infrastructure of roads, other transport facilities, and buildings. Not only does infrastructure attract tourists, tourism is itself a major cause of wilderness fragmentation, particularly as roads and other permanent structures are built for cars and their passengers.

Tourism poses several other threats. Unsustainable levels of hunting and fishing, particularly illegal poaching, are especially worrisome in Russia. Vehicle and foot traffic can damage fragile tundra soils and vegetation. The Icelandic tourism industry, for example, promotes off-road driving in its efforts to increase tourism. Garbage, waste, and pollution are significant problems for many tourism operations, especially as decomposition is slow and waste remains visible atop the permafrost in many Arctic areas. Simply the sheer number of people can disturb both wildlife and local communities.

Different forms of tourism impact the environment in different ways. Snow machine tourists in Kilpisjärvi, Finland (above); Tourists disembarking from a cruise ship on Svalbard (below).

But there is another side to tourism, one that can be beneficial for conservation. Most tourists visit the Arctic to see its natural wonders in pristine condition. In principle, this means that tour operators have a strong interest in protecting the environment. In Svalbard, tour companies have joined with conservation organizations to push for environmental protection. They have prevented construction of a road and spurred the Norwegian government to try to make Svalbard one of the best-managed wilderness areas in the world. Tourists can be effective proponents of conservation as well, spreading awareness of Arctic environmental issues when they return home. The economic benefits of tourism can provide an alternative to potentially more damaging sources of income such as mining, providing jobs to communities without harming their environment.

With both the positive and negative potential of Arctic tourism in mind, the World Wide Fund for Nature (WWF) and its partners in the tourism industry have developed guidelines and incentives for reducing negative impacts on and increasing the benefits for the Arctic environment. The effort is based on ten principles:

1. Make tourism and conservation compatible.
2. Support the preservation of wilderness and biodiversity.
3. Use natural resources in a sustainable way.
4. Minimize consumption, waste, and pollution.
5. Respect local cultures.
6. Respect historic and scientific sites.
7. Communities should benefit from tourism.
8. Trained staff are the key to responsible tourism.
9. Make your trip an opportunity to learn about the Arctic.
10. Follow safety rules.

This approach is finding increasing support among Arctic governments, regions, communities, and private business. With a concerted effort, Arctic tourism can be a long-term asset for conservation.

Peter Prokosch, WWF International Arctic Programme, Oslo, Norway

SOURCE/FURTHER READING

Humphreys, B.H., Å.Ø. Pedersen, P. Prokosch, S. Smith, and B. Stonehouse, eds. 1998. *Linking tourism and conservation in the Arctic.* Meddeleser No. 159. Tromsø: Norsk Polarsinstitutt. 140p.

Ski slopes near Yllästunturi, Finland.

BOX 37. ALIEN INVASIVE SPECIES: A THREAT TO GLOBAL AND ARCTIC BIODIVERSITY

▶ Frequently and for a variety of reasons, foreign or alien species are introduced into new habitats throughout the world. Often, the new habitat is unsuitable for the alien species and they perish. In many cases, however, they not only survive, but also thrive to such an extent that native species are displaced. In these cases, the alien species are said to be invasive.

According to the definition of the World Conservation Union (IUCN), "*Invasive species means an alien species which becomes established in natural or semi-natural ecosystems or habitats, is an agent of change, and threatens native biological diversity*".

Most introductions are accidental, but some are intentional, carried out in the belief that the action is beneficial and that humans are in control. However, actual and potential long-term effects of introductions are extremely hard to predict and quantify.

At present, alien invasive species are considered among the most severe threats to global biodiversity. There are several ways in which they can affect the new environment:

- Native species may be displaced or eliminated.
- Interactions between native species may be disrupted.
- Successful hybridization with native species can result in loss of biological and genetic diversity, which may increase the risk of extinction of the native species due to loss of its genes through diffusion when hybridization occurs.
- Parasites or diseases may accompany the alien species.
- Resource utilization may be affected with socio-economic consequences.

Globally, current evidence indicate that native species are in general declining and are being replaced by a limited number of widespread species that thrive in human-altered environments. Further, the evidence indicate that the declining species tend to represent a non-random selection of higher taxa and ecological groups, which enhances the homogenization of the world's biota.

Some studies have shown that ecosystems with high biodiversity have higher resistance to invasive species than those with low biodiversity. Hence, the Arctic with its few species might be more vulnerable to alien invasions than many other regions. Other studies show that southern ecosystems with high biodiversity also have the highest number of resident alien species. However, compared to many southern areas, the Arctic has had a short history of accidental and intentional species introductions and warming climate may also further intensify the problem.

The last remains of muskoxen at Spitsbergen, Svalbard. Seventeen muskoxen were captured on Greenland and reintroduced to Spitsbergen in 1929. The population peaked around 1960 at approximately 50 animals, but declined steadily after that, probably due to competition for food with the Svalbard reindeer. The last muskox was observed on the island in 1985.

Presently, lack of information makes it difficult to generalize about the actual and potential impacts of alien invasive species in the circumpolar Arctic. However, a few examples from the Norwegian Arctic may provide indication of the magnitude of the problem:

- *Marine environment.* There are many pathways for alien species in the marine environment. These include ship ballast water, sediments carried in tanks or on ship equipment, aquaculture, bait, removal of physical barriers, trade, escapes from research experiments, and fish processing plants. It has recently been confirmed that, from aquaculture and shipping activities alone, six species of phytoplankton, 16 macro algae, and 33 invertebrate species have been introduced to or transferred within the North Sea and North Atlantic. The red king crab, for example, has been invading the Norwegian coast from the Barents Sea where Russian scientists introduced this species some decades ago. The red king crab is at the moment subject to a lucrative fishery, but it also has an ecological impact through competing for food with native benthos and fish, and potentially transferring parasites and pathogens to other marine life.

- *Terrestrial environment.* It is estimated that 48% of vascular plant species, belonging to the most species rich families (i.e. families with more than 100 species) in Norway, have been introduced during the past 1000 years. Prior to 1994, 52 species of birds had been introduced to Norway, of which 17 were introduced intentionally. Some, for example the Canada goose, which was introduced in 1930, have grown substantially in numbers. At the latest count, there were 1,500-2,000 breeding pairs of of this species, adding to the conflict between staging geese and agriculture by inflicting high grazing pressure early in the seed-growing season. About 30 mammal species have been introduced to Norwegian territory in historic times. Among the species introduced to Svalbard are the sibling vole, the Arctic hare and the muskox, the two latter of which have died out. Although poorly studied, a number of invertebrates have also been introduced, some of which have done substantial damage.
- *Freshwater environment.* The freshwater environment offers some of the best-known examples of alien species, due to the socio-economic and recreational impacts they can cause. Nationwide, the most severe case is probably represented by escapees of farmed Atlantic salmon, which threaten to blend with and genetically homogenize native salmon populations. In northern Norway, other species that have received much attention are the vendace, the minnow, and the fish ectoparasite *Gyrodactylus salaris*. All are threats to freshwater fisheries resources of high socio-economic and recreational values, and are therefore monitored and dealt with on a level that is unusual for most alien invasive species.

With some evident exceptions, such as the *Gyrodactylus* parasite, introductions of alien species do not necessarily cause overt local ecological stress. Also, perceived local effects may not necessarily reflect the full extent of the problem, which often is not discovered before the impacts are examined on a regional or global scale.

Although the magnitude of the threat caused by invasive alien species to the Arctic environment is not clear at present, the potential impacts are severe and the problem deserves careful attention.

Dag Vongraven, Norwegian Polar Institute, Tromsø, Norway

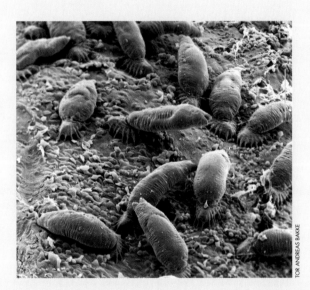

Gyrodactylus salaris *on salmonid skin (scanning electron microscope image). This parasite originates in Central-Asia from where it spread naturally to the Baltic. Its invasion to Scandinavia in the early 1970s was assisted by humans.* ***Gyrodactylus salaris*** *feeds mainly on young salmonids' skin and the fish will ultimately die from fungal or bacterial infections of the wounds. Since its invasion, it has done major damage in fish farms and rivers in Norway.*

SOURCES/FURTHER READING

Gjertz, I. and O. Lønø. 1998. Innførte arter på Svalbard. *Fauna* 51: 58-67. (In Norwegian.)

Hopkins, C. 2001. *Actual and potential effects of introduced marine organisms in Norwegian waters, including Svalbard.* Research report nr. 2001-1, Norwegian Directorate for Nature Management (in press).

Huxel, G.R. 1999. Rapid displacement of native species by invasive species: effects of hybridization. *Biological Conservation* 89: 143-152.

IUCN. 2000. *A guide to designing legal and institutional frameworks on alien invasive species.* IUCN Environmental Policy and Law Paper No. 40.

Levine, J.M. 2000. Species diversity and biological invasions: relating local process to community patterns. *Science* 288: 852-854.

McKinney, M.L. and J.L. Lockwood. 1999. Biotic homogenisation: a few winners replacing many losers in the next mass extinction. *Trends in Ecology and Evolution* 14: 450-453.

Stachowicz, J.J., R.B. Whitlatch and R.W. Osman. 1999. Species diversity and invasion resistance in a marine ecosystem. *Science* 286: 1577-1579.

Tømmerås, B.Å., ed. 1994. *Introduction of alien organisms to Norway.* NINA Utredning 62: 141p.

BOX 38. INTERNATIONAL LEGAL INSTRUMENTS FOR CONSERVATION

▶ OVERARCHING INSTRUMENTS

Three international instruments lay the modern groundwork for the conservation and protection of global biodiversity.
- *The World Charter for Nature* (signed in 1982), a resolution of the United Nations General Assembly, acknowledges that humans are a part of nature and receive benefits from it. Ecosystems and organisms used by humans are to be managed to achieve optimum sustainable productivity, but not at the expense of other species or ecosystems with which they co-exist.
- *Agenda 21* (1992) is a non-binding comprehensive blueprint for environmental action that integrates economic and social considerations into a broader sustainable development response to environmental problems. It defines biodiversity as a capital asset with the potential to yield sustainable benefits if managed properly.
- *The Global Program of Action for Protection of the Marine Environment from Land Based Activities* (GPA) (1995) is a voluntary framework agreement covering pollution abatement based on source categories such as sewage, radionuclides, and nutrients. It also has provisions for the protection of critical habitat.

GLOBAL INSTRUMENTS

Numerous global instruments are designed to assist countries in their efforts to conserve and protect biodiversity, and several apply directly to the Arctic.

Framework Instruments
- *The Convention on Biological Diversity* (CBD) (1992) has three objectives: global conservation of biodiversity, sustainable use of biological resources, and the fair and equitable sharing of benefits arising from the utilization of genetic resources. Its clauses cover a range of topics, from a requirement to establish protected areas to the promotion of indigenous practices and knowledge relevant to conservation and sustainable use.
- *The United Nations Convention on Law of the Sea* (UNCLOS) (1982), defines ocean jurisdiction zones, including the exclusive economic zone (EEZ), establishes rules governing all uses of the oceans and their resources, and extends the rights of non-coastal states to benefit from such uses. It defines fish and other marine biodiversity outside the EEZ as the "common heritage of mankind."

Species Use Instruments
- *The Convention on Trade in Endangered Species* (CITES) (1973) is designed to conserve wildlife species by controlling international trade in endangered flora, fauna, their parts, and derivative products through a system of import and export permits.
- *The Agreement on Straddling Fish Stocks* (1995) builds on the provisions of UNCLOS concerning fish species that move between the exclusive economic zones (EEZs) of countries and the high seas, (i.e. international waters) or migrate over long distances. It aims to optimize use and ensure the long-term conservation and sustainable use of target species while avoiding negative impacts on other species.
- *The Agreement on Fishing Vessels on the High Seas* (1993), which flows from UNCLOS and Agenda 21, recognizes universal rights to fish on the high seas subject to the rules of UNCLOS and aims to prevent non-compliance.
- *The International Convention for the Regulation of Whaling* (1946) seeks to protect whales from overhunting and to regulate the international whale fishery to ensure proper conservation and development of whale stocks. It also provides for the creation of international whale sanctuaries.

Species and Habitat Instruments
- *The Convention on the Conservation of Migratory Species of Wild Animals (Bonn Convention)* (1983) aims to protect endangered migratory species and migratory species with an unfavorable conservation status. It facilitates species agreements among countries within the range of that species.
- *The Convention Concerning the Protection of the World's Cultural and Natural Heritage* (World Heritage Convention – WHC) (1972) calls on parties to designate natural areas and cultural sites of outstanding universal value and to preserve them. There are proposals to establish transboundary world parks under the WHC.
- *The Convention on Wetlands of International Importance especially as Waterfowl Habitat* (Ramsar Convention) (1971) calls on parties to protect migratory stocks of waterbirds and their wetland habitats and to apply the principle of "wise use," as defined by the convention. Under the Ramsar Convention, countries designate wetlands of international importance as Ramsar sites.

Environmental Degradation Instruments
- *The Convention on Prevention of Pollution from Ships* (MARPOL 1973, 1978, and Protocol), restricts vessel discharges of noxious substances and allows for the designation of areas that require special protection from maritime activities. The related London Dumping Convention lists ecologically detrimental substances.

- *The Convention on Environmental Impact Assessment in a Transboundary Context (Espoo)* (1991) establishes a framework to consider environmental factors in domestic decisions concerning large-scale industrial projects and to notify other states of potential impacts.

REGIONAL INSTRUMENTS

Regional biodiversity instruments are geographically focused, and include several among the Arctic countries.

Use Instruments

- *Migratory Bird Conventions* between Canada and the United States (1916) and between the Soviet Union and the United States (1976) regulate hunting and were signed early in the twentieth century to halt the widespread decline of migratory game birds.
- *The Agreement on the Conservation of Polar Bears* (1973), signed by Canada, Denmark, Norway, the Soviet Union, and the United States, prohibits the killing or capture of polar bears except for scientific, conservation, or traditional purposes. The agreement also contains habitat protection provisions.
- *The Porcupine Caribou Herd Management Agreement* (1987) between Canada and the United States obligates the countries to conserve this shared herd and its habitat, and established the International Porcupine Caribou Board as a forum for governments, users, and others to achieve conservation objectives.
- *The North East Atlantic Fisheries Agreement* (1980), signatories to which include Denmark, Finland, Iceland, Norway, Sweden, and Russia, promotes the conservation and optimum use of fisheries resources.
- *Humane Trapping Standards* (1997) are set by agreements signed by Canada, the European Community, Russia, and the United States, and are intended to facilitate the fur trade industry by ensuring that animals are trapped using common standards of animal welfare.

Species and Habitat Instruments

- The Berne *Convention on European Wildlife and Natural Habitats* (1979) is intended to conserve wild flora and fauna and their natural habitats with particular emphasis on rare and endangered species.
- *EU Bird and Habitat Directives* (1979 and 1992) are included in legislation adopted by the European Union to protect birds and habitats by establishing a European Network of Protected Areas and by regulating the hunting, sale and transport of birds.
- *The North American Waterfowl Management Plan* (1998), to which Canada and the United States are party, is designed to conserve habitat for waterfowl through incentives for conservation and management plans for selected waterfowl species.

Environmental Degradation Instruments

- *The Convention for the Protection of the Marine Environment of the North East Atlantic (OSPAR)* (1998), covers the Nordic countries and requires them to take all possible steps to prevent and eliminate pollution, protect the maritime area, conserve marine biodiversity, and use marine resources sustainably.
- *The Nordic Environmental Protection Convention* (1974) defines environmentally harmful activities, specifically discharges of noxious materials and pollutants and sets in place a challenge and compensation system that can be invoked by states or individuals.
- *The North American Agreement on Environmental Cooperation* (1993), signed by Canada and the United States, together with Mexico, established a biodiversity conservation work plan, which includes provisions for species and habitats.
- *The Regional Program of Action for Protection of the Arctic Marine Environment against Land-Based Activities* (1998) has been adopted by the Arctic countries through the Arctic Council, and is a regional instrument under the overarching GPA.
- *Long-Range Transport of Air Pollutants* (LRTAP) (1998) aims to prevent, reduce and control transboundary air pollution from existing and new sources. Fire related protocols address specific pollutants and impacts.

Jeanne Pagnan, IUCN - The World Conservation Union, Ottawa, Canada

BOX 39. U.S./RUSSIA BILATERAL AGREEMENT ON POLAR BEARS BREAKS NEW GROUND IN INTERNATIONAL CONSERVATION

▶ Russia and the United States share the stock of polar bears that inhabit the Chukchi/Bering Seas region and adjacent coastline. Until recently, Russia and the U.S. have managed this population independently. In Russia, polar bear hunting has been banned since 1956, but an unquantified amount of illegal hunting has been reported and protection for the species has dwindled. In the U.S., Alaska Natives are allowed to hunt polar bears for subsistence purposes, as long as populations are not depleted. The current size of the Alaska-Chukotka polar bear population is unknown. Across the Arctic, the most serious threats to polar bears are excessive harvest, habitat loss, pollution, and climate change.

These issues were the motivation for establishing a unified management regime between the two countries. Efforts from a decade of discussions and later negotiations among federal managers and Native users on both sides resulted in development of the *U.S./Russia Bilateral Agreement on the Conservation and Management of the Alaska-Chukotka Polar Bear Population*, signed in October 2000. The primary purpose of the Bilateral Agreement is to assure the long-term conservation of the Chukchi/Bering Seas polar bear population and its habitat by developing a science-based conservation program that can be implemented and enforced in both countries.

The *Bilateral Agreement* is unique in international wildlife management in that it provides a recognized foundation for Native subsistence users' involvement in joint conservation activities. A Native-to-Native agreement between the Alaska Nanuuq Commission in the U.S. and the Association of Traditional Marine Mammal Hunters of Chukotka in Russia is envisaged to complement the government-to-government agreement. Implementation of the *Bilateral Agreement* will be overseen by a Joint Commission, consisting of one federal and one Native representative from each country, and by a Scientific Advisory Group comprised of federal, state or regional, and local participants from Russia and Alaska knowledgeable about polar bear research and management. Conservation and management decisions will be based on the best available information, including traditional ecological knowledge.

Female polar bear **Ursus maritimus** *with cubs.*

The *Bilateral Agreement* recognizes the needs of Native people to harvest polar bears for subsistence purposes. It provides for hunting by Chukotka Natives, as long as harvest monitoring and enforcement programs are in place and comply with domestic law and the terms of the 1973 *Agreement on the Conservation of Polar Bears*. A harvest quota system conservatively based on sustainable yield will result in proactive management, avoiding the reactionary management that occurs in response to over-harvest or severe population declines. Chukotka and Alaska Natives will participate in the collection of data and biological samples, identify important polar bear habitat use areas, and annually exchange information and coordinate future programs.

One of the most remarkable aspects of the *Bilateral Agreement* is the active role that Alaska and Chukotka Natives played in its development and will continue to play in its implementation. It is the hope of all parties to the *Bilateral Agreement* that it will serve as a model for similar agreements for other shared stocks of marine mammals, such as the Pacific walrus.

Susanne Kalxdorff, U.S. Fish and Wildlife Service, Anchorage, Alaska, USA
Scott Schliebe, U.S. Fish and Wildlife Service, Anchorage, Alaska, USA
Stanislav Belikov, All-Russian Institute of Nature Conservation, Moscow, Russia

SOURCES/FURTHER READING

AMAP. 1998. *The AMAP Assessment Report: Arctic pollution issues.* Oslo: Arctic Monitoring and Assessment Programme. xii + 859p.

Bailey, J.L., N.B. Snow, A. Carpenter, and L. Carpenter. 1995. Cooperative wildlife management under the Western Inuvialuit land claim. In: J.A. Bissonette, and P.R. Krausman, eds., *Integrating people and wildlife for a sustainable future.* Bethesda: The Wildlife Society. p. 11-15.

Banks, D., Williams, M., Pearce, J., Springer, A., Hagenstein, R. and D. Olson (eds.) 1999. *Ecoregion-Based Conservation in the Barents Sea: Identifying Important Areas for Biodiversity Conservation.* WWF and The Nature Conservation of Alaska.

Berkes, F. 1999. *Sacred ecology traditional ecological knowledge and resource management.* Philadelphia: Taylor and Francis. 209p.

Bernes, C. 1996. *The Nordic Arctic environment – unspoilt, exploited, polluted?* Copenhagen: The Nordic Council of Ministers.

Björn, L.O., T.V. Callaghan, C. Gehrke, D. Gwynn-Jones, B. Holgren, U. Johanson, and M. Sonesson. 1997. Effects of enhanced UV-B radiation on subarctic vegetation. In: S.J. Woodin and M. Marquiss, eds. *Ecology of arctic environments.* Special Publication 13, British Ecological Society. Oxford: Blackwell Science. p. 241-253.

CAFF (Conservation of Arctic Flora and Fauna). 1994. Habitat Conservation Report No. 1: The State of Protected Areas in the Arctic. Trondheim: Directorate for Nature Management. 163p.

CAFF (Conservation of Arctic Flora and Fauna). 1998. *Technical Report No. 1: Incidental Take of Seabirds in Commercial Fisheries in the Arctic Countries* (Vidar Bakken and Knud Falk, eds.) Akureyri: CAFF. 50p.

CAFF (Conservation of Arctic Flora and Fauna). 1998. *Technical Report No. 2: Human Disturbance at Arctic Seabird Colonies* (John Chardine and Vivian Mendenhall, eds.) Akureyri: CAFF. 18p.

CAFF (Conservation of Arctic Flora and Fauna). 2000 *Technical Report No. 7: Workshop on Seabird Incidental Catch in the Waters of Arctic Countries, 26-28 April 2000. Report and Recommendations* (John W. Chardine, Julie M. Porter, and Kenton D. Wohl, eds.) Akureyri: CAFF. 73p.

Caulfield, R.A. 1992. Alaska's subsistence management regimes. *Polar Record* 28(164): 23-32.

Caulfield, R.A. 1997. *Greenlanders, whales, and whaling: sustainability and self-determination in the Arctic.* Hanover, New Hampshire: University Press of New England.

FAO (Food and Agriculture Organization). 1998. *The state of world fisheries and aquaculture 1998.* Rome: FAO.

Fenge, T. 1993. National Parks in the Canadian Arctic: The Case of the Nunavut Land Claims Agreement. *Environments* 23 (1): 21-36.

Freese, C.H. 2000. *The consumptive use of wild species in the Arctic: challenges and opportunities for biodiversity conservation.* Toronto: World Wildlife Fund-Canada.

Huntington, H.P. 1992. *Wildlife management and subsistence hunting in Alaska.* London: Belhaven Press.

Huntington, H.P. 2000. Using traditional ecological knowledge in science: methods and applications. *Ecological Applications* 10(5).

Nakata, H., S. Tanabe, R. Tatsukawa, Y. Koyama, N. Miyazaki, S. Belikov, and A. Boltunov. 1998. Persistent organochlorine contaminants in ringed seals (*Phoca hispida*) from the Kara Sea, Russian Arctic. *Environmental Toxicology and Chemistry* 17:1745-1755.

National Research Council. 1996. *The Bering Sea ecosystem.* Washington, D.C.: National Academy Press.

Nelleman, C., P. Jordhøy, O. Støen and O. Strand. 2000. Cumulative impacts of tourist resorts on wild reindeer (*Rangifer tarandus tarandus*) during winter. *Arctic* 53:9-17.

Norwegian Polar Institute. 1998. *Linking tourism and conservation in the Arctic.* Proceedings from Arctic Tourism Workshops. Meddelelser No. 159. Tromsø: Norwegian Polar Institute.

Oechel, W.C., A.C. Cook, S.J. Hastings, and G.L. Vourlitis. 1997. Effects of CO_2 and climate change on arctic ecosystems. In: S.J. Woodin and M. Marquiss, eds. *Ecology of arctic environments.* Special Publication 13, British Ecological Society. Oxford: Blackwell Science. p. 255-286.

Usher, P.J. 1997. Common property and regional sovereignty: relations between aboriginal peoples and the Crown in Canada. In: P. Larmour, ed. *The governance of common property in the Pacific Region.* Canberra: National Centre for Development Studies, The Australian National University. p. 103-122.

Walters, C.J., and C.S. Holling. 1990. Large-scale management experiments and learning by doing. *Ecology* 71: 2060-2068.

Woodin, S.J. 1997. Effects of acid deposition on arctic vegetation. In: S.J. Woodin and M. Marquiss, eds. *Ecology of arctic environments.* Special Publication 13, British Ecological Society. Oxford: Blackwell Science. p. 219-239.

5. From forest to tundra

> *The transition zone between the forest and the treeless tundra provides important habitat for many species, and marks a significant climatic boundary. As this chapter describes, the forest-tundra zone is particularly sensitive to disturbance and change.*

The Region

As one moves north from temperate regions, the number of tree species decreases until only a few are left. Moving farther north, the individual trees tend to become smaller and sparser, gradually giving way to large open areas with only a few small, often deformed trees, and finally the treeless tundra. Defining the southern and northern limits of the transition zone between forest and tundra is a problematic task. The circumpolar countries often use different terms to describe this area, many of which also include adjacent areas. This chapter refers to the transition zone from the boreal forest to the tundra, or forest-tundra for short.

More precisely, the forest-tundra is the belt between the forest-line (the northern limit of the continuous forest) and the treeline (the limit for the occurrence of tree species in tree-size form, i.e., 2-3 meters tall) or tree-species line (the limit for the occurrence of a tree-species without regard to size or form). This transition, where the protective forest encounters an open windy landscape, often indicates a boundary between extreme, harsh conditions and more moderate ones. The forest-tundra is sometimes called a combat zone, where trees struggle for existence by trying to adapt their ways of regeneration, growth forms, stature, and physiological and genetic

properties to extreme circumstances in order to survive.

The species that form the treeline are different across the Arctic, and include spruce, pine, birch, poplar, and larch. Over large areas and mostly in discontinuous form, permafrost extends through the forest-tundra and widely into the boreal forest. Some species, such as spruce and larch, have shallow root systems and are not greatly affected by frozen ground below them. Others, such as birch and poplar, have a deep tap root that requires unfrozen ground.

The transition zone is hundreds of kilometers wide in the region of Taimyr and eastern Siberia, in the interior of Labrador, and in parts of central Canada. In eastern coastal Labrador and parts of Fennoscandia, it is only a few kilometers wide. Elevation also affects the transition zone. In Alaska, the crest of the Brooks Range runs across the area in which the treeline would be expected to occur, making the effects of altitude and latitude indistinguishable. Similar effects are found in northwestern Europe. The Taimyr Peninsula in Russia is one of the few places where the gradient of latitude is not complicated by other factors.

In addition to the trees and other plants of the region, the forest-tundra is home to many animal species. It provides relatively open summer habitat to forest birds and mammals and fairly protected winter habitat to species that summer on the tundra. Many insects thrive in the forest-tundra, living on trees and other plants and finding refuge from wind and storms. These insects support insectivorous birds such as warblers and woodpeckers. Accordingly, large numbers of birds are found in the forest-tundra during summer.

Key characteristics

The presence of trees distinguishes the forest-tundra from the tundra. The presence of large open areas and the loss of continuous tree cover separate the forest-tundra from the boreal forest. The forest-tundra ranges from sparse trees to a mosaic of tundra and forest. The proportion of tundra increases northwards, and the northernmost trees are as a rule short (less than 2 or 3 meters), stunted, and malformed.

Several factors help determine, over the long and short term, how far north trees will grow. Climate is the major factor over the long term. The 10°C sum-

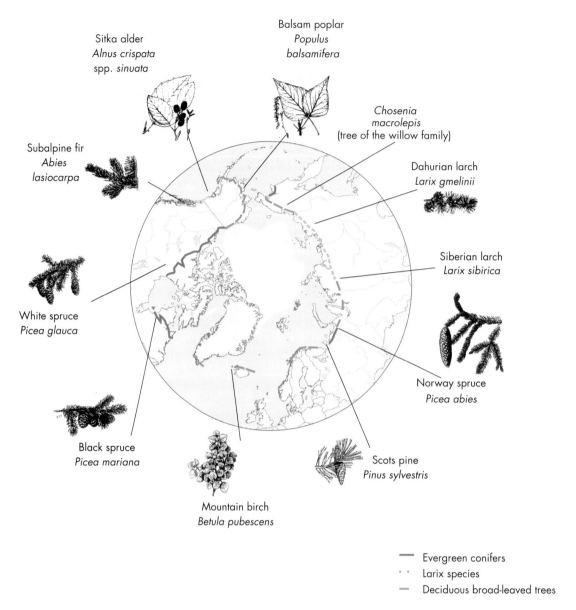

Figure 37. The Arctic treeline, with characteristic species for various areas.

mer isotherm approximates the potential treeline, although wind and moisture are also important. Several other factors affect the treeline on the scale of tens of kilometers. In many places, the tree- and forest-lines have been pushed farther south by human activities such as reindeer grazing or logging. Similarly, forest fires, either natural or started by humans, may also affect the tree- and forest-lines, particularly if they coincide with simultaneous climatic deterioration. Other factors, like insect pests such as the autumn moth in northern Fennoscandia, have caused the treeline to retreat. Furthermore, soil conditions and topographic factors may considerably affect the location of the treeline.

Surprisingly, perhaps, the occurrence of permafrost correlates only vaguely either with the northern treeline or with boreal vegetation zones. As a generalization, the permafrost is discontinuous in the forest-tundra and continuous under the tundra. The treeline

*Treeline species from different parts of the Arctic: Siberian larch **Larix sibirica**, Northern Ural Mountains, Russia – Mountain birch **Betula pubescens** ssp. **czerepanovii**, Fennoscandia. – Norway spruce **Picea abies**, Finnish Lapland.*

Typical species of the forest tundra: Moose **Alces alces**, *whooper swan* **Cygnus cygnus** *and Common crane* **Grus grus**.

in northwestern Canada coincides fairly closely with the southern limit of continuous permafrost. In the farthest inland areas of Eurasia, continuous permafrost extends deep into the boreal forest.

The forest-tundra has the curious feature of having fewer plant species than either the tundra on one side or the boreal forest on the other. This is likely due to climate factors, among them the average position of air-mass fronts. The forest-tundra transition coincides with the mean summer position of the Arctic front. The diversity of Arctic species is greater north of the forest-tundra, where arctic air-masses almost always dominate, and the diversity of boreal species increases to the south. The species that are found in the transition zone tend to be ones that are also found on both sides, trees being the obvious exception. Thus, the true tundra species and true forest species do poorly in the intermediate zone, while more ecologically flexible species survive.

In human terms, the treeline forms a rough cultural boundary. In North America, Inuit, Inuvialuit, Iñupiat, and Yupiit generally live in the tundra areas, while other indigenous peoples such as the Athabascans and Dene live within the boreal forest. Similar divisions are found in Russia, for example between the tundra-dwelling Nenets and the forest-dwelling Komi on the western side of the Ural Mountains.

Functionality

The basic ecological functions of the forest-tundra are similar to those of other terrestrial areas, including the tundra, which is described in more detail in the following chapter. Because trees are the distinctive flora of the forest-tundra, this section describes the processes that affect trees or the treeline specifically.

The plants, and especially the trees, that make up an area of the forest-tundra change over time. Storms, fire, insect outbreaks, or human activity may change the vegetation, hence altering or initiating the steps of plant succession. The first species to establish themselves are those that reproduce easily and thrive in sunny conditions. They include deciduous trees such as birch and poplar. As these trees mature, others such

Norway spruce **Picea abies** *battered by snow, wind and frost at the treeline.*

as spruce that can tolerate the shade of the leafy trees start to grow. Mosses begin to dominate the ground, covering most of the area. Moisture and decaying plant leaves and twigs have to pass through the moss before they reach the roots of other plants. With a covering of moss, the soil tends to become cooler, favoring the mosses and spruce, which grow less quickly than many other plants. Eventually, coniferous trees and mosses dominate the area, until fire returns at an interval of up to hundreds of years.

Fungi play an especially important role in the forest-tundra. Stresses such as a cold climate and nutrient-poor soils make trees more dependent on mycorrhizae, the symbiotic union of fungi and plant roots. Without the boost provided by mycorrhizae, the treeline would be much farther south. Various fungi depend on all stages of the life cycle of trees. Their diversity indicates the amount of time that has passed since the last disturbance.

Mammalian population cycles are a pronounced feature of both the forest-tundra and the tundra. The three-year cycle for several species of lemming and the approximately ten-year cycle for Canada lynx and snowshoe hare in North America are important examples. There is strong evidence that herbivore population cy-

BOX 40. MOUNTAIN BIRCH AND THE NORDIC TREELINE

▶ The treeline in the Nordic region is formed by the mountain birch. This species complex, composed of a variety of forms, subspecies, and hybrids, is found in Fennoscandia and on the Kola Peninsula. Mountain birch and its close relatives extend farther east in Russia and westward to Iceland and Greenland.

A number of factors may explain why the Nordic treeline is formed by a broadleaf tree, unlike the coniferous treeline found across much of the rest of the Arctic. The climate of Fennoscandia, in particular the influence of the ocean, is a major contributor. Higher humidity helps accelerate the bud break of birch to a greater degree than for most other deciduous trees. More importantly, the birch grows at lower summer temperatures than spruce and pine in Fennoscandia. The average-daily-maximum isotherm of 13.2°C for the four warmest summer months has been found recently to be the figure of highest significance in association with the mountain birch forest-line. In most areas of Fennoscandia, this places the upper limit of birch some 150-200 meters above the treeline for conifers, and also farther north.

While climate determines the potential treeline, the actual limit of mountain birch in the Nordic countries has often been depressed by grazing. In northern Finland, northernmost Norway and Sweden, and in some subalpine areas of southern Norway, reindeer are the main grazing animal. In other parts of Fennoscandia, such as the subalpine areas of western Norway and particularly in Iceland, grazing by sheep has been very extensive and has strongly influenced the growth of mountain birch. Grazing by invertebrates, particularly the cyclic outbreaks of geometrid moths, is an important cause of mortality for mountain birch.

In the last 50 to 80 years, many of the outfarms established over the course of several centuries for summer grazing in the mountain birch belt were abandoned, and there may now be a succession towards new mountain birch growth at those locations. Because of the harsh climate, however, the succession is slower than might have been expected, indicating that the position of the current tree line reflects more the influence of past climate than that of current conditions. The slow response may explain why the Nordic mountain birch treeline has only weakly advanced in recent years, despite higher average temperatures. Human influences today include increased tourism and the building of cottages and ski lifts. The slow recovery rate in the region causes the impacts of such activities to be visible for several years.

Frans E. Wielgolaski, University of Oslo, Oslo, Norway

---

cles are caused by plant-herbivore interactions, rather than by other factors such as the predator-prey relationship. Small mammals, especially some rodents such as voles and lemmings, play a key role in tundra and forest-tundra ecosystems, forming a major link between primary producers and secondary consumers.

Interactions with other regions

Ecologically, the forest-tundra functions in ways similar to other terrestrial areas. Freshwater areas are spread throughout the forest-tundra in the form of lakes and ponds, rivers and streams, and wetlands such as bogs, swamps, and marshes. Runoff from the land carries nutrients to lakes and rivers, helping aquatic plants and animals to grow. Many birds and mammals eat water plants and fish, bringing nutrients back onto land. Insects that breed in wetlands are a significant portion of the diet of many small birds that nest among the trees and shrubs.

As a typical ecotone or transition zone, the forest-tundra acts as a buffer between the boreal forest and the open tundra. Some tundra species take refuge and find protection in the trees. Forest species search for open areas for feeding or to avoid predators. Few if any species are found exclusively in the forest-tundra. Many migratory species such as reindeer spend part of the year in the forest-tundra, or at least pass through it as they move from the tundra to the closed forest. The species of the forest-tundra include trees and other plants that are also found far to the south. As the marginal range of many of these species, the forest-tundra is a refuge in case of catastrophe and an evolutionary area for adaptation to the Arctic.

BOX 41. PALUDIFICATION, FIRE, AND TOPOGRAPHY STRUCTURE THE NORTHERN FOREST

▶ From the air, the forested hills of interior Alaska show repeating patterns: black spruce grows on north-facing slopes, aspen on south-facing hillsides, birch on east- and west-facing slopes, and white spruce sprinkled everywhere except amid the black spruce. In river valleys, white spruce, cottonwood, and birch line the riverbanks, while black spruce and shrub vegetation grow in the peatlands that cover flats beyond the active floodplain. These patterns are formed by three basic processes:

Fire in a spruce forest.

- *Paludification* is the natural process by which forest becomes peatland. Plant litter can accumulate rapidly in high latitudes, because decomposition is slow. Amid the litter, mosses flourish, helping retain water and insulate the soil. Tree seedlings have difficulty establishing themselves in the vegetative mat, and permafrost rises as the soil cools. Over the course of centuries or millennia, the forest becomes peatland.
- *Wildfires* can be viewed as the antagonist of paludification. Fires destroy the developing mat of vegetation, exposing mineral soil, which can then warm up, allowing trees to colonize the barren areas. For instance, birch, aspen, and white spruce all establish themselves most effectively on bare mineral substrates rather than on top of plant litter or mosses. Through fire, the feedbacks that lead to peatland are broken.
- *Topography* affects both paludification and fire behavior. Steeper slopes are better drained, preventing the buildup of water that aids paludification. The direction a slope faces determines incident solar radiation and thus the amount of solar heating. North-facing slopes are cooler and wetter than south-facing slopes. Topography can help channel fires, especially in concert with wind.

An example of the interaction of these processes can be seen near the entrance of Denali National Park in Alaska. Riley Creek has downcut through several hundred feet of river and glacier sediments, creating a descending series of gravel terraces over the course of 10,000 years. The terraces face south and are well-drained due to their gravel substrate and position on a slope. The older terraces are forested, and fires recurring every 40-60 years are the dominant disturbance. Plant species able to resprout from roots following fires increase in importance with terrace age. A forest of white spruce and cottonwood trees covers terraces 1,000-4,000 years old.

On the terraces of Riley Creek that are older than about 4,000 years, soils and vegetation reach a stable state. Aspen, white spruce, and ericaceous shrubs grow on well-drained, poorly developed soils, and fires occur at frequent intervals. We infer that this ecosystem maintains itself through positive feedbacks involving clonal reproduction, the vegetation's effect on the fire regime, and fire's effects on halting soil development and preventing paludification and the invasion of peatland. Topographic position is critical in making it possible for root-sprouting, fire-prone vegetation to dominate this upland site.

Dan Mann, Institute of Arctic Biology, University of Alaska, Fairbanks, Alaska, USA

Skiing hunter, northern Finland.

Natural variability

In the absence of other external disturbances, treeline vegetation is in a dynamic equilibrium with climate. The location of the treeline and of the forest-tundra is largely determined by climate, and reacts even to relatively short-term climate fluctuations. In places where the land is still recovering from the glaciers of the last Ice Age, the treeline is moving northward as the processes of soil formation and plant colonization continue.

Since the last glaciation, the treeline, as well as the forest-tundra, in most regions has advanced northward and retreated southward several times. Deforestation often appears to involve interactions between a cooling climate and fire, as shown in northern Quebec. In the time known as the Little Ice Age, between the late 16th and late 19th Centuries, the pine and birch limits retreated in northern Scandinavia. A more favorable climate in the 1930s caused treelines to advance in many parts of the northern circumpolar zone, only to retreat again in the 1950s and 1960s. Recent reports from around the Arctic indicate that treelines continue to recede in some areas, whereas in other regions the retreat has stopped and the treeline is even advancing.

Severe climate restricts tree reproduction more than tree growth. Mature trees can live for hundreds of years without regenerating, and several treeline species can reproduce vegetatively. To mature, seeds need more warmth than that necessary to keep the tree alive and growing. In an average year in the forest-tundra, the amount of heat is insufficient for tree seeds to mature and germinate or for seedlings to go through their early juvenile stages. Thus, generative reproduction is possible only after a series of favorable years. Research in central and eastern Canada suggests that the treeline there should be regarded as relict, in that the trees are not capable of generative reproduction because they cannot produce mature seeds.

Periodic insect outbreaks are a major disturbance in the forest-tundra. In some areas, these insects can devastate a region as drastically as fire, which itself may spread through the dry, dead timber left by the insects. In Alaska, for example, spruce bark beetles have in recent years killed a significant portion of the spruce forests throughout much of the state, leading to an increase in forest fires. In the mountain birch woodlands of northern Fennoscandia, geometrid moths cause large-scale defoliation at intervals of about a decade. Although the effects of such outbreaks will be felt for decades to centuries as the forest regenerates, the general characteristics of the forest-tundra will remain, provided that there is no climatic deterioration.

Human uses and impacts

Humans have disturbed the forest-tundra for centuries, and today there are few areas at or near the treeline that have not been affected. In Iceland, the early

The great grey owl **Strix nebulosa** *is a taiga species that preys mainly on voles* **Microtus** *spp.*

*Forest damaged by the autumn moth **Epirrita autumnata**.*

settlers cut down most of the trees that were on the island. As a result, the land eroded more easily and dried out in many areas. Livestock grazing added to the disturbance by removing even more of the vegetation that held the soils in place. Similar patterns have been seen in Fennoscandia, where regeneration of birch forests is suppressed in areas heavily grazed by reindeer or sheep. As sheep farming has declined in Norway, young and dense birch forests have replaced pastures.

Fur trapping has been commonly practiced for centuries in the forest-tundra, in pursuit of bear, lynx, wolverine, wolf, and smaller mammals. In Russia, the majority of small communities engaged in reindeer herding, hunting, and fishing are located in the forest-tundra and the boreal forest, with seasonal migrations to the tundra. Some traditional activities, such

*Autumn moth **Epirrita autumnata** on a birch tree.*

BOX 42. ARY-MAS: AN ISLAND OF REMNANT FOREST

▶ The northernmost forest in the world is Ary-Mas, an isolated "forest island" on the Taimyr Peninsula in northern Siberia. It is likely that Ary-Mas is a remnant forest, left from a warmer period, or several warmer periods, when the tree line extended much farther north than it does today, a phenomenon well-documented in the Taimyr and elsewhere. Thus, the Ary-Mas system, preserved by favorable local conditions, is actually much older than the surrounding tundra system.

Ary-Mas was first described by scientists in the 1930s. A research station of the Botanic Institute of the Russian Academy of Sciences was established there in 1969, and a monograph on the area was published in 1987. Today, a year-round monitoring program is carried out by the staff of the biosphere reserve Taimyrsky, including studies of the larch biology, bird populations, microclimate, and soil thawing trends.

The trees in Ary-Mas are all Dahurian larch, spread over 60 square kilometers along the upper terrace of an alluvial plain of the Novaya River. The area has many bogs and lakes, but the tree canopy is relatively dense, between 30 and 50 percent closed. Other forest islands are found elsewhere in the Arctic, but composed of different tree species. One of the fascinating features of Ary-Mas is that it is home to many species that are not found in the surrounding tundra, but only in the boreal forest that begins some 30 kilometers to the south. Among these species are water plants, shrubs, the redbacked vole, wren, Naumann's thrush, and many others.

Elena Pospelova, Moscow State University, Moscow, Russia

---

as reindeer grazing and gathering firewood, push the treeline south in many areas.

The forest-tundra is typically less exploited than the boreal forest to the south, largely because the trees of the forest-tundra are smaller and thus less desirable for wood products. However, in recent decades, commercial logging operations have advanced closer to the forest-tundra across much of the boreal region. In Fennoscandia and northwestern Russia, for example, logging operations were extended close to the forest-tundra in the 1960s-1990s. Regeneration has been difficult in many areas. In Norway, the Norwegian spruce has been planted extensively for logging, replacing up to 30% of the native birch forest.

Dams, roads, mines, smelters, oil and gas production, pipelines, and electric transmission lines are the result of increasing industrial development. These activities affect not only species but entire landscapes through fragmentation and the disruption of basic processes such as permafrost and water flow. The development of non-renewable resources is increasing. In Russia, mining of non-ferrous metals has been a major industry in forest-tundra areas such as Norilsk and the Kola Peninsula. In the Russian Far East, there is substantial gold mining.

In the Fennoscandian forest-tundra, wolf and wolverine populations are low because of hunting as well as habitat loss. During the last few decades, protection has led to increases in their numbers. The arctic fox, however, remains rare in the region despite total protection for the past 60 years. Although the golden eagle is persecuted in reindeer management areas because it preys on reindeer calves, its population is stable, in part because of recruitment from areas outside the Arctic. Other birds of prey, such as the peregrine falcon, have had high concentrations of contaminants, which lowered their fecundity. The gyrfalcon population is low in Fennoscandia, and probably across Eurasia, and requires conservation effort.

Tourism, including sport hunting and fishing, attracts moderate though increasing numbers of visitors to forest-tundra areas. This places additional pressure on the region's resources, sometimes leading to conflicts between local and visiting hunters. The forest-tundra in general has a low tolerance for trampling. Even the temporary presence of humans often leaves a lasting impact.

BOX 43. THE INTRODUCED NOOTKA LUPINE IN ICELAND: BENEFIT OR THREAT?

▶ Species often expand into new areas naturally, taking advantage of favorable conditions to compete with the plants or animals already there. Humans frequently introduce species, intentionally or otherwise. Most introductions are unable to thrive in new areas, but some take hold too well. The Nootka lupine was brought to Iceland from Alaska to help revegetate areas of bare ground. Its success has spurred a debate about the perceived benefits of introducing species.

Of the perhaps 10,000 plant species that humans have introduced to Iceland, most of them for gardens, only a handful have been able to spread naturally. The Nootka lupine was introduced in 1945 and planted in sparsely vegetated reforestation areas in the lowlands. The lupine grew vigorously, spreading to bare and infertile soils. Foresters, gardeners, and others helped spread the lupine widely, and the plant is now found in many lowland areas throughout the country.

Lupine plants are very productive, and their accumulating litter helps enrich the soil biota and gradually increase soil organic matter. In sparsely vegetated areas and heathlands, the lupine is very successful in competition with the native plants that are generally of lower stature and pro-

Nootka lupine **Lupinus nootkatensis** *in lowland heath, northern Iceland.*

Meadow vegetation in northern Iceland.

ductivity. As the years pass, the dominant low shrubs, mosses, and lichens decline in the shade underneath the lupine canopy, and species richness among plants declines. In some areas the lupine starts to degenerate after 15-20 years, replaced by grassland vegetation and tall herbs that thrive in the rich soil created by the lupine. Lupine patches produce a complete change in vegetation composition and thus a considerable change in the landscape.

But is the lupine a threat in Iceland? Soil erosion and loss of vegetation cover is a serious problem. Efforts to combat the erosion and revegetate some of the barren areas have been going on for almost a century. The Nootka lupine has proven to be the most efficient plant for reclamation in many barren lowland areas. Once the plant has been seeded or established, it will look after itself and spread without needing costly fertilizer application or other treatment. For these reasons, the plant continues to be attractive for reclamation efforts.

On the other hand, it is clear that the utmost care should be taken in the use and distribution of the plant. The native plant communities are already disturbed after a prolonged history of over-grazing. Some of them, such as low-shrub heaths, have little resistance against invasion by the lupine. Once the lupine has become established, it is very difficult to control its spread. Efforts to do so, for example, in the Skaftafell National Park in southeast Iceland, have proven ineffective. Grazing sheep prevent the spread of the lupine by eating the seedlings, but the overall decline in sheep farming in Iceland has left many large lowland areas free of grazing. As vegetation recovers in these areas, the lupine is likely to thrive at the expense of native plants.

The story of the Nootka lupine in Iceland is not long, and the consequences of its introduction cannot be foreseen in detail. Lupine colonization on barren areas is welcomed and considered a benefit by most Icelanders, but its invasion of heathlands is looked upon as undesirable. A useful tool or a dangerous weapon, the Nootka lupine must be used wisely and with care.

Borgthór Magnússon, Agricultural Research Institute, Reykjavik, Iceland

SOURCES/FURTHER READING

Bjarnason, H. 1978. Erosion, tree growth and land regeneration in Iceland. In: M.W. Holdgate and M.J. Woodman, eds. *The breakdown and Restoration of Ecosystems.* New York: Plenum Press. p. 241-248.

Einarsson, E. 1997. Aðfluttar plöntutegundir á Íslandi (Introduced plant species in Iceland). In: A. Ottesen, ed. Nýgræðingar í flórunni. *Innfluttar plöntur – saga, áhrif, framtíð (Introduced plants in the Icelandic Flora).* Ráðstefna Félags garyrkjumanna 21. og 22. febrúar 1997. Reykjavík: Félag garyrkjumanna. p. 11-15. (In Icelandic).

Magnússon, B., ed. 1995. Biological studies of Nootka lupine (*Lupinus nootkatensis*) in Iceland. Growth, seed set, chemical content and effect of cutting. *RALA Report No. 178.* Reykjavík: Agricultural Research Institute. 82 pp. (In Icelandic with an English summary and subtitles).

Magnússon, B. 1999. Biology and utilization of Nootka lupin (*Lupinus nootkatensis*) in Iceland. In: G.D. Hill, ed. *Towards the 21st century.* Proceedings of the 8th International Lupin Conference, Asilomar, California, 11-16 May 1996. Canterbury, New Zealand: International Lupin Association. p. 42-48.

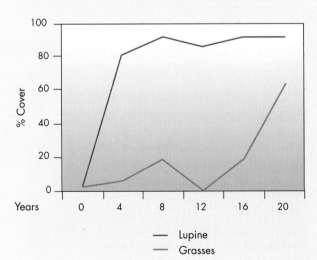
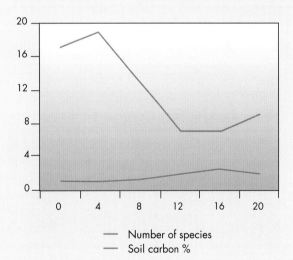

*Figure 38. Plant successional changes in a Nootka lupine **Lupinus nootkatensis** patch in southwestern Iceland. The site was eroded and had sparse vegetation cover when it was invaded by the lupine.*

BOX 44. MANAGEMENT AND BIODIVERSITY IN FENNOSCANDIAN TIMBERLINE FORESTS

▶ Northern forests grow slowly, and many of the species and ecological relationships found in them take time to develop. In Fennoscandia, economic pressures have pushed the logging industry northward into the timberline forests. The impacts of this expansion are extensive and long-lasting, but new management approaches may help.

In Finland, for example, 41% of all threatened species are rare because of forestry operations. Loss of habitats and fragmentation of forests together with hunting have caused declines in the number of large predators. Species requiring old-growth coniferous forests, such as Siberian tit, Siberian jay, and capercaillie, have declined in number as the area of virgin forest has shrunk.

Intensive forestry has changed the forest structure in many ways:

- Tree species composition has changed, because coniferous trees are favored over broadleaf trees in forest regeneration. Many invertebrate species, such as butterflies, moths, and beetles, that live on deciduous trees have declined.
- As a result of suppressing forest fires, pyrophilous (fire-dependent) species have become rare.
- Plowing of soil and ditch digging have increased the leaching of nutrients from soil and increased humus in waters.
- More silt has accumulated on river and lake bottoms, destroying the spawning areas of fish species. Salmonids, which are especially sensitive to changes in water quality, have been the first to disappear.

A decrease in the amount of decaying wood or coarse woody debris is another of the impacts of intensive forestry. A large number of invertebrates and lower plants such as mosses, lichens, and wood-decomposing fungi have declined in number or become threatened. These species play an important role in the decomposition of woody material. Many of these species require large-diameter logs to live on, and cannot survive on the small branches left after logging.

New forest management guidelines in many Fennoscandian countries attempt to address these problems. These methods include:
- Smaller regeneration areas and favoring of natural regeneration whenever possible
- Protective measures for wetlands and freshwater systems
- Forestry management at the landscape level.

Landscape ecological management planning, or LEMP, takes into account larger units than single stands, and considers the balance between cut and uncut stands as well as connections with other areas. Under this approach, areas of special ecological significance are left untouched, such as key biotopes with specific species compositions, corridors that allow movements of animals between uncut stands and protection zones around waters and other sensitive areas. Dead and living trees are also left in logged areas to maintain species that depend on decaying wood.

Anna-Liisa Sippola, Arctic Centre, Rovaniemi, Finland

Clear-cut and plowed forest, Inari, Finnish Lapland.

Forest regeneration with seed-tree cutting, Inari, Finnish Lapland.

*Capercaillie **Tetrao urogallus**, Siberian tit **Parus cinctus** and Siberian jay **Perisoreus infaustus** are old-growth forest species that have suffered from large-scale forestry.*

SOURCES/FURTHER READING

Hallman, E., M. Hokkanen, H. Juntunen, K-M. Korhonen, S. Raivio, O. Savela, P. Siitonen, A. Tolonen, and M. Vainio. 1996. Alue-ekologinen suunnittelu (Landscape ecological planning). Metsähallituksen metsätalouden julkaisuja 3. Helsinki. 59 p. (In Finnish)

Niemelä, J. 1999. Management in relation to disturbance in boreal forest. *Forest Ecology and Management* 115:127-134.

Rassi, P., H. Kaipiainen, I. Mannerkoski, and G. Ståhls. 1992. *Report on the monitoring of threatened animals and plants in Finland.* Helsinki: Ministry of Environment. Komiteanmietintö 1991:30. Helsinki. 323 p. (In Finnish with English summary.)

*The status of the golden eagle **Aquila chrysaëtos** in Finnish Lapland has improved because reindeer herders are compensated for calves on the basis of nesting pairs in their herding areas.*

The forest-tundra is vulnerable to diffuse disturbances, such as pollution, acidification, and increased ultraviolet radiation. While most of the pollution reaching the Arctic comes from distant sources, there are also significant pollution sources within the Arctic. In northern Russia, for example, heavy metals and sulfur dioxide pollution from the large mining complexes have killed trees, setting off a chain of erosion and further environmental damage. The damage caused by the mining areas of the Kola Peninsula extends into northern Finland. Acid rain caused by emissions of sulfur dioxide causes needle loss on Scots pine and other species, and has severely damaged lichens in Finland's Inari region.

The spread of diseases, fungi, and insects can be enhanced by human disturbance, which can lower the natural resistance of plants and animals. The organisms that create widespread destruction of a forest may always be present, but only become able to overwhelm trees and other plants when they are weakened by disturbance or pollution.

Climate change

Climate change is likely to affect the forest-tundra because the treeline is so greatly influenced by summer temperature. Climatic variability since the last Ice Age has led to great swings in the position of the

*The peregrine falcon **Falco peregrinus** has suffered from pollutants. The populations are slowly recovering due to active conservation efforts.*

treeline. As noted above, the current limit of trees in some areas may be as much a reflection of former climates as it is a result of current conditions. Other changes have been observed that are consistent with the predictions of global warming. In recent decades, nine species of birds have extended their breeding ranges northward to the Thelon Game Sanctuary, an outlying forest-tundra area in the Northwest Territories of Canada. In addition, moose and red squirrels have established breeding populations in the Thelon Valley.

The impacts from climate change are difficult to predict but are likely to include melting of discontinuous permafrost, which in turn will lead to slumping and other deformations of the ground. Climate change may increase the frequency and severity of forest fires, which will further accelerate permafrost thawing. The most profound change is likely to be simply the location of the forest-tundra: if the treeline moves north, the forest-line is likely to do the same, taking the transition zone with them. The impacts of this northward creep are hard to predict and are likely to depend on the specific physical conditions that are encountered.

Climate may also affect the distribution of insects and diseases that damage trees, altering outbreak cycles. Shifts in human patterns, including reindeer herding, commercial logging, and other types of de-

Destroyed forest near smelters, Montchegorsk, Kola Peninsula, Russia.

velopment, are also likely to change with climate shifts. The direction of change and its consequent impacts, however, cannot be predicted with confidence.

Conservation

The conservation measures currently in widespread use in the forest-tundra include those typically used on land. Protected areas encompass significant portions of the forest-tundra in Alaska and modest portions in Canada. In Fennoscandia, relatively large timberline forests are protected as national parks, nature reserves, or wilderness areas. In Iceland, some of the remaining forest areas are protected. Other areas have been replanted with trees, often with exotic species such as the Sitka spruce, Siberian larch, and lodgepole pine. These are intended partly to regenerate the forest and partly to develop a local timber industry. Key habitats within the forest-tundra – those areas with high species richness and rare and vulnerable species – require special consideration for conservation. River and stream valleys, cliff walls, and bogs are particularly important.

Most mammals and birds found in the forest-tundra spend much of their time each year in either the forest or the tundra. Members of such species may depend on the forest-tundra for critical parts of their life cycles, or may merely pass through during their migrations. Loss or degradation of this habitat through indirect or direct disturbance will affect animals that spend a significant or critical portion of their time in the forest-tundra. Destruction of migratory corridors, including the creation of physical barriers to movement, will affect the transient species. The forest-tundra, as the zone of transition, is a critical area for the interaction of the tundra and the forest.

SOURCES/FURTHER READING

Andreyev, V.N. 1966. Osobennosti zonal´nogo raspredeleniya nadzemnoy fitomassy na Vostochnoevropeyskom Severe (In Russian). *Botanicheski Zhurnal* 51(10):1401-1411.

Arnalds, A. 1987. Ecosystem disturbance in Iceland. *Arctic and Alpine Research* 19(4):508-513.

Blöndal, S. 1987. Afforestation and reforestation in Iceland. *Arctic and Alpine Research* 19(4):526-529.

Böcher, T.W. 1954. Oceanic and continental vegetational complexes in southwest Greenland. *Meddelelser om Grønland* 148 (1):1-336.

Briggs, D., P. Smithson, K. Addison, and K. Atkinson 1997. *Fundamentals of the physical environment.* London: Routledge. 557p.

Brown, R.J.E. 1970. Permafrost as an ecological factor in the Subarctic. In: Ecology of the subarctic regions. UNESCO, Proceedings of the Helsinki Symposium. *Ecological Conservation* 1:129-140.

Bryson, R.A. 1966. Air masses, streamlines and the boreal forest. *Geographical Bulletin* (Ottawa) 8:129-140.

Bull, J. and J. Farrand, Jr. 1977. *The Audubon Society field guide to North American birds. Eastern region.* New York: Alfred A. Knopf. 775p.

Chapin, F.S. III, and C. Korner, eds. 1995. *Arctic and alpine biodiversity: patterns, causes and ecosystem consequences.* Berlin: Springer Verlag. 332p.

Eronen, M. 1979. The retreat of pine forest in Finnish Lapland since the Holocene climatic optimum: a general discussion with radiocarbon evidence from subfossil pines. *Fennia* 157(2):93-114.

Harrison, C. 1982. *An atlas of the birds of the western Palearctic.* London: Collins. 322p.

Heikkinen, O. 1984. The timber-line problem. *Nordia* 18(2):105-114.

Heikkinen, R.K., and S. Neuvonen. 1997. Species richness of vascular plants in the subarctic landscape of northern Finland: modelling relationships to the environment. *Biodiversity and Conservation* 6:1181-1201.

Hosie, R.C. 1979. *Native trees of Canada.* Don Mills, Ontario: Fitzhenry and Whiteside. 380p.

Hultén, E. 1958. The amphi-Atlantic plants and their phytogeographical connection. *Kungliga Svenska Vetenskapsakademiens Handlingar* 7(1):1-340.

Hultén, E. 1964. The circumpolar plants. I. Vascular cryptogams, conifers, monocotyledons. *Kungliga Svenska Vetenskapsakademiens Handlingar* 8(5):1-290.

Hultén, E. 1971. The circumpolar plants. II. Dicotyledons. *Kungliga Svenska Vetenskapsakademiens Handlingar* 13(1):1-463.

Hustich, I. 1958. On the recent expansion of the Scotch pine in Northern Europe. *Fennia* 82(3):1-25.

Hustich, I. 1966. On the forest-tundra and the northern tree-lines. A preliminary synthesis. *Reports from the Kevo Subarctic Research Station* 3:7-47.

Jobbágy, E.G., and R.B. Jackson. 2000. Global controls of forest line elevation in the northern and southern hemispheres. *Global Ecology and Biogeography* 9:253-268.

Kallio, P., and J. Lehtonen. 1973. Birch forest damage caused by *Oporinia autumnata* (Bkh.) in 1965-66 in Utsjoki, N. Finland. *Reports from the Kevo Subarctic Research Station* 10:55-69.

Kankaanpää, S., T. Tasanen, and M-L. Sutinen, eds. 1999. *Sustainable development in northern timberline forests.* Proceedings of the Timberline Workshop, May 10-11, 1998 in Whitehorse, Canada. Finnish Forest Research Institute, Research Papers 734, 187p.

Kihlman, A.O. 1890. Pflanzenbiologische Studien aus Russisch Lappland (In German). *Acta Societatis pro Fauna et Flora Fennica* 6(3):1-256.

Krebs, J.S., and R.G. Barry 1970. The arctic front and the tundra-taiga boundary in Eurasia. *Geographical Review* 60:548-554.

Kullman, L. 1988. Short-term dynamic approach to tree-limit and thermal climate: evidence from *Pinus sylvestris* in the Swedish Scandes. *Annales Botanici Fennici* 25:219-227.

Kullman, L. 1993. Dynamism of the altitudinal margin of the boreal forest in Sweden. In: B. Frenzel, ed. Oscillations of the alpine and polar tree limits in the Holocene. *Paläoklimaforschung* 9:41-55.

Kullman, L. 2000. Tree-limit rise and recent warming: a geoecological case study from the Swedish Scandes. *Norsk Geografisk Tidsskrift* 54:49-59.

Larsen, J.A. 1980. The boreal ecosystem. New York: Academic Press. xvi + 500p.

Mattsson, J. 1995. Human impact on the timberline in the far north of Europe. In: O. Heikkinen, B. Obrebska-Starkel, and S. Tuhkanen, eds: *Environmental aspects of the timberline in Finland and in the Polish Carphatians.* Zeszyty Naukowe Uniwersitetu Jagielloskiego 956, Prace Geograficzne, Zeszyt 98:41-55.

Mikola, P. 1970. Forests and forestry in subarctic regions. In: Ecology of the subarctic regions. UNESCO, Proceedings of the Helsinki symposium. *Ecological Conservation* 1:295-302.

Mitchell, V.L. 1973. A theoretical tree line in central Canada. *Annals of the Association of American Geographers* 63: 296-301.

Moser, M.M. 1967. Die ektotrophe Ernährungsweise an der Waldgrenze (In German). In: *Ökologie der alpinen Walgrenze.* Mitteilungen der Forstliche Bundes-Versuchsanstalt Wien 75: 357-380.

Myklestad, A., and H.J.B. Birks. 1993. A numerical analysis of the distribution patterns of *Salix* L. species in Europe. *Journal of Biogeography* 20:1-32.

Müller, M.J. 1982. *Selected climatic data for a global set of standard stations for vegetation science.* Tasks for vegetation science 5. The Hague: Dr. W. Junk. 306p.

Nichols, H. 1976. Historical aspects of the northern Canadian timberline. *Arctic* 29:28-47.

Norment, C.J., A. Hall, and P. Hendricks. 1999. Important bird and mammal records in the Thelon River Valley, Northwest Territories: range expansions and possible causes. *Canadian Field-Naturalist.* 113(3):375-385.

Payette, S. 1983. The forest tundra and present tree-lines of the northern Québec-Labrador peninsula. *Collection Nordicana* 47:3-23.

Payette, S. and R. Gagnon 1985. Late Holocene deforestation and tree regeneration in the forest-tundra of Québec. *Nature* 303(6003):570-572.

Payette, S. and C. Morneau 1993. Holocene relict woodlands at the eastern Canadian treeline. *Quaternary Research* 39:84-89.

Payette, S., L. Filion, L. Gauthier, and Y. Boutin 1985. Secular climate change in old-growth tree-line vegetation of northern Québec. *Nature* 315(6015):135-138.

Regel, C. 1950. Klimaänderung und Vegetationsentwicklung in eurasiatischen Norden (In German). *Österreichishe Botanische Zeitschrift* 96:369-398.

Tenow, O. 1972. *The outbreaks of* Oporinia autumnata *Bkh. and* Operophthera *spp. (Lep., Geometridae) in the Scandinavian mountain chain and northern Finland 1862-1968.* Zoologiska Bidrag från Uppsala, Supplement 2, 107p.

Tikhomirov, B.A. 1961. The changes in biogeographical boundaries in the north of U.S.S.R. as related to climatic fluctuations and activity of man. *Botaniska Notiser* 56:285-292.

Tikhomirov, B.A. 1970. Forest limits as the most important biogeographical boundary in the North. In: Ecology of the subarctic regions. UNESCO, Proceedings of the Helsinki symposium. *Ecological Conservation* 1:35-40.

Tikkanen, E., and I. Niemelä, eds. 1995. *Kola Peninsula pollutants and forest ecosystems in Lapland.* Ministry of Agriculture of Finland and the Finnish Forest Research Institute. 82p.

Tuhkanen, S. 1980. Climatic parameters and indices in plant geography. *Acta Phytogeographica Suecica* 67, 110p.

Tuhkanen, S. 1982. Trädgränsen och klimatet (In Swedish). *Naturen* 106(3):82-94.

Tuhkanen, S. 1999. The northern timberline in relation to climate. In: S. Kankaanpää, T. Tasanen, and M-L. Sutinen, eds. *Sustainable development in northern timberline forests.* Proceedings of the Timberline Workshop, May 10-11, 1998 in Whitehorse, Canada., Finnish Forest Research Institute, Research Papers 734:29-61.

Ukkola, R. 1995. Trampling tolerance of plants and ground cover in Finnish Lappland, with an example from the Pyhätunturi national park. In: O. Heikkinen, B. Obrebska-Starkel, and S. Tuhkanen, eds: *Environmental aspects of the timberline in Finland and in the Polish Carphatians.* Zeszyty Naukowe Uniwersitetu Jagielloskiego 956, Prace Geograficzne, Zeszyt 98: 91-110.

Van Cleve, K., F.S. Chapin III, P.W. Flanagan, L.A. Viereck, and C.T. Dyrness, eds. 1986. *Forest ecosystems in the Alaskan taiga.* Ecological Studies 57. New York: Springer Verlag. x + 230p.

Veijola, P. 1998. *The northern timberline and timberline forests in Fennoscandia.* The Finnish Forest Research Institute, Research Papers 672, 242p.

Viereck, L.A., and E.L. Little, Jr. 1972. *Alaska trees and shrubs.* Agriculture handbook 410. Washington, D.C.: U.S. Department of Agriculture, U.S. Forest Service. 265p.

Virtanen, T., and S. Neuvonen. 1999. Climate change and Macrolepidopteran biodiversity in Finland. Chemosphere - Global Change Science 1(4):439-448.

6. The Tundra and the Polar Desert

The tundra and polar desert are shaped by climate and influenced by soil conditions and moisture. They are sensitive to human actions. This chapter describes these characteristic Arctic regions.

The Region

Beyond the forest-tundra lies the tundra and the polar desert. Several ways of classifying the tundra and polar desert have been proposed, typically according to vegetation characteristics. Here, we will distinguish between tundra, which is largely covered by plants, and polar desert, which is largely bare ground or rock. While areas of alpine tundra exist south of the Arctic, they are beyond the scope of this report.

Broadly, tundra exists where there is adequate moisture and summer temperatures remain substantially above freezing. Along the treeline, Arctic tundra vegetation forms a narrow belt of tall shrubs, mostly erect willows, which gradually changes to low-shrub tundra with semi-erect willows, alder, and dwarf birch. Farther from the treeline, the shrubs become more and more sporadic, though prostrate forms of willows are found even in the polar desert.

The polar desert typically occurs farther north or at higher elevations than the tundra, in areas lacking sufficient moisture and warmth to sustain extensive plant cover. Geological factors also play a role in determining the distribution of tundra and polar desert. Plants take longer to colonize rugged rocky areas like the Canadian Archipelago than lowland plains such as the Taimyr Peninsula. Polar deserts are highly susceptible to climatic disturbances such as the Little Ice Age, during which extensive areas of vegetation were destroyed by the growth of glaciers and ice caps. The plant communities in polar deserts, therefore, are of very recent origin. As the climate warms across much of the Arctic, more and more areas are going through this stage of primary succession.

Arctic soils, named cryosols because of the frozen layer at their base, are young, generally infertile, and poorly developed. The active layer, the depth to which soil thaws in summer, varies from more than a meter in the low Arctic to only a few centimeters in some polar deserts. A thick layer of organic matter or peat characterizes tundra soils. They are often saturated with water, becoming bog soils. In contrast, polar desert soils often consist of sand and gravel with only traces of organic material. The most developed and fertile soil is Arctic brown soil, found on warm, well-drained sites with a deep active layer.

In both tundra and polar desert, permafrost strongly influences the landscape and vegetation. First, the frozen subsoil supplies no nutrients to plants, nor does it allow deep roots to grow, thus limiting the size and types of plants. It also prevents water from draining away. Second, the freezing and thawing action of the soil and its water moves and sorts material, making a physically unstable habitat for plants. On slopes, soils may creep downhill from the action of freezing and thawing, and from the presence of water and ice, which reduce friction.

Polar semi-desert with flowering arctic poppies, **Papaver radicatum**. *Northeast Greenland National Park.*

BOX 45. THE TUNDRA AND POLAR DESERT: DISSIMILARITIES BETWEEN NORTH AMERICA AND EURASIA

▶ Patterns of climate, geography, and geology result in pronounced differences in vegetation and ecosystem characteristics between and within the Arctic tundra and polar desert of Eurasia and North America. The northern treeline runs fairly high above the Arctic Circle in Eurasia and in western North America, but descends far south in Beringia and in central and eastern Canada. Similarly, the boundary separating the Low and High Arctic swings deep below the Arctic Circle in the Canadian continental mainland of the Keewatin and Hudson Bay regions. The differences are explained largely by the Gulf Stream bringing warm water to the Arctic on the Scandinavian side, and by cold currents of the Davis Strait and ice-dominated waters of the Bering Sea on the North American side.

In North America, the Low Arctic vegetation complexes consist mostly of sedge-dominated mires and tussock tundra, while elevated drier landscapes support heath communities. In Eurasia, low willow and birch shrub tundra forms a wide transition zone from the forest-tundra ecotone, often extending to the ocean shore. While Eurasia and North America share the circumpolar Low Arctic more or less equally, the poorly vegetated High Arctic landscapes called polar deserts are mostly located in North America. Even the concept of "polar desert" is diversely defined and classified in scientific literature describing the Old and New World tundra systems.

The North American polar deserts have mostly formed on coarse granitic or carbonate rock substrates present in the Keewatin uplands and Arctic Island plateaus. They are much drier and poorer in the abundance of cryptogams (i.e., mosses, ferns, fungi), scattered herbs, and occasional prostrate shrubs, with a handful of vascular species and a maximum of 5% vegetation cover. Small isolated sedge and herb meadows around embryonic bogs and shallow lakes are also common. These "barrens" represent the largest area of the Canadian High Arctic. In contrast, the Eurasian polar deserts are almost exclusively confined to the Arctic Ocean islands. They have developed on finer, more humid soils, are richer in species diversity, support larger moss and lichen biomass, and have a relatively high herb and grass cover. Sheltered and drier areas in the North American High Arctic are occupied by cushion plants, mosses, and lichens with higher ground cover, and are classified as polar semi-deserts. In the Eurasian High Arctic, these complexes are widespread and are placed in the polar desert category.

Josef Svoboda, Department of Botany, University of Toronto, Mississauga, Ontario, Canada

SOURCES/FURTHER READING
Alexandrova, V.D. 1988. *Vegetation of the Soviet polar deserts*. New York: Cambridge University Press. 228p.
Bliss, L.C., J. Svoboda, and D.I. Bliss. 1984. Polar deserts, their plant cover and plant production in the Canadian High Arctic. *Holarctic Ecology* 7:305-324.
Bliss, L.C., and N.V. Matveyeva. 1992. Circumpolar arctic vegetation. In: F.S. Chapin III, R.L. Jefferies, J.F. Raynolds, G.R. Shaver, and J. Svoboda, eds. *Arctic ecosystems in a changing climate*. San Diego: Academic Press. p.59-90.
Bliss, L.C. 1979. Vascular plant vegetation of the southern circumpolar region in relation to the antarctic, alpine and arctic vegetation. *Canadian Journal of Botany* 57:2167-2178.

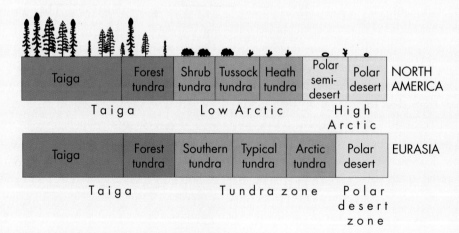

Figure 39. Classification of northern ecozones according to North American and Eurasian usage. Note the differences within the tundra zone (Source: Bliss & Matveyeva 1992).

Wet hummocky tundra, Greenland.

The result of frost action is a landscape that contains a wide range of local habitats at scales ranging from a few meters to several kilometers. Such habitats range from relatively deep, water-filled trenches with lush vegetation, to dry ridges with few plants, to raised mounds used by arctic foxes for dens or by birds for nests. They support diverse species and ecological processes.

Key characteristics

The dominant characteristic of tundra and polar deserts is the cold, short summer. On the Taimyr Peninsula, the number of vascular plants decreases from 250 in the south to 50 in the north, closely following the decrease in average July temperature. Most tundra plants grow slowly, and thus are capable of supporting only a limited amount of grazing. While caribou and reindeer herds may be large, their ranges cover vast areas. Thus, the overall density of animals is far lower than in the grasslands of the temperate and tropical regions, the other ecosystems dominated by grazing animals.

For much of the year, the tundra is covered in snow and is relatively inactive. Migratory animals have moved to other areas. Some mammals, such as the marmot and brown bear, hibernate. Most if not all insects are dormant during winter. Resident non-hibernating species – lemmings, ptarmigan, arctic foxes, and in some areas caribou and muskoxen – typically spend the winter trying to survive until the following summer, living off fat reserves and the little

BOX 46. DECOMPOSITION ENSURES RECYCLING OF NUTRIENTS

▶ Life on Earth depends on two complementary biological processes: *photosynthesis* (fixation of atmospheric carbon by plants) and *decomposition* (release of carbon back to the atmosphere). The balance between photosynthesis and decomposition determines the amount of organic matter accumulated in the soil. Thus decomposition is an essential process in the circulation of carbon, as well as other nutrients, in all ecosystems. In the Arctic, as elsewhere, decomposition is the product of three interacting sets of factors:

Soil organisms
- Compared with temperate latitudes, the diversity of soil macro-organisms is low in the Arctic, but there is no evidence of a reduced range of bacterial and fungal processes; organisms show physiological flexibility
- Large populations of mites, collembola (springtails), enchytraeid worms, and insect larvae assist in decomposition; earthworms are curiously absent from the North American tundra, but represented by at least five species in the Eurasian tundra
- Adaptation to low temperature allows decomposition and nutrient mineralization to proceed even when temperatures are around 0°C

Physico-chemical environment
Some aspects of the Arctic environment aid decomposition:
- While sub-zero temperatures limit the activities of soil organisms, solar radiation raises surface temperatures to 30°C or more during summer
- Freeze-thaw cycles fragment litter (dead plant matter), helping to replace the function of earthworms

Other factors inhibit decomposition:
- Lack of moisture in deserts and rocky areas limits decomposition
- Waterlogging causes anaerobic conditions, inhibiting decomposition and causing peat accumulation
- Permafrost maintains low temperatures and reduces activity within the soil
- Acid conditions on granite reduce decomposition, although higher pH is widespread over limestone

Organic matter quality
Quality of plant litter is highly variable:
- Herbs such as cloudberry produce litter that decomposes rapidly because of its low lignin content and high nutrient content (low carbon:nitrogen ratio)
- Leaves and needles from shrubs and heaths decompose more slowly; wood and mosses very slowly (high carbon:nitrogen ratio)
- Soluble carbohydrates are quickly decomposed and the more resistant fractions gradually accumulate as soil organic matter, some of which is hundreds of years old

The balance between these three factors determines rates of decomposition and nutrient mobilization. Rapid decomposition encourages plants that are more productive and that demand more nutrients, which in turn produces nutrient-rich herbage and litter. Slow decomposition selects for plants which, while less productive, also retain their nutrients and produce low quality litter.

In the Arctic, annual plant production varies from about 5 to 500g dry weight per square meter from the polar semi-deserts to the more sheltered moist meadows in the low Arctic. But, because of the slower rate of decomposition due to the soil microclimate, the amount of accumulated organic matter varies from about 5 to 300 kg per square meter, some of it locked in permafrost. The largest amounts accumulate in waterlogged mires and muskeg.

Up to a quarter of the world's soil carbon is stored in northern soils; the tundra has been a sink for atmospheric carbon through accumulation in soil over thousands of years. Climatic warming will increase both plant production and decomposition, but the balance will be variable and current estimates indicate that the tundra is likely to fluctuate between being a sink and a source of atmospheric carbon. One recent model also estimates that the tundra will decrease to half its current area by 2100, mainly due to forests expanding in response to a combination of factors.

Bill Heal, University of Edinburgh, Edinburgh, Scotland

SOURCES/FURTHER READING

Nadelhoffer, K.J., A.E. Giblin, G.R. Shaver, and A.E. Linkens. 1992. Microbial processes and plant nutrient availability in Arctic soils. In: F.S. Chapin, R,L. Jefferies, J.F. Reynolds, G.R. Shaver, and J. Svoboda, eds. *Arctic Ecosystems in a Changing Climate.* Academic Press, San Diego. p. 281-300.

Robinson, C.H. and P.A. Wookey. 1997. Microbial ecology, decomposition and nutrient cycling. In: S.J. Woodin and M. Marquiss, eds. *Ecology of Arctic Environments.* Blackwell Science, Oxford. p. 41-68.

White, A., M.G.R. Cannell, and A.D. Friend. 2000. The high-latitude terrestrial carbon sink: a model analysis. *Global Change Biology* 6: 227-245.

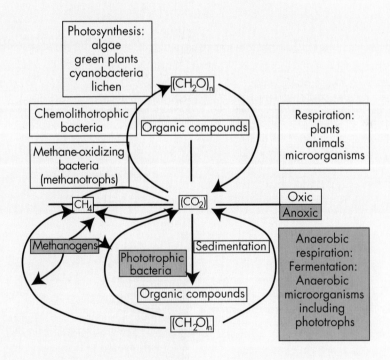

Figure 40. The carbon cycle. Microrganisms play a vital role as primary producers and decomposers.

BOX 47. MIRES AND WETLANDS

▶ A mire can be defined as an ecosystem sustained by a humid climate and a high water table, due to which partially decomposed organic matter is accumulated as peat. In the Arctic, such systems are complex and biologically rich.

Mires can be broadly classified in two groups. Ombrotrophic peatlands, or bogs, receive water and chemical elements from atmospheric deposition only, that is, from rain and snow. In minerotrophic peatlands, including fens and swamps, additional water and nutrients flow from a larger drainage area, passing through mineral soils on their way to the mire.

While mires are found throughout the Arctic, there are distinct differences between climatic zones. Aapa mires, or patterned fens, are distributed in the southernmost parts of the Arctic. These mires are minerotrophic, and have a shallow peat layer and a depressed center. In late spring floods, the centers of aapa mires are often covered in water, which may prevent *Sphagnum* moss from becoming the dominant species. The surface of an aapa mire is distinctive, consisting of high strings and low pools. These patterns are unstable, and the strings may move several tens of centimeters a

*In addition to climate, the gradient of the mineral bottom affects the surface patterns of aapa mires. Where the bottom slopes, the surface flow of meltwater is strong, creating a clear pattern on the surface. The strings often form dams, creating pools. The vertical difference between the top and bottom pools can be over 50 m (above). Aapa mires with a level mineral base may have an even surface dominated by sedges and **Sphagnum** mosses (right).*

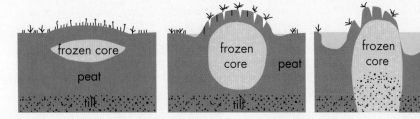

Figure 41. Development of a palsa mound. In the first stage (left), a frozen peat core develops. The frozen peat core grows and the palsa starts to rise (middle). Finally, the outer peat layer of the palsa begins to dry and crack (right). Then, the palsa breaks down and collapses.

year, likely pushed by hydrostatic pressure from the pools.

Farther north are palsa mires. Palsas are mounds with ice cores covered in peat. The vegetation of palsa mires is similar to that of aapa mires, but the string-and-pool surface patterning is absent. Instead, the palsas, which can grow as high as seven meters, rise above the otherwise flat surface of the mire. In some areas, *pounikkos* are found instead of palsas, the difference being that the center of a *pounikko* is frozen seasonally rather than permanently. *Pounikko* is a Finnish word meaning a high hummock covered by *Sphagnum fuscum*, or an extensive hummock area in an aapa or a palsa mire.

The northernmost mires are polygonal mires. Like palsa mires, polygonal mires form only in areas where the mean annual temperature is below -1°C. Frost action in the surface layer of these mires forces coarse-grained soil to the edge of the polygon while fine-grained soil remains in the center. The constant movement makes such polygons a difficult environment for plants. Between polygons are ice wedges, 0.1-3 meters wide at the top and extending downwards for up to several meters.

In addition to their significance for plants, mires are excellent habitat for insects and, consequently, for nesting birds such as waders and geese. In most systems, the number of species of plants and animals decreases when moving northwards, but the numbers of birds and bird species nesting on open mires increases in Finland the farther north one goes. Because such birds migrate from the Arctic throughout the world, Arctic mires play a major role in global ecology.

Mires in the Arctic are sensitive to changes in temperature and hydrology. Palsas, for example, grow until the covering peat layer dries and cracks open, exposing the ice core to warmth, whereupon it melts and the palsa collapses. New palsas form as ice cores develop in the mire. But if the temperature rises, new palsas may no longer form, and palsa mires will disappear. Similarly, if precipitation patterns change, the water balance of a mire might change, turning it into a dry area or a pond. Human disturbance, in the form of roads, pipelines, mines, and so on, can also shift the hydrology of an area, with drastic consequences for mires. Because of their importance and sensitivity, protecting mires is an essential component of conservation in the Arctic.

Harri Vasander, Department of Forest Ecology, University of Helsinki, Finland

SOURCES/FURTHER READING

Botch, M.S., and V.V. Masing. 1979. *Ekosistemy bolot SSSR. (Mire ecosystems of the USSR)* (In Russian). Leningrad: Nauka. 188 pp.

Botch, M.S., and V.V. Masing. 1983. Mire ecosystems in the U.S.S.R. In: A.J.P.Gore, ed. *Ecosystems of the World 4B. Mires: Swamp, Fen and Moor. Regional Studies.* Amsterdam: Elsevier. p. 95-152.

Järvinen, O., J. Kouki, J., and U. Häyrinen. 1983. Reversed latitudinal gradients in total density and species richness of birds breeding on Finnish mires. *Ornis. Fennica* 64: 67-73.

Moen, A. 1999. *National Atlas of Norway: Vegetation.* Hønefoss: Norwegian Mapping Authority. 200 p.

Ruuhijärvi, R. 1983. The Finnish mire types and their regional distribution. In: A.J.P. Gore, ed. *Ecosystems of the World 4B. Mires: Swamp, Bog, Fen and Moor. Regional Studies.* Amsterdam: Elsevier. p. 47-67.

Seppälä, M. 1988. Palsas and related forms. In: M.J. Clark, ed. *Advances in Periglacial Geomorphology.* Chichester: John Wiley & Sons. p. 247-272.

Seppälä, M., and L. Koutaniemi. 1985. Formation of a string and pool topography as expressed by morphology, stratigraphy and current processes on a mire in Kuusamo, Finland. *Boreas* 14: 287-309.

Seppä, H. 1996. The morphological features of the Finnish peatlands. In: H. Vasander, ed. *Peatlands in Finland.* Jyväskylä: Finnish Peatland Society. p. 27-33.

Zoltai, S.C., and F.C. Pollett. 1983. Wetlands in Canada: Their classification, distribution, and use. In: A.J.P. Gore, ed. *Ecosystems of the World 4B. Mires: Swamp, Bog, Fen and Moor. Regional Studies.* Amsterdam: Elsevier. p. 245-268.

Polygon tundra, Bylot Island, Canada.

Palsa wetlands are examples of the effects of frost action on tundra soils.

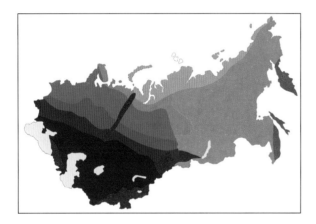

Zones
- Polygon mires
- Palsa mires
- Discontinous zone of aapa mires
- Raised bogs
- Pine bogs and fens
- Reed and sedge fens
- Fresh- and saltwater marshes

Provinces
- Continental provinces of Siberia and the far East
- Maritime Provinces of the far East
- High mountain Provinces

Figure 42. Mire zones and provinces in Russia, showing the change of mire types when moving northwards.

food that is available. In summer, however, the tundra bursts to life: migratory birds arrive in huge numbers from far away; some of the largest barren-ground caribou herds return from wintering in the boreal forest and forest-tundra; insects abound; and plants take advantage of the extended daylight to put out new shoots, flowers, and berries.

Summers, however, are highly variable from year to year. Species must be able to survive poor years and take advantage of good years. One strategy among plants is to recycle nutrients or to store nutrients and energy in one year for the next year's growth. This helps to spread the risk of poor conditions over a longer period. Cushion plants, such as the white mountain-avens and the moss campion, grow a new layer of leaves each year, covering and incorporating the previous year's layer into the cushion. The old layers slowly decompose, releasing nutrients into the cushion for recycling, while creating a buffer against drying winds and temperature changes. Tundra plants tend to stay small and compact to avoid wind damage.

Many insects spend most of their life cycle underground or underwater, and emerge only for a brief mating period, letting environmental conditions trigger life cycle changes. Others take a long time to mature, such as the woolly-bear caterpillar, which takes 7-14 years to develop from a fertilized egg to a ma-

BOX 48. MOSQUITOES

▶ Everyone who has put his or her nose out-of-doors in the northern summer knows what a mosquito is. In Finland alone, 38 species of mosquitoes in five genera have been recorded, the most plentiful of which are the *Aedes* species, which include the most painful bloodsuckers. The species that attack humans vary from south to north, and can be recognized by their gray colour and black-and-white-striped abdomen. When they attack, however, we normally just kill them with a quick slap. Do we really know who they are?

Mosquitoes are spread all over the world, and are especially abundant in northern latitudes close to the treeline. A curious exception is Iceland where mosquitoes have never been recorded. Some species have a circumpolar distribution, while others occur either in Eurasia or North America. Not all attack humans – some specialize in other mammals or birds. Reindeer and caribou suffer from these flies so badly that they migrate to open areas that have fewer mosquitoes. This behavior allows them to avoid mosquito-borne diseases and parasites, which include tularemia, arboviruses, and various encephalitises.

Only female mosquitoes are bloodthirsty. Males sit in vegetation feeding on the nectar of flowers. After having a proper blood meal, the female seeks a suitable place for egg-laying. Depending on the species, this may be either a small pond or just moist ground where there is likely to be shallow water the following spring. The eggs are resistant to dry and cold conditions and hatch as the snow melts. A few species overwinter as adults and emerge early in summer. In the water, the tiny larvae catch small plankton and grow rapidly. Many aquatic invertebrates, like diving beetles, prey upon those larvae. Within two to three weeks of hatching, the larvae change into pupae, which also swim. After a few days, the pupae swim to the surface where it sheds its skin. The resulting imago sits for a short period on the empty pupal skin, drying out and hardening its wings. It then flies to nearby vegetation to hide. Adult mosquitoes are an important part of the diet of many northern insectivorous birds.

Juhani Itämies, Zoological Museum, Oulu University, Finland

SOURCES/FURTHER READING

Hackman, W. 1980. A check list of the Finnish Diptera. I. Nematocera and Brachycera (*s.str.*). *Notulae Entomologicae* 60:17-48.

Itämies, J. and I. Lumiaho. 1982. Mosquitoes (Diptera, Culicidae) attracted by man in SW and NE Finland. *Aquilo Seriologica Zooligica* 21: 1-5.

Natvig, L.R. 1948. Contributions to the knowledge of the Danish and Fennoscandian mosquitoes. Culicini. Norsk Entomologisk Tidskrift Suppl I:1-567.

Wood, D.M., P.T. Dang, and R.A. Ellis. 1979. The mosquitoes of Canada. Diptera: Culicidae. *The Insects and Arachnids of Canada*, Part 6. p. 1-389.

---

ture adult. Similarly, a long life span allows many species of birds to select favorable years for reproduction rather than trying to produce eggs and rear chicks in poor summers. Mammal populations are often highly cyclical. Small mammals, such as lemmings, are able to reproduce rapidly when conditions are good, sending their populations soaring. When this happens, the populations of their predators – ermines, least weasels, arctic foxes, snowy owls, and jaegers (skuas) – rise rapidly as well.

Functionality

Many Arctic plants are good at colonizing bare ground, relying on nutrients taken from the exposed rocks and soils. Lichens can grow, though extremely slowly, on surfaces that are not disturbed. Mosses compete with vascular plants for favored habitats. Some vascular plants are able to thrive even in undeveloped polar desert soils, provided there is adequate moisture. Once plants are able to flourish, they can alter

The mountain burnet **Zygaena exulans** *inhabits the tundra of Russia and Fennoscandia and is also found in mountainous areas of central Europe. Its hairiness is a characteristic Arctic/alpine adaptation.*

burst of spring and summer feeds herbivores, which return nutrients in the form of feces. The grazing animals also remove dead plant litter, helping to stimulate further plant growth. Some Arctic plants are adapted to grazing, and respond by re-growth even under intense grazing pressure.

Animals are affected by climate directly and through its effects on plants. Early winters may place enough stress on animals to affect their physical condition and subsequent reproductive success. Resident birds and mammals, such as ptarmigan and muskoxen, can be sensitive to even brief delays in the arrival of spring with its fresh forage. Migratory species such as caribou follow the melting snow, often seeking areas where they can graze on exposed vegetation and then take refuge from insects by moving to snow fields. Changes in vegetation structure and composition alter the productivity of the landscape, and thus its ability to support herbivores. One result of this variability is the fluctuating populations of many species, especially small mammals and their predators.

Several studies in the Arctic areas have detected regular variations in lemming population dynamics, often called population cycles. The abundance of lemming predators, such as weasels, ermines, arctic foxes, raptors and owls, follows the lemming cycles. In northeastern Greenland, for instance, ermine is the main predator of lemming, and its abundance fol-

the local environment and, indeed, the entire landscape. Plants help retain water that might otherwise drain away or evaporate. They also insulate the soil and underlying permafrost.

Overall, the composition of plants across a landscape may remain roughly constant, while the local distribution of plants, insects, birds, and mammals varies widely. Physical changes, including fires, erosion, thawing, and lake drainage, bring areas back to early successional stages. Where the plant cover is broken, for example by overgrazing or industrial activity, the soils and vegetation may change, perhaps permanently.

The weather and its variability are the primary factors that control plant growth in a given year. The

There are many Fritillary butterflies in the Arctic. Freija's fritillary **Clossiana freija** *is found in the Eurasian lowland tundra.*

Caloplaga elegans, *a crustaceous lichen.*

lows the lemming cycles. When the lemming population increases from a low point of the cycle, the population of ermines, also at a low level at the start of the cycle, increases as well, but lags behind the lemming population because of the time required for the ermines to reproduce in response to the increase in available food. Eventually, the lemming population crashes. Because the ermine population is high at this point of the cycle, ermine predation on the remaining lemmings is intense, further depressing the lemming population. Next, the ermines run out of food and decline as well. Here, the cycle begins again.

The lemming cycle drives the behavior of other species, both directly and indirectly. Snowy owls and jaegers nest only in places and years when lemmings are abundant, since prey availability is essential for laying and incubating eggs, and for feeding chicks. Eiders nesting in areas with snowy owls and jaegers have much higher success rearing chicks when lemmings are abundant, even though they do not themselves eat lemmings. Rather, the eiders nest close enough to the larger birds' nests to benefit from their ability to drive away foxes, but not so close as to attract predation by the owls or jaegers. In years when lemmings are not abundant, the owls and jaegers do not drive the foxes away, and the eiders' nests have no such protection.

The tundra and certain oases in polar desert areas, including the freshwater areas within them, are critical breeding, feeding, and molting habitat for the majority of the world's geese and for many species of waders and other migratory birds. Many areas within the Arctic are either protected or sufficiently remote to ensure

BOX 49. LEMMINGS: KEY ACTORS IN THE TUNDRA FOOD WEB

▶ Arctic lemmings include true lemmings, of the genus *Lemmus*, and collared lemmings, of the genus *Dicrostonyx*. Although both have a circumpolar – though not identical – distribution, the two genera differ in many respects. *Dicrostonyx* is much more resistant to extreme low temperatures than *Lemmus*. Consequently, its distribution extends farther north: collared lemmings are found on the northernmost islands of the Canadian Arctic and in northern Greenland, whereas the range of true lemmings reaches down to the boreal zone and does not include Greenland or the northern Canadian islands. *Lemmus* lemmings eat mosses, supplemented by grasses and sedges. *Dicrostonyx* lemmings prefer forbs and shrubs like avens and willows. This distinction is reflected in habitat use. In the Arctic tundra, *Lemmus* is usually found on wet lowlands or moist patches. *Dicrostonyx* lives almost exclusively on dry and sandy hills and ridges.

Lemmings are well-known for their violent population fluctuations, the impacts of which are reflected in various ways. High numbers of lemmings have a strong impact on the vegetation and affect the nutrient cycle. Shallow permafrost slows down soil processes, but when lemmings at peak densities graze away much of the vegetation cover, the summer thaw can penetrate deeper into the soil,

The Norwegian lemming **Lemmus lemmus**.

making more nutrients available. Lemmings also have a strong impact on the biomass and species composition of the vegetation.

The numbers of both mammalian and avian predators depend on the lemming fluctuations. In low lemming years, resident mammalian predators, including the arctic fox, ermine, and least weasel, hardly breed. During peak years,

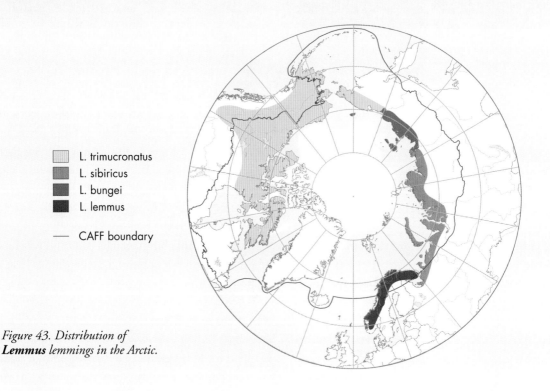

Figure 43. Distribution of **Lemmus** *lemmings in the Arctic.*

- L. trimucronatus
- L. sibiricus
- L. bungei
- L. lemmus
- CAFF boundary

*The collared lemming **Dicrostonyx groenlandicus**.*

arctic foxes can have litters of up to 20 kits, and ermine and weasel numbers increase rapidly. Many species of birds of prey move nomadically, or briefly visit their breeding grounds in the Arctic in the search of lemming peaks. Snowy owls and jaegers (or skuas) are particularly well-known for their dependence on lemmings, breeding only in years when lemmings are abundant.

The causes of these population cycles have long been a topic of research and often heated discussion. Food, predation, and intrinsic physiological factors are possible explanations, each of which has some scientific support. It may well be that in different areas different factors play the critical role. In the Canadian Arctic, for example, the long-term low density of *Dicrostonyx* is explained by heavy predation. In Fennoscandia, on the other hand, lemming crashes in alpine areas seem to be caused by shortage of food, but when lemmings migrate into the boreal forests, they are regulated by predation.

In Northern Fennoscandia, the famous migrations of the Norwegian lemming (*Lemmus lemmus*) can temporarily extend its range some 200 kilometers into the boreal forest, though they only occur three times a century. Norwegian lemmings have both spring and fall movements. Spring movements are caused by snow melt and last only two or three weeks, whereas fall migrations are density-dependent, and may last two to three months. Such remarkable movements in the fall have not been reported for other lemming species.

Heikki Henttonen, Vantaa Research Centre, Finnish Forest Research Institute, Finland

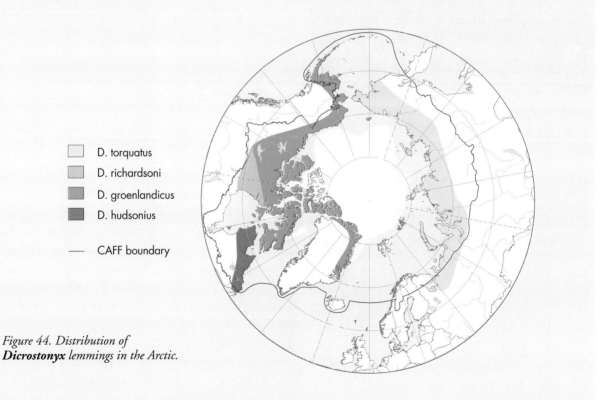

*Figure 44. Distribution of **Dicrostonyx** lemmings in the Arctic.*

relatively healthy habitat. Environmental changes and hunting pressure, both within and outside the Arctic, have greatly affected the numbers of many species. Some, such as the Steller's eider, have become endangered, while others, such as the snow goose and barnacle goose, have prospered. While the quality of the tundra may generally be good for the birds that depend on it, climate change and other large-scale alterations are likely to have long-term effects that will greatly reduce nesting and molting habitat.

Interactions with other regions

The tundra and polar desert cover most of the land of the Arctic region. Many species are shared with, or migrate to, the forest-tundra, which also provides refuge from extreme weather conditions. Freshwater is found through much of the tundra, in the form of lakes, ponds, rivers, streams, and wetlands, the boundaries of which can be hard to distinguish in some low-lying areas. In the polar desert, freshwater is less common and highly seasonal. It shapes the landscape through erosion, shaping the ground surface, and providing moisture to plants. Most of the Arctic coastline is a boundary between tundra or polar desert and the ocean. In winter, shorefast sea ice – the ice attached to land – can blur the boundary, as land animals may wander onto the frozen ocean. Many land mammals have reached islands in this way, and the solid surface can also help plant seeds spread to islands.

The tundra and polar desert interact functionally with other parts of the Arctic system. Sea birds feed from the sea but nest on land, and may be affected by land animals and conditions. Birds and mammals migrate from the tundra to the forest-tundra and the forest. Floods deposit soils and nutrients on land. Runoff brings nutrients to freshwater and to the ocean. Animals that feed in water but live on land return organic matter in the form of feces and, when they die, their bodies.

Many migratory species, especially birds, link the tundra and polar desert with temperate and tropical areas. Migrations of ducks, geese, and waders are a substantial transfer and net export of biomass each

Lemming predators:
Long-tailed jaeger **Stercorarius longicaudus**
Ermine **Mustela erminea**
Arctic fox **Alopex lagopus**

BOX 50. PARASITISM: AN UNDERESTIMATED STRESSOR OF ARCTIC FAUNA?

▶ Parasitism is perhaps the most successful form of life and occurs everywhere. Every animal species is the natural host to a variety of parasites. A parasite can infect the host with little or no effect, or can cause derangements to the function of the animal, resulting in disease or death. The effects of parasites and disease on the long-term health of wildlife populations are unclear and controversial, often because not enough is known. Parasites have caused short-term crashes in some wildlife populations. Recent discoveries point to more subtle effects as well. In Svalbard reindeer, for example, gastro-intestinal nematodes reduce pregnancy rates.

Some Arctic parasites can infect humans. Sarcocystis and toxoplasma (protozoal parasites), and trichinella (a nematode) encyst in the skeletal muscle of a variety of animals. They generally do not harm the host, but can cause disease in people via the consumption of improperly prepared meat. Most cases of clinical trichinosis in people in the Arctic, for example, have been caused by eating uncooked polar bear or walrus meat. Another parasite, the fox tapeworm, *Echinococcus multilocularis*, can infect humans but is usually treatable.

Microparasites, such as bacteria, viruses, and protozoa, can multiply rapidly in the host and tend to exhibit boom-and-bust behavior (epidemics) in the host populations. Antibodies to morbilliviruses have been found in marine mammals in the Arctic. These viruses have the potential to cause epidemic disease similar to the phocine distemper that caused mass mortality of common seals in the North Sea in the 1980s. Rabies causes regular epidemics in arctic and red fox, often leading to transmission to sled dogs and potential human contact. Tularemia, a bacterial disease, occurs in several Arctic species, including arctic hares and rodents, and can cause epidemic and endemic disease.

Microparasites can also be present endemically, or occur at a high rate in the population normally. These forms are thought to be adapted to the host and most likely not to affect the population, unless they affect the reproductive ability of the host animal. Brucellosis is common in most of the Arctic populations of caribou and reindeer, and has the potential to cause abortion. It may be more pathogenic in moose, and can be transmitted to humans.

Macroparasites, such as worms and flukes, are multicellular, tend to be enzootic (to remain at fairly constant levels through time), and often have complicated life cycles. Although the eggs or larvae of some macroparasites can pass directly from one definitive host (the animal in which the parasite reproduces) to another, many macroparasites go through one or more intermediate hosts – animals in which the parasite lives but does not reproduce.

The impact of the parasite on the intermediate host can be very different from its effect on the definitive host. The intermediate host is often harmed, enhancing the likelihood that the parasite will be transferred to the definitive host, for example, by making prey animals easier to catch so that predators will eat the parasite along with the prey. In the definitive host, on the other hand, it is usually not advantageous for the parasite to be too harmful, because the death of the definitive host would mean the death of the parasite.

Because of the time taken to develop at each stage, the parasite's response to changes in host populations is typically slow, and is further retarded by the harsh conditions of the Arctic. The periodicity of host population fluctuations can also affect the parasite's capacity to track host densities. The four-year cycles of voles in Finnish Lapland seem to be too short for helminths (intestinal worms) to catch up with the changes in host densities. On the other hand, in the ten-year cycles of snowshoe hares in Canada, the abundance of similar kinds of helminths clearly follows the host cycle.

The successful life cycle and infective stages of many parasites depend on weather. This strongly suggests that climate change will shift the rates and intensity of infection. Experimental work with the recently discovered muskox lungworm, *Umingmakstrongylus pallikuukensis*, revealed the importance of summer temperatures for its development. It is notable that the lungworm infections have become conspicuous during a time when summer temperatures have been increasing. Early summer precipitation is essential for the survival of intermediate stages of many helminths in Arctic rodents. Changing climatic patterns are likely to affect the abundance and other dynamics of these parasites.

Heikki Henttonen, Vantaa Research Centre, Finnish Forest Research Institute, Finland
Kathleen A. Burek, Alaska Veterinary Pathology Services, Eagle River, Alaska, USA

SOURCES/FURTHER READING
Haukisalmi, V., and H. Henttonen. 1990. The impact of climatic factors and host density on the long-term population dynamics of vole helminths. *Oecologia (Berlin)* 83:309-315.
Haukisalmi, V., H. Henttonen, and F. Tenora. 1988. Population dynamics of common and rare helminths in cyclic vole populations. *Journal of Animal Ecology* 57:807-826.
Hoberg, E.P., L. Polley, A. Gunn, and J.S. Nishi. 1995. *Umingmakstrongylus pallikuukensis* gen.nov. et sp. nov. (Nematoda: Protostrongylidae) from muskoxen, *Ovibos moschatus*, in the central Canadian Arctic, with comments on biology and biogeography. *Canadian Journal of Zoology* 73:2266-2282.
Irvine, R.J., A. Stein, O. Halvorsen, R. Langvatn, and S.D. Albon. In press. Life-history strategies and population dynamics of abomasal nematodes of the high arctic Svalbard reindeer (*Rangifer tarandus platyrhynchus*). Submitted to *Parasitology*.
Kutz, S. 1999. The biology of *Umingmakstrongylus pallikuukensis*, a lung nematode of muskoxen in the Canadian Arctic: field and laboratory studies. Unpublished doctoral dissertation, University of Sasketchewan, Saskatoon, Sasketchewan.
Schantz, P.M., B. Gottstein, T. Ammann, and A. Lanier. 1991. Hydatic and the Arctic. *Parasitology Today* 7:35-36.
Thompson, R.C.A., and A.J. Lymbery, eds. 1995. *Echinococcus and hydatid disease.* Oxford: CAB International. 477p.

BOX 51. SANDPIPERS: BIRDS OF THE TUNDRA

▶ While the diversity and abundance of most taxonomic groups declines as one moves north, the sandpiper family is a dramatic exception. Of the 24 species of sandpipers, nearly all have their breeding grounds substantially or entirely in the Arctic. Of the estimated 19 million birds in this family, 94% breed in the Arctic.

An analysis of the species richness of Calidrid sandpipers (the eighteen species of sandpipers in the genus *Calidris*, plus two closely-related species, the spoon-billed sandpiper and the stilt sandpiper) shows an area of high diversity in the Bering Sea region, reaching a peak in easternmost Chukotka where 11 species are found. By contrast, the area from 80°W (west of Baffin Island) eastward to 80°E (east of the Yenisei River) has no area with more than four species of calidrids. This phenomenon has been found for other taxa, for example geese, but has not been fully explained.

The area with more than three species coincides precisely with the area defined as Arctic according to botanical criteria. The boundary of this area is approximately the treeline, which appears to be the southern boundary in summer for most breeding sandpipers. In addition, the area of highest diversity is the area containing the highest number of rare plant species, a connection that requires more examination.

Sandpipers flock to the Arctic in summer for the innumerable insects on the tundra, which provide an ideal food source for rearing young, and which are always available in the perpetual daylight. In winter, sandpipers head south, reaching virtually all the world's wetlands on their migrations and in their winter habitats. Despite the attractions of the Arctic, breeding conditions can still be harsh, and sandpipers have evolved special breeding strategies.

One strategy particularly common among Arctic sandpipers, including the dotterel, sanderling, and the phalaropes, is that the males care for the young. The female dotterel and phalarope has more colorful plumage. After laying her eggs, she may breed again with a different male, producing two or more clutches of eggs in a good summer.

For many other sandpiper species, such as the curlew sandpiper, the female provides most of the care for the hatchlings, whereas the broad-billed sandpiper evolved an intermediate strategy, under which the female incubates the eggs but departs once the eggs have hatched.

These strategies appear to be closely linked with migration patterns. Small changes in these conditions, either inside or outside the Arctic, could have a dramatic effect on sandpiper populations by throwing their migratory and breeding patterns out of balance. Conservation efforts need to consider the entire flyway and the breeding habits of sandpipers.

A final observation about sandpipers is that many Arctic wading birds, as well as some species of other bird families, have brownish-red coloration, especially on the breast and neck. Why this is so is not clear, though the color is always the same shade. Perhaps it is camouflage for birds sitting on the nest on the summer tundra, where low-angled sunlight gives a russet hue to many objects.

Christoph Zöckler, UNEP – World Conservation Monitoring Centre, Cambridge, U.K.

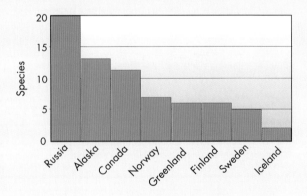

*Figure 45. The number of **Calidris** species breeding in each Arctic country.*

year. The presence of huge numbers of these species can transform areas of tundra through grazing and via the deposit of large quantities of fecal matter. Events and changes in the winter range of these species can have substantial impacts on the tundra and polar desert. The example of snow geese in North America described on page 87 shows how improved food supplies on the winter range of the geese can lead to overgrazing in the Arctic. By contrast, destruction of winter habitat for other species has led to great declines in the numbers of birds reaching the tundra, with consequences for predators, including humans.

SOURCES/FURTHER READING

Piersma, T. 1996. Scolopacidae. In: J. del Hoyo, A. Elliot, and J. Sargatal, eds. *Handbook of the Birds of the World*. Vol. 3: Hoatzins to Auks. Barcelona: Lynx. p. 444-487

Talbot, S., B.A. Yurtsev, D.F. Murray, G.W. Argus, C. Bay, and A. Elvebakk. 1999. *Atlas of Rare Endemic Vascular Plants of the Arctic*. Conservation of the Arctic Flora and Fauna (CAFF) Technical Report No. 3. Anchorage, Alaska: U.S. Fish and Wildlife Service. iv + 75p.

Tomkovich, P. S. 1996. *Calidris*-sandpipers of north-eastern Siberia. *Dutch Birding* 18: 11-12.

Yurtsev, B.A. 1994. The floristic division of the Arctic. *Journal of Vegetation Science* 5(6): 765-776.

Zöckler, C. 1998. *Patterns in Biodiversity in Arctic Birds*. WCMC Bulletin No.3. Cambridge, U.K.: World Conservation Monitoring Centre. 15 p.

Table 5. Population size and distribution of sandpipers in the Arctic, with the percentage of their breeding range that is in protected areas. Species in bold breed entirely in the Arctic. Figures in brackets indicate percentage in areas of high breeding density.

COMMON NAME	SCIENTIFIC NAME	POPULATION (1000 IND.) TOTAL	ARCTIC	BREEDING AREA (Mill. km^2)	AREA PROTECTED IN ARCTIC (%)
Great Knot	*Calidris tenuirostris*	330	280	1.05	0.2
Red Knot	***C. canutus***	1261	1261	1.19	18
Sanderling	***C. alba***	478	478	1.23	22
Semi-palmated Sandpiper	***C. pusilla***	3500	3500	2.53	11
Western Sandpiper	***C. mauri***	2500	2500	0.15	52
Red-necked Stint	***C. ruficollis***	471	471	0.46	2.6
Little Stint	***C. minuta***	1411	1411	1.07	12.8
Temminck's Stint	*C. temminckii*	300	285	2.01	4.3
Long-toed Stint	*C. subminuta*	65	6	2.59	–
Least Sandpiper	*C. minutilla*	150	100	4.4	10
White-rumped Sandpiper	***C. fuscicollis***	75	75	1.31	6.2
Baird's Sandpiper	***C. bairdii***	75	75	2.31	12.3
Pectoral Sandpiper	***C. melanotos***	150	150	2.63	12.6
Sharp-tailed Sandpiper	***C. acuminata***	166	166	0.13	18.4 (3.4)
Curlew Sandpiper	***C. ferruginea***	1096	1096	0.84	16 (30.8)
Purple Sandpiper	*C. maritima*	180	175	0.98	13
Rock Sandpiper	*C. ptilocnemis*	5	1	0.12	33.7
Dunlin	*C. alpina*	2834	2300	2.65	13 (19.8)
Spoon-billed Sandpiper	***Eurynorhynchus pygmeus***	2	2	0.068	4.02
Stilt Sandpiper	***Micropalama himantopus***	75	75	0.527	12.11
Surfbird	*Aphriza virgata*	50–70	16	n/a	n/a
Broad-billed Sandpiper	***Limicola falcinellus***	13–22	13–22	n/a	n/a
Buff-breasted Sandpiper	***Tryngites subruficollis***	25	25	n/a	n/a
Ruff	*Philomachus pugnax*	3600	3000	n/a	n/a

Natural variability

The tundra and polar desert are constantly changing on small and large scales, driven by short- and long-term climatic variation. Longer-term changes are still unfolding, including the development of mature soils following the last Ice Age. Plants are colonizing areas exposed by retreating glaciers in the past century or two. These changes alter the landscape substantially over time, for example, by the movement of the treeline as described in the previous chapter. Small-scale changes also take place, includ-

Temminck's stint **Calidris temminckii**

Curlew sandpiper **C. ferruginea**

Little stint **C. minuta**

Sanderling **C. alba**

Pectoral sandpiper **C. melanotos**

Red-necked stint **C. ruficollis**

Dunlin **C. alpina**

Western sandpiper **C. mauri**

Purple sandpiper **C. maritima**

American golden plover **Pluvialis dominica**

Turnstone **Arenaria interpres**

Ringed plover **Charadrius hiaticula**

Red phalarope **Phalaropus fulicarius**

Grey plover **Pluvialis squatarola**

ing the patterns of landscape change and plant succession.

Animal populations, too, often go through alternating periods of abundance and scarcity. The dynamics of these cycles are not entirely understood, though predation, the quality of forage, and climate play important roles. The cycles also affect habitat and other species. Overgrazing can cause substantial changes in vegetation and the underlying soil. Abundant prey causes predator populations to rise, with related effects such as the improved nesting success of tundra birds when lemming populations are high.

Human uses and impacts

From the earliest times, humans have hunted and gathered tundra plants and animals for food and material, uses that continue today in much of the Arctic. Many animals are hunted for meat or fur. Plants are gathered for food, medicines, clothing, baskets, and other purposes.

In recent centuries, the tundra has been used in increasingly intensive ways. Reindeer herding spread across Eurasia, supporting larger human populations while affecting rangelands, the position of the treeline, and the abundance and distribution of wild reindeer and other grazing animals. Trapping for the fur trade brought the profit motive to remote areas, allowing people to acquire manufactured goods and spurring significant harvests in many regions. The hunting of caribou and muskoxen to sell meat and hides developed in the 20th Century. Sport hunting for polar bears, caribou, and muskoxen is another economically significant use. The impacts of these types of use vary widely, depending largely on how effectively they are regulated.

The 20th Century also saw the beginning of large-scale industrial use of tundra and polar desert regions, particularly for mining and for oil and gas development. Such activities have taken place throughout the Arctic, and many continue to expand. There is considerable interest in exploring for oil and gas in Alaska, Russia, and Canada, and new mines are being developed in Greenland. The construction of roads and pipelines across the tundra introduces a variety of impacts to the surrounding area, the long-term effects of which may be severe. Tourism is another form of economic development on the rise, although its environmental impacts are poorly understood.

The vegetation of the tundra and polar desert is vulnerable to physical disturbance. Vehicle tracks can last for years or even decades, especially in areas of permafrost with high concentrations of ice. Disturbances to the vegetation cover can alter freeze-thaw patterns, leading to the thawing and collapse of the ground and the development of deep cavities and erosion channels. In areas with lower concentrations of ice, vehicle tracks can affect species richness and total plant biomass. Such impacts can affect basic ecosystem functions such as nutrient cycling. They can also act in a cumulative way with other impacts, making predictions more complicated. Such effects are not limited to summer travel, but can be caused by vehicles traveling across frozen ground in winter.

Roads and pipelines can fragment the landscape by creating physical and biological barriers, especially if drainage patterns are affected. Such disturbances are most common near human settlements and in areas of industrial development. An example is the railway that was built across the migratory route of wild reindeer on the Taimyr Peninsula in the 1960s. Trains killed some animals, and many more were deflected or delayed by the prospect of crossing the tracks. In 1969, a gas pipeline was built parallel to the railway. Some 20,000 reindeer did not cross it, leaving them stranded south of their usual calving and summer range. Eventually, a 60-kilometer fence was built to direct the reindeer away from the transportation corridor.

Many areas of the Arctic have been explored for minerals and for oil and gas deposits, including the North Slope of Alaska. The National Petroleum Reserve-Alaska (NPRA) was established in 1924 and explored extensively in the 1940s and 1950s through a program of seismic testing and test wells. While the tundra is largely intact functionally, there are visual signs of local disturbance in the form of vehicle tracks, rusting oil drums, and waste pits from the test wells. The oil development sites currently in operation on

Vezdekhod, the Russian all-terrain vehicle goes anywhere, but can cause long-term damage to the tundra.

the North Slope of Alaska, centered at Prudhoe Bay, affect the surrounding tundra. The construction of roads and gravel pads, dust plumes from vehicle traffic on gravel roads, and alterations to drainage patterns across the tundra have left their mark.

In Russia, oil and gas development on the Yamal Peninsula has, over the course of three decades, destroyed some areas of tundra and degraded others. The region is now less able to support reindeer herds. In addition to changing the vegetation cover, greater hunting pressure from the increased number of people brought to the region by the development has reduced populations of fur-bearing animals. These activities plus the effects of grazing by reindeer have considerably changed the ecology of the region. An additional consequence of such development is the spread of species new to the Arctic, which have colonized disturbed areas in the tundra. In addition to industrial impacts, reindeer grazing has had an impact on an estimated 20% of the Russian tundra.

Livestock herding has significant impacts in some areas. Sheep have caused widespread erosion in Iceland. Predator control efforts have removed most large carnivores from Fennoscandia. Wild reindeer fared poorly in competition with domesticated herds in Russia, though the wild herds have increased following the decline of reindeer farms over the past decade. Changes in reindeer herding and marketing practices in Fennoscandia led to a great increase in domestic reindeer populations in the past twenty years. In Norway, the reindeer population has declined in the last decade as a result of government policies, poor conditions, and predation. Along fence lines and migration routes, trampling and overgrazing have degraded the plant cover. In Russia, reindeer herds used to be spread widely across northern areas, dispersing the impact of grazing. This was possible because the collective farms of the Soviet era had helicopters and other means of supporting the distant herders. Today, herders stay closer to the settlements, and the herds are concentrated, leading to localized overgrazing.

Overhunting has eradicated species in some areas. Muskoxen disappeared from Alaska in the 19th Century, largely as a result of the use of rifles to provide meat for commercial whalers who wintered in the Arctic aboard their ships. In the 20th Century, they were reintroduced in several places in the state, which has been a controversial move. Muskoxen are often thought to compete with caribou, and they may introduce diseases or other unexpected problems. Efforts to restore bears, wolves, and wolverines in Fennoscandia have met with objections on similar grounds, particularly from reindeer herders. On the other hand, the return of such species can enhance opportunities for viewing and hunting, and can restore natural ecosystem functions and patterns.

Diffuse disturbances in the form of contaminants, acid rain, and ultraviolet radiation pose an additional threat to Arctic ecosystems. Radionuclides from the atmospheric testing of nuclear weapons between the 1940s and the early 1960s, and from industrial accidents, most notably at Chernobyl in 1986, have spread across the entire Arctic. In the tundra and polar desert, radioactive contaminants accumulate in lichens, mosses, and cushion plants that draw heavily on air and precipitation for their nutrients. Caribou and reindeer, in turn, are exposed through the lichens that make up most of their winter diet. Thus, they have some of the highest concentrations of radionuclides among Arctic animals. The radionuclide burden from atmospheric testing has diminished greatly from its peak in the 1960s, but areas that were heavily affected by Chernobyl still suffer elevated levels.

Heavy metals, such as lead, cadmium, and mercury, can be toxic to many forms of life if the concentrations of certain chemical forms of the metals are high enough. Many of these metals are transported long distances by wind and deposited with precipitation in the Arctic. The metals are also found in the rocks and soils of many parts of the Arctic. Plants and animals in these areas can have high concentrations of heavy metals as a result of natural processes. Human activities can add to natural levels. In areas near large mining and smelting operations, particularly in Russia, the forest and the tundra have been affected over large areas. Persistent organic pollutants (POPs) originate largely in the temperate and tropical latitudes and are transported to the Arctic by air and water. Although the concentrations are typically higher in the marine food web than in the terrestrial one, these substances are found in tundra plants and animals.

Acid rain is primarily a product of industrial activity, and may occur hundreds of kilometers from the source of pollution. Particularly in the European Arctic, acid rain poses a significant threat to tundra systems. Tundra and polar deserts are sometimes classified by the acidity or alkalinity of their soils, which influences the types of plants and microbes found in the area. Changes in the acid balance of the soils can lead to changes in the type and extent of vegetation. Acid rain can also dissolve components of soils and rocks, affecting both the soils and the freshwater that passes through them, as is discussed in the next chapter.

Ultraviolet radiation (UV) has increased at the Earth's surface in the Arctic due to thinning of the ozone layer. It is potentially a source of significant impacts to the tundra and polar desert. UV is capable of causing changes to the chemical structure of plant and animal cells and genes. It can also harm the eyes of animals. Although not enough is known about the effects of increased UV on tundra plants to be able to predict impacts, the measured increases in UV are typically greatest at higher latitudes. The Arctic could be affected to a greater degree than temperate and tropical regions.

BOX 52. EROSION IN ICELAND

▶ Iceland's soils, the so-called *andosoils* that form in volcanic deposits, are fertile but sensitive to disturbance. Once, such soils covered most of the island, supporting lush vegetation such as birch and willow woodlands. The Vikings arrived in 875, and the pressures brought by human settlement, such as grazing and the cutting and burning of woodlands, have combined with the effects of a cooling climate to degrade much of Iceland's land cover. When humans arrived, Icelandic deserts covered 5,000 to 15,000 square kilometers. Today, more than 37,000 square kilometers are barren, and an additional 10,000 to 15,000 have limited plant cover and production. Much of the degradation occurred early, leading to social unrest and poverty, possibly contributing to Iceland's loss of independence in the 13th Century.

The current state of Iceland's terrestrial ecosystems is generally poor. In 1900, only a few small native birch stands remained. Today, birch trees cover only 1% of the island. The remaining vegetation is often very degraded, dominated by species that can tolerate or avoid grazing by sheep. Some areas, particularly lowlands, as well as the highlands in the west and northwest, have retained their soil cover and are the backbone of Icelandic agriculture. Erosion remains an enormous environmental problem, however, and despite a moist climate, the sandy deserts are probably the largest outside the world's arid regions.

Icelanders have long recognized the poor state of their lands. The Icelandic Soil Conservation Service is the oldest such agency in the world, having been formed in 1907 to combat sand encroachment and soil erosion. The Icelandic Forest Service was established around the same time, to preserve the remaining stands of trees on the island. Today, the Icelandic government, farmers, non-governmental organizations, and the general public are fighting to restore Iceland's ecosystems through extensive reclamation work.

The task ahead is immense, and cannot be accomplished without a much larger effort than is currently underway. More resources must be devoted to restoration, and research and teaching must be directed towards an understanding of how Iceland's ecosystems work and how they can be restored economically and ecologically.

Ólafur Arnalds, Agricultural Research Institute and Soil Conservation Service, Reykjavik, Iceland

Erosion in northeast Iceland.

The effects of overgrazing by reindeer in Finnmark, Norway.

Climate change

Climate change is already affecting the distribution of moisture and the northward expansion of the forest. It is likely to affect the distribution of most plants and animals of the tundra and polar desert. Climate change models predict a great reduction in the tundra as the forest moves towards the Arctic Ocean, although the process of treeline shift is slow. Some estimates suggest that by 2100, half of the tundra could be lost. For many bird species, the loss of breeding and molting habitat would have severe consequences. Melting of permafrost due to climate warming can lead to widespread disturbances in vegetation through landslides, sinking and deformation of soil, and other physical changes.

Such changes can disrupt entire tundra ecosystems through changes to vegetation cover, water flow, and migration routes. It is essential, however, to consider each of these changes at an appropriate time scale, which may vary from the decadal to the millennial. In the meantime, other factors will modify the course of changes in the Arctic and beyond. The relationships between various changing environmental factors will complicate the short- and long-term impacts. Increases in snowfall, for example, may lead to cooler climates on local and regional scales. If the snow lasts longer in spring, it will reflect more sunlight away from the surface.

Furthermore, different species will be affected to different degrees. Species that are adapted to the cold may fare poorly when their competitors no longer

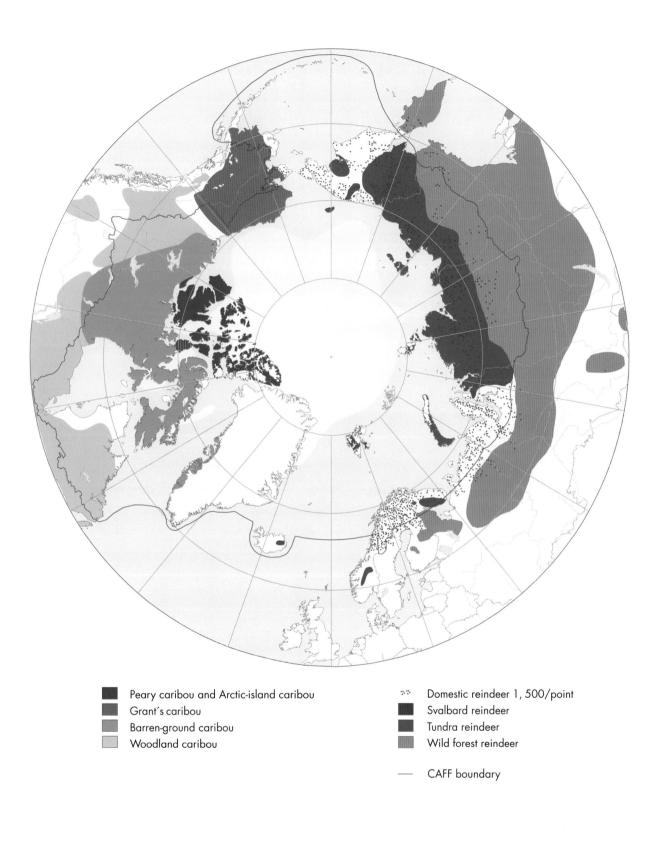

Figure 46. Distribution of wild reindeer, caribou, and domestic reindeer in the Arctic.

BOX 53. PEARY CARIBOU: EXTREME DEMANDS ON AN ENDANGERED CARIBOU

▶ The Peary caribou is the smallest of the forms of North American caribou and it faces the most extreme conditions of any caribou. It is only found on the Queen Elizabeth Islands of the Canadian High Arctic, where brief summers limit plant growth and large areas are scarcely vegetated polar desert or unvegetated gravel, barrens, rock slopes, and icefields.

Wide population swings among Peary caribou have probably been common over time. These swings tend to require long periods for a population to increase, whereas even one exceptionally severe winter can cause sudden crashes. The first count of Peary caribou was in 1961, when the total number was estimated at 25,845. The number fell through the 1960s, 1970s, and mid-1980s to 2,100 animals, before increasing again. Where there was hunting, Inuit voluntarily reduced it. From 1994-97, however, the number of caribou on the western Queen Elizabeth Islands dropped to 1,100, the lowest ever recorded. In the Bathurst Island complex, the population declined from 3,600 in 1961 to 300 in 1974, took 20 years to climb back to 3,000, and then crashed to fewer than 100 animals between 1995 and 1997.

Peary caribou are well-adapted to extreme conditions, and have the ability to find alternatives in all but the worst years. Even when Peary caribou were at their highest number, there was no evidence that they were limited by the total amount of food present. Instead, the main factor influencing rates of births and deaths is climate, specifically the type and duration of snow cover. Freezing rain, strong winds that create compacted drifts, crusting snow, and ground-fast ice in spring lock in much of the vegetation and make grazing difficult or impossible in winter and spring. The resulting malnutrition can reduce the ability of the caribou to reproduce and survive. The winters of 1994-97 were

Peary caribou *Rangifer tarandus pearyi*.

especially severe in this regard, and climate change models predict more of the same.

Currently listed as endangered in Canada, the long-term survival of the Peary caribou depends primarily on whether climatic conditions will allow the populations to recover. Other factors may play a role now that the populations are so low. Wolf predation may have a serious impact on the remnant herds, a warmer climate could lead to greater exposure to disease and parasites, and human impacts from resource development or contaminants may hinder the recovery of the Peary caribou.

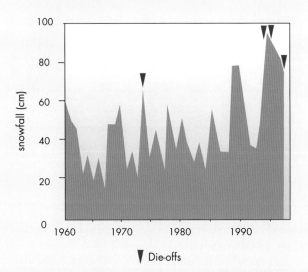

*Figure 47. Die-offs of Peary caribou **Rangifer tarandus pearyi** from the 1960s to 1990s.*

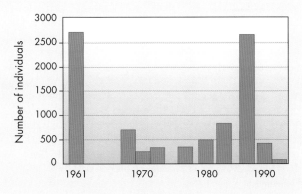

*Figure 48. The Peary caribou **Rangifer tarandus pearyi** population on Bathurst Island from the 1960s to 1990s.*

Frank Miller, Canadian Wildlife Service, Edmonton, Canada
Anne Gunn, Department of Resources, Wildlife and Economic Development, Government of the Northwest Territories, Yellowknife, Canada

SOURCES/FURTHER READING

Gunn, A., F.L. Miller, and D.C. Thomas. 1981. The current status and future of Peary caribou (*Rangifer tarandus pearyi*) on the Arctic Islands of Canada. *Biological Conservation* 19(1980-81):283-296.

Maxwell, B. 1997. *Responding to global climate change in Canada's Arctic.* Downsview, Ontario: Environment Canada. 82 pp.

Miller, F.L. 1990. Peary caribou status report. Environment Canada report prepared for the Committee on the Status of Endangered Wildlife in Canada, Edmonton, Alberta. 64 pp.

Miller, F.L. 1992. Peary caribou calving and postcalving periods, Bathurst Island complex, Northwest Territories, Canada, 1990. *Technical Report Series* No. 151:1-87. Edmonton, Alberta: Canadian Wildlife Service, Western & Northern Region.

Miller, F.L. 1995. Peary caribou studies, Bathurst Island complex, Northwest Territories, July-August 1993. *Technical Report Series* No. 230:1-76. Edmonton, Alberta: Canadian Wildlife Service, Western & Northern Region.

Miller, F.L. 1998. Status of Peary caribou and muskox populations within the Bathurst Island complex, south-central Queen Elizabeth Islands, Northwest Territories, July 1996. *Technical Report Series* No. 317:1-147. Edmonton, Alberta: Canadian Wildlife Service, Western & Northern Region.

Miller, F.L., R.H. Russell, and A. Gunn. 1977. Distributions, movements and numbers of Peary caribou and muskoxen on western Queen Elizabeth Islands, Northwest Territories, 1972-74. *Canadian Wildlife Service Report Series* No. 40:1-55.

Miller, F.L., E.J. Edmonds, and A. Gunn. 1982. Foraging behaviour of Peary caribou in response to springtime snow and ice conditions. *Canadian Wildlife Service Occasional Paper* No. 48:1-41.

Tener, J.S. 1963. Queen Elizabeth Islands game survey, 1961. *Canadian Wildlife Service Occasional Paper* No. 4:1-50.

BOX 54. RARE ENDEMIC VASCULAR PLANTS OF THE ARCTIC

▶ Rare plants are of aesthetic, ecological, educational, historical, and scientific value. Studies of rare plants address the immediate practical concerns of species conservation, assist in identifying and delimiting habitats that warrant protection, and contribute to a better understanding of the ecological and evolutionary processes that are fundamental to all of life's diversity. Rare plants are often ecological specialists that occupy microhabitats where competition with other species is absent or rare. Certain rare tundra species are presumed to be relicts of late-glacial steppe environments. These minor elements in contemporary tundra are well-adapted to recapture their former area should there be a return to the environmental conditions under which they had once been more widespread. Under such changing conditions, these rarities could become the "seeds" for widespread vegetational changes.

In light of the above, botanists from the eight Arctic countries have worked cooperatively to identify rare vascular plants endemic to the Arctic, to establish an annotated list of these plants, and to perform a gap analysis on the results. The recently compiled CAFF *Atlas of Rare Endemic Vascular Plants of the Arctic* identifies and provides an annotated list of ninety-six rare endemic plants and plots their distribution. It is hoped that the Atlas will be used to guide program decisions, especially the allocation of resources for protection and additional research. Preservation of habitat is the only logical strategy to save rare plants, hence it is necessary to learn as much as possible about the ecological requirements of the plants in question. For large areas of the Arctic, adequate data on the distribution of plants

Purple wormwood **Artemisia globularia** *var.* **lutea** *is a distinct yellow-flowered race of a rare vascular plant that occurs only on the islands of the Bering Sea, Alaska.*

are still lacking. The Atlas is expected to stimulate greater interest in these plants, and possibly the discovery of additional localities where they are found.

Gap analysis uses a rapid appraisal technique to identify gaps in the protection of biodiversity by determining whether target species and ecosystems are adequately represented within the existing network of protected areas. Once gaps are identified, the challenge will be to fill them through new reserves or changes in management practices. Using gap analysis, the relationship of rare plants to areas of protected habitats was assessed. Plants were then grouped into three categories: 1) unprotected (no localities are within protected areas), 2) partially protected (some localities are within protected areas), and 3) protected (all localities are within protected areas). Results indicate that 47% of the rare endemics are unprotected, 23% partially protected, and 30% fully protected. Following the IUCN (World Conservation Union) Red List threat categories, it was determined that 1% of the plants are Endangered, 19% Vulnerable, 29% Lower Risk Near Threatened, 26% Lower Risk Least Concern, and 24% Data Deficient.

Stephen S. Talbot, U.S. Fish and Wildlife Service, Anchorage, Alaska, USA

Drummond's bluebells **Mertensia drummondii** *is a rare plant found on sandy sites in the tundra of Alaska and Canada.*

SOURCES/FURTHER READING

Falk, D.A., and K.E. Holsinger, eds. 1991. *Genetics and conservation of rare plants.* New York: Oxford University Press. 302p.

Holmgren, A.H. 1979. Strategies for the preservation of rare plants. *Great Basin Naturalist Memoirs* 3:95-99.

IUCN. 1994. *IUCN red list categories.* Gland, Switzerland : IUCN (The World Conservation Union). 21p.

Lysenko, I.G., M.J.B. Green, R.A. Luxmoore, C.L. Carey-Noble, and S. Kaitala. 1996. *Gaps in habitat protection in the circumpolar Arctic: a preliminary analysis.* CAFF Habitat Conservation Report No. 5. Moscow: Ministry of Environmental Protection and Natural Resources, Russian Federation. 19p.

Rowe, J.S. 1988. On saving flora and fauna. *Canadian Plant Conservation Newsletter* 3(2):2-4.

CAFF (Conservation of Arctic Flora and Fauna). 1999. *Technical Report No. 3: Atlas of rare endemic vascular plants of the Arctic* (S.S. Talbot, B.A. Yurtsev, D.F. Murray, G.W. Argus, C. Bay, and A. Elvebakk, eds.) Anchorage: U.S. Fish and Wildlife Service. iv + 73p.

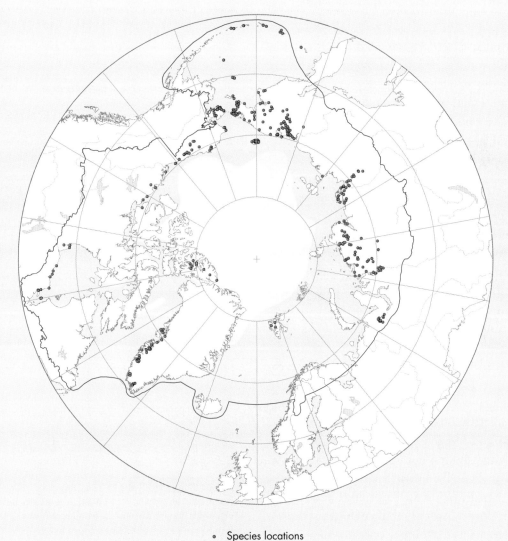

• Species locations

Figure 49. The distribution of rare endemic vascular plants of the Arctic.

face such severe conditions. Increases in storms and other extreme weather events may destroy tall plants or those living in areas that are easily flooded. Pest outbreaks will become more likely. New plant communities will develop over time, changing the patterns of rarity and abundance that are seen today. While the precise effects are unpredictable, the overall changes are likely to be substantial and readily observed.

Conservation

A variety of measures are used to conserve the tundra and polar desert regions of the Arctic. Chief among these is the designation of protected areas. Although the overall percentage of protection is relatively high, some habitat types and some geographic regions are poorly represented. Protected areas, though valuable, are neither sufficient by themselves nor the only means of protecting habitats and species. Land-use regulations offer degrees of protection for plants and animals. Hunting regulations help prevent over-exploitation. To achieve conservation aims, however, such regulations must be followed. Compliance and enforcement are lacking in some areas, particularly in Russia. Environmental impact assessments are widely used to identify and help prevent serious problems from industrial activities, though cumulative impacts are often ignored. Once development begins, further monitoring and regulation are needed, especially to make sure that preventive measures are working adequately.

For migratory species and for issues of global significance, international agreements can help achieve particular conservation goals. For migratory birds, several treaties address research and management issues across national borders. Several shared populations of animals, such as the Porcupine Caribou Herd in North America, are the subject of bilateral conservation agreements. Environmental contaminants and climate change are currently the subjects of global negotiations, (see box 38).

SOURCES/FURTHER READING

AMAP. 1998. *The AMAP Assessment Report: Arctic pollution issues.* Oslo: Arctic Monitoring and Assessment Program. xii +859p.

Behnke, R.H. 2000. Equilibrium and non-equilibrium models of livestock population dynamics in pastoral Africa: their relevance to Arctic grazing systems. *Rangifer* 20:141-152.

Bernes, C. 1996. *The Nordic Arctic environment-unspoilt, exploited, polluted?* Copenhagen: Nordic Council of Ministers. 240p.

Billings, W.D. 1992. Phytogeographic and evolutionary potential of the Arctic flora and vegetation in a changing climate. In: F.S. Chapin III, R.L. Jefferies, J.F. Reynolds, G.R. Shaver, and J. Svoboda, eds. *Arctic ecosystems in a changing climate: an ecophysiological perspective.* San Diego: Academic Press. p. 91-109.

Bliss, L.C. 1997. Arctic ecosystems of North America. In: F.E. Wielgolaski, ed. *Polar and alpine tundra. Ecosystems of the world, Volume 3.* Amsterdam: Elsevier Science. pp. 551-684.

Bliss, L.C., and N.V. Matveyeva. 1992. Circumpolar arctic vegetation. In: F.S. Chapin III, R.L. Jefferies, J.F. Reynolds, G.R. Shaver, and J. Svoboda, eds. *Arctic ecosystems in a changing climate.* San Diego: Academic Press. p.59-90.

Bliss, L.C., and K.M. Peterson. 1992. Plant succession, competition, and the physiological constraints of species in the Arctic. In: F.S. Chapin III, R.L. Jefferies, J.F. Reynolds, G.R. Shaver, and J. Svoboda, eds. *Arctic ecosystems in a changing climate.* San Diego: Academic Press. pp. 111-136.

Bliss, L.C., J. Svoboda, and D.I. Bliss. 1984. Polar deserts, their plant cover and plant production in the Canadian High Arctic. *Holarctic Ecology* 7: 305-324.

Briggs, D., P. Smithson, K. Addison, and K. Atkinson 1997. *Fundamentals of the physical environment.* London: Routledge. 557p.

Cargill, S.M. R.L. and Jefferies. 1984. The effects of grazing by lesser snow geese on the vegetation of a subarctic salt marsh. *Journal of Applied Ecology* 21: 669-686.Chapin, F.S. III, and C. Körner, eds. 1995. *Arctic and alpine biodiversity.* Ecological Studies 113. Berlin: Springer-Verlag. xviii + 332p.

Chapin, F.S. III, R.L. Jefferies, J.F. Reynolds, G.R. Shaver, and J. Svoboda, eds. 1992. *Arctic ecosystems in a changing climate.* San Diego: Academic Press. xvii + 469p.

Dyke, A.S. 1978. Indications of neoglacierization on Somerset Island, District of Franklin. Current Research, Part B: Geological Survey of Canada, Scientific and Technical Notes. Paper 780-1B, p. 215-217.

Fahselt, D., P.F. Maycock, and J. Svoboda. 1988. Initial establishment of Saxicolous lichens following recent glacial recession in Sverdrup Pass, Ellesmere Island, Canada. *Lichenologist* 20(3): 253-268.

Ferguson, S.H., M.K. Taylor, A. Rosing-Asvid, E. Born, and F. Messier. 2000. Relationship between denning of polar bears and conditions of sea ice. *Journal of Mammalogy* 81:1118-1127.

Forbes, B.C. 1992. Tundra disturbance studies. I. Long-term effects of vehicles on species richness and biomass. *Environmental Conservation* 19:48-58.

Forbes, B.C. 1993. Aspects of natural recovery of soils, hydrology and vegetation at an abandoned high arctic settlement, Baffin Island, Canada. *Proceedings of the VI International Conference on Permafrost, Beijing, China, July 1993.* South China Institute of Technology Press. Vol. 1, p. 176-181.

Forbes, B.C. 1996. Plant communities of archaeological sites, abandoned dwellings, and trampled tundra in the eastern Canadian Arctic: a multivariate analysis. *Arctic* 49(2): 141-154.

Forbes, B.C. 1997. Species establishment on anthropogenic primary surfaces, Yamal Peninsula, northwest Siberia, Russia. *Polar Geography* 21(1): 79-100.

Forbes, B.C. 1998. Cumulative impacts of vehicle traffic on high arctic tundra: soil temperature, plant biomass, species richness and mineral nutrition. *Nordicana* 57:269-274.

Forbes, B.C. 1999. Reindeer herding and petroleum development on Poluostrov Yamal: sustainable or mutually incompatible uses? *Polar Record* 35(195): 317-322.

Fox, J.L. 1998. Finnmarksvidda: reindeer carrying capacity and exploitation in a changing pastoral ecosystem – a range ecology perspective on the reindeer ecosystem in Finnmark. In: S. Jentoft, ed. *Commons in a cold climate*. Tromsø: University of Tromsø. p. 17-39.

Gauthier, G., L. Rochefort, and A. Reed. 1997. The exploitation of wetland ecosystems by herbivores on Bylot Island. *Geoscience Canada* 23(4): 253-259.

Henry, G.H.R., and A. Gunn. 1991. Recovery of tundra vegetation after overgrazing by caribou in Arctic Canada. *Arctic* 44(1): 38-42.

Henry, G.H.R., and J. Svoboda. 1986. Dinitrogen fixation (acetylene reduction) in high arctic sedge meadow communities. *Arctic and Alpine Research* 18: 181-187.

Henry, G.H.R., B. Freedman, and J. Svoboda. 1986. Survey of vegetated areas and muskox populations in east central Ellesmere Island. *Arctic* 39:78-81.

Jefferies, R.L. F.L. Gadallah, D.S. Srivastava, and D.J. Wilson. 1994. Desertification and trophic cascades in arctic coastal ecosystems: a potential climatic change scenario? In: T.V. Callaghan, ed. *Global changes and Arctic terrestrial ecosystems*. Ecosystem Report no. 10. Luxemburg: European Community. p. 206-210.

Kevan, P.G., B.C. Forbes, V. Behan-Pelletier, and S. Kevan. 1995. Vehicle tracks on high arctic tundra: their effects on soil, vegetation and soil arthropods. *Journal of Applied Ecology* 32:656-669.

Krupnik, I.I. 1993. *Arctic adaptations*. Hanover, New Hampshire: University Press of New England. xvii + 355p.

Kukal, O., and P.G. Kevan. 1987. The influence of parasitism on the life history of a high arctic insect, *Gynaephora groenlandica* (Wocke) (Lepidoptera: Lymantriidae). *Canadian Journal of Zoology* 65: 156-163.

Levesque, E., and J. Svoboda. 1999. Vegetation re-establishment in polar "lichen-kill" landscapes: a case study of the Little Ice Age impact. *Polar Research* 18(2): 221-228.

Lundmark, L. 1989. The rise of reindeer pastoralism. In: N. Broadbent, ed. *Readings in Saami history, culture, and language*. Miscellaneous Publications 7. Umeå, Sweden: Umeå University. pp. 31-44.

Matveyeva, N.V. 1998. *Zonation in plant cover of the Arctic*. Russian Academy of Sciences, Proceedings of the Komarov Botanical Instutite, Issue 21. 220 p. (In Russian.)

Mech, D.L. 2000. Lack of reproduction in muskoxen and Arctic hares caused by early winter? *Arctic* 53(1): 69-71.

Muc, M., J. Svoboda, and B. Freedman.1994. Soils of and extensively vegetated polar desert oasis, Alexandra Fiord, Ellesmere Island. In: J. Svoboda, and B. Freedman, eds. *Ecology of polar oasis, Alexandra Fiord, Ellesmere Island, Canada*. Toronto: Captus University Publications. p. 41-50.

Raillard, M., and J. Svoboda. 1989. Muskox winter feeding strategies at Sverdrup Pass, Ellesmere Island, NWT. *Musk-ox* 37: 86-92.

Raillard, M.C., and J. Svoboda. 1999. Exact growth and increased nitrogen compensation by the arctic sedge *Carex aquatilis var. stans* after simulated grazing. *Arctic, Antarctic, and Alpine Research* 31:21-26.

Reynolds, J.F., and J.D. Tenhunen. 1996. *Landscape function and disturbance in arctic tundra*. Ecological Studies 120. Berlin: Springer-Verlag. xx + 437p.

Skrobov,V.D. 1984. Human intervention and wild reindeer. In: E.E. Syroechkovskii, ed. *Wild reindeer of the Soviet Union*. New Delhi: Amerind Publishing Co. p. 90-94.

Sittler, B., and T.B. Berg. 1995. Karupelv Valley project 1988-1998: an ongoing long-term research of lemming cycles. *Northern Sciences Network Newsletter* 17:11-12.

State Committee of the Russian Federation for Environment Protection. 1997. *Biodiversity conservation in Russia*. Moscow: Biodiversity Conservation Center. 170p.

Svoboda, J., and G.H.R. Henry. 1987. Succession in marginal arctic environments. *Arctic and Alpine Research* 19(4): 373-384.

Syroechkovskii, E.E. 2000. Wild and semi-domesticated reindeer in Russia: status, population dynamics and trends under the present social and economic conditions. *Rangifer* 20:113-126.

Treshnikov, A.F., ed. 1985. *Atlas of Arctic*. GUGK under Cabinet of Ministers, Moscow. 204p. (In Russian).

Walker, M.D. 1995. Patterns and causes of Arctic plant community diversity. In: F.S. Chapin III and C. Körner, eds. *Arctic and alpine biodiversity*. Ecological Studies 113. Berlin: Springer-Verlag. pp. 3-20.

Wielgolaski, F.E., ed. 1997. *Polar and alpine tundra. Ecosystems of the World*, Volume 3. Amsterdam: Elsevier Science.

Wolfe, S.A., B. Griffith, and C.A. Gray Wolfe. 2000. Response of reindeer and caribou to human activities. *Polar Research* 19:1-11.

Zöckler, C., and I. Lysenko. 2000. *Water birds on the edge: first circumpolar assessment of climate change impact on arctic breeding water birds*. WCMC Biodiversity Series 11, 25p. + annex.

7. Rivers, lakes, and wetlands

The seasonal patterns of freshwater have a great influence not only on aquatic species but also on the surrounding land and the ocean into which rivers drain. This chapter describes the role of freshwater and its sensitivities.

The Region

Freshwater begins as precipitation. Snow and rain, dew and rime ice bring water from the air to the ground, along with dust and gases. Once on the ground, there are many paths that the water drops and ice crystals may follow. Some will soon evaporate, though relatively little will do so in the cool Arctic climate. Snow and ice may remain frozen, in the case of glaciers for centuries or millennia. Water from rain or melted snow flows across the ground, where it either is absorbed by plants and soils or collects in ponds and streams. As it moves, the water releases some of the substances it contained when it fell from the air, and picks up others from whatever it flows through and across. Where it goes and what it contains are the essential features of freshwater. They are critical to its impact on the surrounding land and on the microbes, plants, and animals that use freshwater areas and their products.

In the Arctic, freshwater is found in many forms. Glaciers and ice caps hold a substantial portion of the world's freshwater, nearly ten percent of which is held in the Greenland Ice Cap alone. In mountainous regions of the tundra, glaciers and accumulated snow are the source of rivers. Rapid warming in spring results in a surge of meltwater that can lead to substantial erosion. The sediment load carried in the spring runoff is considerable and is deposited downstream on flood plains or in riverbeds when the current slows down. The melt water and sediments carry both nutrients and contaminants.

Freshwater ecosystems are a major component of the Arctic landscape. Wetlands dominate vast stretches of the tundra and are found throughout much of the forest-tundra, from peat bogs to marshes. Ponds and lakes are common in many regions of the Arctic. In summer, some areas of the tundra appear to be nearly as much water as land. From small streams to some of the world's largest rivers, running water is also common in the Arctic. The runoff from these rivers helps to keep the top layers of the Arctic Ocean far less salty than most ocean water. Migrations of fish from

Glacier in Svalbard.

Rapa River delta in Sarek National Park, Sweden.

Tundra-peatland mosaic in east-European Russia near Naryan-Mar.

the ocean to freshwater bring significant quantities of nutrients from the sea.

Key characteristics

As with other parts of the Arctic environment, freshwater goes through drastic cycles each year. Lakes and ponds are typically covered with ice for more than half the year, and often longer. Ice thickness can reach 2-3 meters, so that shallow ponds and streams may be frozen solid. In deeper ponds and lakes, unfrozen water remains beneath the ice, where biological activity can continue through the winter, though slowed by cold temperatures and the lack of sunlight. Rivers keep flowing, but many braided rivers consist of deep water holes joined by shallow flow beneath, within,

The yellow-billed loon (white-billed diver) **Gavia adamsii** *nests in Siberia, Alaska, and Arctic Canada. It winters along the coast of Norway, and along the Pacific coasts of Alaska and Canada.*

The harlequin duck **Histrionicus histrionicus** *occurs in two disjunct breeding areas in eastern and western North America, and its range extends to western and southeastern Greenland, Iceland, and eastern Siberia.*

or over ice. The deep water holes are essential habitat for wintering fish. In spring, melting snow produces a period of high runoff with water flowing over still-frozen ground. The period of high flow in rivers can be destructive through the formation and break-up of ice dams, as well as through erosion and flooding. It is also essential to many smaller ponds that are isolated at other times in the year.

In areas that are predominantly covered by vegetation, freshwater often carries high concentrations of dissolved organic carbon, the result of decaying plant matter. Water can pick up much of this material from the vegetation and soils through which it flows. A portion of the dissolved organic carbon, in particular humic and fulvic acids, colors the water. Dark waters in the Arctic are typically rich in these materials, which absorb sunlight and act as a natural screen against ultraviolet radiation. The Lena River in Siberia, for example, is a black-water system because of its high concentrations of dissolved organic carbon. When it reaches the ocean, the Lena creates a carbon-rich plume that extends more than a hundred kilometers offshore.

Freshwater systems in the Arctic support an array of flora and fauna, including a great number of species of bacteria, protists, algae, and micro-invertebrates. Some of these live in sediments, while others float in the water column. Many of these small-cell

At the end of summer, ponds dry out. Few plants (here **Carex saxatilis**) *can survive the daily freeze-thaw cycles.*

RIVERS, LAKES AND WETLANDS 165

BOX 55. CONSERVING ARCTIC CHAR

▶ Arctic char come in many forms, as described in box 14. This variety, as well as the other characteristics of char, present a number of conservation challenges.

Arctic char are long-lived, and in lake populations in northern areas, individuals may live over 30 years. Anadromous populations typically live 20 years or more, which is still long in comparison with other fishes. Arctic char reach sexual maturity as late as age 10, and in northern areas where productivity is low, may reproduce only every second year or even less frequently. Despite their long lives, the total reproductive output of arctic char, and thus their potential for sustainability, may be quite limited.

All arctic char spend the first few years of life in freshwater. After that, anadromous fish begin the process of transition to the marine environment, and eventually spend each summer feeding in the sea and each winter in freshwater. In theory, this adaptation allows the fish access to the productive ocean, and in fact anadromous char are larger at the same age than their freshwater counterparts and produce more and better quality eggs. For these reasons, anadromous char are typically regarded as more productive and thus capable of supporting greater levels of exploitation. This may not be true, however, especially as anadromous char require multiple habitats widely separated in space.

Spawning habitat is usually in short supply, and for char inhabiting rivers, overwintering habitat can be very limited. In northern Alaska and Canada, for example, spawning, rearing, and overwintering sites all correspond to the same perennial springs which feed the mountain rivers used by Dolly Varden char. Downstream of these areas, the rivers freeze to the bottom. Thus, the entire population of char must survive in relatively small volumes of water for eight or nine months each year. While lake populations of char likely lead relatively stable lives, anadromous fish are susceptible to disturbances in their various habitats. These factors combine to reduce the ability of char to withstand disturbances, whether from humans or from natural causes, as each successive impact takes a significant toll on the fish's ability to survive and reproduce.

Char are important in the Arctic for their evolutionary significance as the only freshwater fish in high latitudes, for their ecological role as a top predator and intermediate prey, and for the extent of their use by humans. This last factor can greatly complicate conservation efforts because fishing for sea-run char may take individuals from several stocks without being able to determine relative levels of exploitation. Fishing in rivers and lakes can mean that the same stock is targeted more than once during the year. The importance of char to household, sport, and commercial uses makes effective management more complicated and, hence, more essential.

Our knowledge of char is still limited in relation to the great diversity of types of char, their population characteristics and dynamics, and the effects of fishing and of indirect impacts such as pollution and climate change. To conserve char effectively, more research is needed in these areas, and in particular, greater attention to the methods used in the management of such a complex species group.

James D. Reist, Fisheries and Oceans Canada, Winnipeg, Canada

SOURCES/FURTHER READING

Craig, P.C. 1989. An introduction to anadromous fishes in the Alaskan Arctic. *Biological Papers of the University of Alaska* 24: 27-54.

Gross, M.R. 1987. Evolution of diadromy in fishes. *American Fisheries Society Symposium* 1: 14-25.

Johnson, L. 1980. The Arctic char, *Salvelinus alpinus*. In: E.K. Balon ed. *Chars, Salmonid fishes of the Genus Salvelinus*. The Hague: Dr. W. Junk. p. 15-98.

Johnson, L. 1983. Homeostatic characteristics of single species fish stocks in Arctic lakes. *Canadian Journal of Fisheries and Aquatic Sciences* 40: 987-1024.

Johnson, L. 1989. The anadromous Arctic char, *Salvelinus alpinus*, of Nauyuk Lake, N.W.T., Canada. *Phys. Ecol. Japan*, Spec. Vol. 1: 201-227.

Reist, J.D. 1997. Potential cumulative effects of human activities on broad whitefish populations in the lower Mackenzie River basin. In: R.F. Tallman and J.D. Reist, eds. Proceedings of the workshop on the biology, traditional knowledge, and scientific fisheries management of the broad whitefish, *Coregonus nasus*, in the lower Mackenzie River. *Canadian Technical Report on Fisheries Aquatic Sciences* 2193: 179-197.

---

species form a complex microbial food web that provides the biomass and energy to support larger animals. The web also recycles carbon, nitrogen, phosphorus, and other nutrients.

In addition to resident fish species, freshwater is essential for the life cycle of anadromous fish, which spawn in freshwater but live part of their lives in the ocean. Atlantic salmon and Pacific salmon are the best known, but a number of whitefish, char, and other species follow similar patterns. Not only do anadromous fishes require freshwater habitat, they often bring substantial quantities of nutrients from the ocean to

Figure 50. Circumpolar distribution of arctic char species complex, **Salvelinus alpinus**, *and related species.*

Red-throated loon (diver) **Gavia stellata** *is a circumpolar bird which breeds in the freshwaters of the tundra and forest-tundra zones.*

freshwater in the form of their bodies, which decompose in freshwater or are eaten by predators. The open wetlands of the subarctic and Arctic are the primary breeding habitat for some species of waterfowl, such as loons, the tundra swan, black brant, and several other species of geese, greater scaup, eiders, and sandpipers. Some mammals, such as the beaver, live largely in freshwater, while others, such as moose and brown bear, feed on freshwater plants and animals.

There are many extreme habitats for freshwater life in the Arctic: permanently ice-covered lakes, saline lakes and ponds, perennial springs, naturally acid lakes, and meltwater on and within glaciers and ice shelves. These environments often contain complex communities of microscopic life-forms such as viruses, bacteria, cyanobacteria, micro-algae, protists, nematodes, rotifers, and tardigrades that are highly tolerant of the severe environmental conditions. The biodiversity and functioning of these so-called "extremophiles" is attracting increasing interest among ecologists because of their biotechnological potential and their importance for understanding fundamental biological processes.

The diversity of crustaceans and fishes, which have been better surveyed than other groups, is low. Very few freshwater species are restricted to the Arctic, although many fishes have genetic and behavioral adaptations for life in Arctic waters. Dwarf forms of whitefish, for example, are genetically distinct from southern strains of the same species. Among stream insects, there is a general decrease in diversity from south to north, with running waters in the highest latitudes dominated exclusively by cold-tolerant dipterans (flies). Reptiles and amphibians are all but missing. In Russia, only the common lizard, the Siberian newt, and three species of frogs are found in the tundra region.

Functionality

Most nutrients in freshwater come from the rocks and soils of riverbeds and uplands, and from decaying vegetation from the land and at the water's edge. The characteristics of the surrounding land are, therefore, significant factors in determining the quality of the water and its ability to support life. Drainage patterns are critical. Rivers can carry nutrients over great

BOX 56. FRESHWATER FOOD WEBS

▶ Although Arctic lakes and rivers are characterized by a lower biodiversity than temperate and tropical latitude waters, these ecosystems contain rich communities of plants, animals and microbes that make up two interconnected food webs.

The lower food web contains the greatest diversity and biomass of aquatic species, and consists of interlinked populations of micro-organisms: many types of bacteria that degrade organic material, minute picocyanobacteria (blue-green algae) that efficiently capture light for photosynthesis, larger cell species with various modes of nutrition (nanoflagellates and other phytoplankton), small-bodied zooplankton such as ciliates and rotifers that graze the small cells, and viruses that attack each microbial component. A liter of lake water will typically contain thousands to millions of these various microscopic life forms. Organic carbon plays a central role in the dynamics of the microbial food web and enters the system from photosynthesis in the lake (autochthonous carbon), as well as in the form of coarse organic matter from the soils and vegetation in the surrounding catchment (allochthonous carbon).

The lower food web contributes to breaking down the organic matter to fine particles, which then are used by larger filter feeders, especially insects such as mosquitoes and blackflies. Carbon and energy from the microbial food web also enter the upper food web via grazing by larger zooplankton, or by bottom-dwelling animals such as snails. Certain fish species may play a keystone role in the flow of carbon to higher trophic levels. In Toolik Lake, Alaska, lake trout exert this effect by controlling the abundance of sculpins, graylings, burbot, snails, and zooplankton.

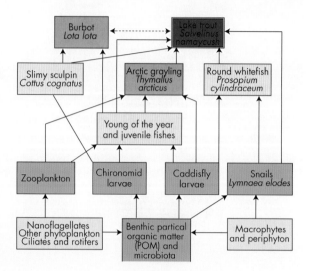

Figure 51. The upper food web of Toolik Lake, Alaska.

Warwick F. Vincent, Laval University, Sainte Foy, Quebec, Canada

SOURCE/FURTHER READING
Vincent, W.F., and J.A. Hobbie. 2000. Ecology of Arctic lakes and rivers. In: M. Nuttall and T. V Callaghan, eds. *The Arctic: environment, people, policies*. Amsterdam: Harwood Academic Publishers. p. 197-231.

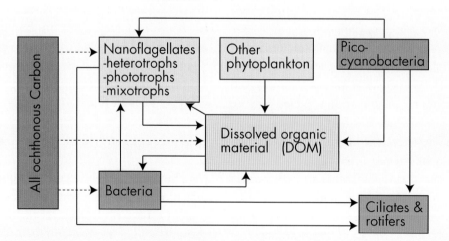

Figure 52. The microbial food web of Toolik Lake, Alaska.

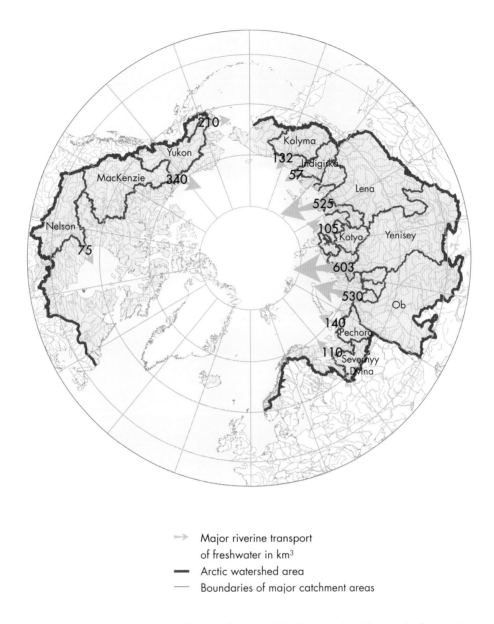

Figure 53. Catchment area of the Arctic Ocean, showing the annual discharge (cubic kilometers) of major rivers.

distances. The ways in which water flows from land to lakes and streams determines the extent to which it will pick up materials. Wetlands, where water sits for long periods will produce water with much higher levels of carbon than slopes, down which the water flows quickly. Water that flows underground may go through several steps of absorbing and depositing different elements. The deltas of large rivers, where waters often slow down and deposit what they have carried from far upstream, may be affected by what occurs in the entire drainage basin.

The nutrients that flow into freshwater produce a variety of floating plants such as algae, as well as larger plants that grow along the shores of ponds and streams. Primary production in tundra ponds can be high, and a great deal of it takes place under the ice. Once the ice goes out, primary production by algae in the water column peaks, typically limited by the

Blackflies can make field work a challenge. Lake Myvatn, Iceland.

availability of nutrients. Bottom-dwelling micro-algae (especially mats and films of cyanobacteria), moss, and larger plants dominate production in many Arctic lakes, streams, and wetlands. These benthic communities provide an important habitat and food source for insects and other invertebrates, which in turn are foraged upon by fish and waterfowl.

Low temperatures, low nutrient availability, and the brief growing season mean that primary production rates are typically low in Arctic freshwaters, thereby restricting the amount of animal production that can be supported. An experiment in a stream in northern Alaska added phosphorus for several years. There was a shift in the photosynthetic community from periphyton (algae growing over the surface of rocks) to large standing stocks of moss, with accompanying increases in insect and fish production.

Grazing invertebrates can limit the amount of algae that are in a pond at any time. In ponds with no fish, these invertebrates may be up to two centimeters long. In ponds with fish, which prey on invertebrates, only smaller species of invertebrates are found. Freshwater fishes in the Arctic include species that graze on plants and small invertebrates as well as larger predatory fish such as lake trout and burbot. As noted above, anadromous species are common in many areas.

Arctic freshwater fishes have two basic strategies for spawning and surviving the winter. Spawning in fall allows the parent fish to produce larger eggs, after feeding during summer, but requires the eggs to survive the winter. Spawning in spring means the parent fish must produce eggs after the long winter, but allows the eggs to hatch and become small fish before they, in turn, must survive the winter. Both strategies place a substantial burden on the parent fish. Some, such as Pacific salmon, do not survive after spawning. Others, such as the whitefish, are capable of breeding annually.

Some marine mammals use freshwater at certain times of the year. Beluga whales use river mouths and estuaries for molting and calving, and in rare instances have been seen hundreds of kilometers up the Yukon River. Young male bearded seals may spend much of the summer in rivers.

Interactions with other regions

In the tundra and forest-tundra, freshwater is typically a major force in shaping the physical and biological landscape of the region. The transition from solid ground to wetlands to rivers and lakes is often a gradual one. Many species of plants are found in wet areas, from moist soils to the edges of lakes and ponds. Many birds and some mammals live on land but feed in freshwater at least part of the time. Freshwater is also intimately linked with the marine environment. Rivers flow eventually to the ocean, bringing substantial quantities of soil and other mineral and biological runoff. Freshwater flowing into the ocean can create a temporary freshwater corridor between rivers through which freshwater species can spread from one watershed to another. Several fish species use both the marine and freshwater environments.

Nutrients in freshwater come from runoff from land, as well as from the decay of anadromous fishes and resident organisms. Freshwater flora and fauna provide food for a variety of birds and mammals, which spread feces on land, adding nutrients to soils. Anadromous fishes depend on freshwater for spawning and, in some cases, for overwintering. Terrestrial predators, such as brown bears, may depend heavily on such fish, making a strong link between the ocean and land, via freshwater. Many insects lay eggs and live their first life stages in freshwater. Their abundance supports not only freshwater species but also many of the birds found in the Arctic in summer.

Rivers deliver great quantities of nutrients to the oceans, and the estuaries and river mouths of large Arctic rivers are often highly productive environments. Physically, the rivers also transport heat. The treeline extends farther north near the mouths of large rivers such as the Mackenzie in North America. Through flooding, rivers carry sediments and nutrients onto their riverbanks and across their floodplains. This action can be destructive through erosion and the uprooting of plants, but it can also be beneficial by providing rich new material for plant growth.

Freshwater areas in the Arctic typically have less physical interaction with areas outside the Arctic than do the land and marine regions, although some large rivers flow from temperate latitudes to the Arctic. The massive flow of freshwater into the Arctic Ocean has a major influence on the salinity and heat content of the sea. This in turn affects the production of North Atlantic Deep Water, a driving force in the oceanic circulation system. Biologically, many of the large populations of Arctic waterbirds migrate far to the south during winter. The birds use various freshwater habitats for staging, molting, breeding, nesting, and feeding. Anadromous fishes, such as salmon, swim far into the ocean and often well to the south during their life cycle.

Natural variability

The patterns of freshwater in the Arctic, including precipitation, ground water, and surface water, vary over time. The amount and timing of precipitation differ greatly from year to year. Streams and rivers change course. Braided streams move back and forth across a gravel plain. Meandering streams cut into riverbanks making oxbows, isolating old loops. Thaw lakes form as the ice in permafrost melts over time, creating depressions that fill with water and eventually join to become a lake. The stream flowing out of a thaw lake may melt the ground around it, cutting a deep channel through which the lake suddenly empties. The newly bare ground freezes again as permafrost returns, perhaps producing a pingo as the expanding ice underground forces its way upward into a small hill. In the now-dry lakebed, the tundra grows again, and the cycle repeats.

These changes, to the course of rivers and the existence of certain lakes, affect the flora and fauna found

A sport fisher in Finnish Lapland with a brown trout ***Salmo trutta****.*

BOX 57. THE ATLANTIC SALMON: A SPECIES IN TROUBLE

▶ Often regarded as the King of Fish, the Atlantic salmon is one of the world's great voyagers. To complete its journey from gravel spawning beds in the upper reaches of a river, down to the ocean where it travels thousands of kilometers, and back to its natal stream, the salmon has to avoid predators and compete for food. It must also survive the dangers created by humans.

Once, the Atlantic salmon was found from northern Spain to the Arctic, and from North America to northern Russia, but in the twentieth century its reign over the piscatorial world began to decline. Damming of rivers, fishing, pollution, and other human activities have contributed to the demise of the salmon in many areas. Fish farming, too, has threatened wild stocks by introducing diseases and weakening the genetic strains of wild stocks as farmed fish escape to mingle and compete with wild salmon. In Norway, for example, salmon stock imported from the Baltic Sea for farming brought with them a freshwater parasite to which only the Baltic stocks are immune. Now, one third of Norway's rivers are devoid of life, having been deliberately poisoned in an effort to eradicate the parasite. Even this effort has failed in some of Norway's larger rivers. Norway's total salmon stock has dropped four-fold in the last thirty years.

Overfishing is another threat to wild salmon. Salmon from a particular river tend to stay together in the ocean. Improvements in fishing technology have enabled a single fishing vessel to catch the entire run of a salmon river in just a few hours. The effect of such practices is that the salmon population as a whole has declined drastically, and total fish catches have followed. Because the total catches are likely to include escaped farm fish, the catch of wild salmon has probably dropped even further.

Today, Iceland and northern Russia are the only places where the wild salmon remains relatively healthy. Rivers on the Kola Peninsula have been protected by their remoteness while in Iceland a ban on virtually all salmon netting, strong conservation measures, and active restoration programs have together protected wild spawning stocks and helped them recover from natural disasters such as volcanic eruptions that destroy spawning populations. Such efforts are not cheap, but recreational salmon fishing can attract tourist revenue up to one hundred times the market value of the fish itself. With catch-and-release fishing, the salmon need not even die to benefit the local economy.

The fight for control of the dwindling salmon population continues. Organizations such as the North Atlantic Salmon Fund promote the elimination of commercial fishing through the purchase of salmon quotas and the retraining of fishermen for other species or other jobs. The long-term prospects for the wild Atlantic salmon are unclear. If the King of Fish is to continue its reign, fisheries and habitat management will need to change.

Orri Vigfusson, North Atlantic Salmon Fund, Reykjavik, Iceland

---

in them as well as the variety of birds and mammals that use freshwater areas. Typically, such changes do not affect the overall character of the landscape. Individual ponds may come and go, but some ponds will always be found on the tundra plain. Rivers may move back and forth but are likely to continue to contain deep water holes in which fish can overwinter.

Interannual changes in habitat characteristics can have a major influence on biological activity. The breeding success of lesser snow geese along Hudson Bay, for example, is highly variable and is strongly correlated with climatic variables, especially early season air temperature. Snow and ice are dominant features of the Arctic, and small year-to-year variations in climate can result in a striking difference in the extent, depth, and duration of open water for aquatic birds and other species.

Human uses and impacts

Rivers and lakes provide drinking water for Arctic communities, especially as groundwater is scarce in permafrost areas. The volume of water used is typically low, as are the impacts. In some areas, however, using water from lakes may affect the overwintering habitat of fish and other species. Human communities also use rivers and lakes for wastewater, including sewage. In certain cases, this practice can lead to highly

Salmon fisher in Utsjoki, Finnish Lapland.

polluted areas and even the spread of diseases. As shown by long-term experiments at the Toolik Lake site in northern Alaska, prolonged nutrient enrichment of Arctic lakes and rivers can lead to major shifts in plant and animal species composition.

Fishing is the most widespread use of freshwater resources in the Arctic. From subsistence fishing to sport fishing to commercial fishing, Arctic waters provide substantial quantities of fish each year. The number of species is relatively low, but many, such as Atlantic and Pacific salmon, whitefish, and sturgeon, are an abundant and valued source of food and income. Others, such as lake trout, char, and grayling, attract sport fishers from around the world while also providing for commercial fisheries. The most common impacts from fishing are local declines in some species through overharvesting.

Large rivers can be major travel and transportation routes as well as significant barriers to the movements of animals and people. In summer, the Mackenzie and Yukon rivers in North America and the Lena, Ob, and Yenisey rivers in Siberia see a great volume of boat and barge traffic. In winter, they are used as roadways. Reindeer herders use winter ice to get from one side of a river to the other, though this is disrupted if ice-breaking ships are used to keep barges moving. The use of rivers as major transportation corridors inevitably produces pollution. Riverbank

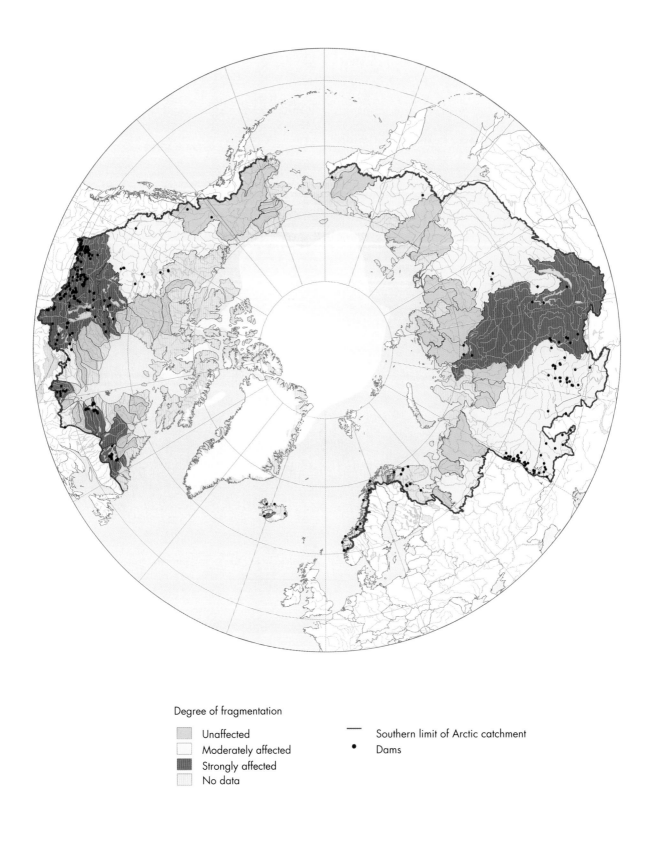

Figure 54. River channel fragmentation and major river dams in the Arctic catchment area. (Source: Dynesius and Nilsson 1994).

The Lokka Reservoir in Finnish Lapland was constructed in the 1960s. Many reindeer drowned while trying to reach their old grazing and breeding areas.

development contributes to erosion and increases the amount of dirt washed into the river.

The size of many rivers in the Arctic, and the lack of alternative local energy sources in some areas, has led to the development of hydroelectric dams, particularly in the Nordic region and northern Quebec. Such dams produce tremendous amounts of electricity, but they affect the surrounding landscape, as well as the river system. Restricted water flow and a physical barrier across the river severely limit fish movements. Salmon blocked from their spawning streams are the most widely known example, but other fish move to different parts of a river at different times of the year or through their life cycle. Dammed rivers may not have the high water flows in spring that can erode riverbanks, spread nutrients, and clear debris. The lakes created behind dams flood large areas and can create barriers to migrating land mammals. These artificial lakes may flood wetlands and can also release contaminants such as mercury that were bound to soil particles. This causes problems, particularly for predators such as white-tailed eagles and osprey, as well as for humans who eat fish from these watersheds.

RIVERS, LAKES AND WETLANDS 175

The diversion of large, north-flowing rivers in North America and Eurasia to provide freshwater to southern areas would likely have a considerable impact on the Arctic. The reduction in freshwater reaching the Arctic Ocean might increase the salinity of the surface layer of the ocean, lowering its freezing point, and thus reducing the extent of sea ice, with further impacts on marine life and climate. The dissolved organic matter carried by these rivers influences carbon availability in the nearshore area, the amount of ultraviolet radiation penetrating the water, and the amount of light available for photosynthesis. Thus, human activities that affect the export of these materials can have an impact on the Arctic Ocean food web.

Another serious human impact is pollution dumped into rivers and lakes. This type of disturbance is most common near industrial operations, where contaminants are released. It is also a concern in the larger rivers that flow from more heavily developed areas south of the Arctic, especially in Russia. Mining operations, logging that leaves bare ground, and other physical disturbances on land and in the water can turn clear water muddy and radically change the habitat offered by the water body. The use of gravel from riverbeds, for road building and other purposes, can affect freshwater habitat, in particular the overwintering areas available to fish. Use of the water itself, to make ice pads for drilling operations or to build roads, can reduce overwintering habitat. By reducing the amount of water left, such extraction can also affect the oxygen content and other qualities of the remaining water.

In addition to physical disturbances, biological disturbances have occurred through the introduction of species into areas where they do not naturally occur. Some of these introductions are intentional, often with the idea of improving fishing opportunities. In Fennoscandia, the brook and rainbow trout have been introduced to benefit sport fishers. In the 1950s, vendace was introduced to two small lakes in the Lake Inari watershed in northern Finland. By the 1980s, it had spread throughout the watershed and become especially abundant in Lake Inari. This abundance led to a commercial fishery. After ten years, however, the vendace population was greatly reduced and no longer commercially viable.

In Sweden and Norway, electric companies introduced the opossum shrimp to mountain lakes to replace invertebrates that had been lost to changes in water levels from damming. The shrimp were intended to provide prey for fish and thus support fish populations. Some fish such as burbot and brown trout have benefited. But the opossum shrimp eats the zooplankton that arctic char and whitefish feed on, and those species have declined, leading to an overall decline in fish production.

Other introductions are accidental. The parasite *Gyrodactylus salaris* was introduced to Norway through a delivery of salmon from Sweden. From a hatchery in southern Norway, the parasite spread to several rivers where the salmon were released to increase local populations. The parasite is devastating, killing all the salmon in the river. The treatment is no less severe. Poison is used to kill all the fish, leaving the parasite with no hosts so that it cannot survive. After the parasite is gone, the river can be re-stocked. The method has been effective in some rivers but not in others.

Biological disturbances include diminishing the genetic diversity of species. If a discrete population of a species is lost, its distinctive genetic characteristics may be lost as well. Such problems are especially prominent with species such as salmon, which have separate stocks in each river system. Genetic contamination is another problem, for example through programs to stock salmon streams by releasing smolt taken from other places. In some spawning areas in Norway, 70% of the salmon were found to be escaped farm fish. If the genetic identity of the original stock is lost, so are the special adpatations it has developed for that particular stream.

Human activities also cause diffuse disturbances, including acid rain, contamination, and increased ultraviolet radiation. While the chemicals that produce acid rain have natural as well as human-caused origins, the levels of acid rain associated with ecological damage stem largely from industrial sources. In the Arctic, acid rain is a problem primarily in the European Arctic, where pollution from central and eastern Europe is

BOX 58. FOOD WEB CONTAMINANTS IN ARCTIC FRESHWATERS

▶ Heavy metals, pesticides, and other contaminants are now accumulating in lake and river biota throughout the circumpolar Arctic. They enter the region via long range transport from lower latitudes and are then biomagnified in concentration through the aquatic food web. Mercury, for example, now occurs in much higher concentrations in surface lake sediments in remote lakes in subarctic Quebec than it did in the 19th Century, and mercury concentrations per unit biomass have been found to increase with trophic position in the food web.

Local effects may also control the level of contamination in high latitude aquatic biota. A substantial mercury release, for example, accompanied the flooding of land in subarctic Quebec for hydroelectric development. The mercury accumulated rapidly in fish that feed on zooplankton, such as lake whitefish, but more slowly and to higher levels in northern pike, a top predator that feeds on other fish.

Warwick F. Vincent, Laval University, Sainte Foy, Quebec, Canada

SOURCE/FURTHER READING
Vincent, W.F., and J.A. Hobbie. 2000. Ecology of Arctic lakes and rivers. In: M. Nuttall and T. V Callaghan, eds. *The Arctic: environment, people, policies*. Reading, United Kingdom: Harwood Academic Publishers. p. 197-231.

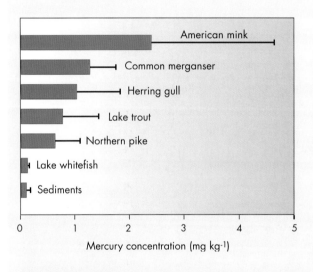

Figure 55. Mercury accumulation in food chains of subarctic lakes in Quebec, Canada.

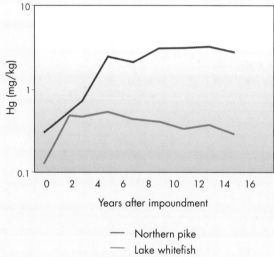

Figure 56. Trends in mercury level in fish muscle after damming of rivers in Quebec, Canada.

carried north. The main culprit for acidification is sulfur dioxide, which can become sulfuric acid in water. Within the Arctic, the largest sources of sulfur are metal smelters, primarily on the Kola Peninsula and at Norilsk in Russia, where sulfur deposition is high. When acid rain falls, it can change the chemical balance of soils and water. In some areas, the minerals in the soil are able to neutralize acidity effectively, but in others the acid remains and becomes toxic to many freshwater plants. Fennoscandia typically has low tolerance for acid rain. When the acid is stored in snow, the spring runoff can carry a concentrated pulse of acidic water through a system, with severe impacts.

Other contaminants, such as heavy metals and persistent organic pollutants, are also found throughout the Arctic, often carried by wind and deposited di-

BOX 59. THE LAKE MÝVATN ECOSYSTEM: MANY USES, MANY THREATS

▶ Lake Mývatn is one of the largest and most remarkable lakes in the volcanic zone of Iceland. A combination of cold, phosphate-rich spring water and silica-rich warm water creates a luxuriant growth of diatoms, which in turn are utilized by a phenomenal abundance of midge larvae living on the shallow lake bottom. The midges are the main food source for a diverse and dense community of migratory waterfowl, including 15 species of ducks that regularly breed on the shores of the lake. Most birds migrate from Europe, but among the North American elements is Barrow's goldeneye, which is characteristic of the area. The Laxá River, which flows out from the lake, ranks among the most fertile of rivers, supporting a massive population of blackflies. These form the main food source for fish and waterfowl, including substantial populations of brown trout and Atlantic salmon.

The shore of Lake Mývatn is dominated by volcanic landforms, a multitude of bizarre lava formations and a variety of beautifully shaped craters. Holes and crevices in the lava serve as nesting habitat for Barrow's goldeneye, replacing the tree holes that are the normal nest sites in North America. Water-filled caves in the surrounding lava are home to a dwarf variety of arctic char. Among the most interesting features of Lake Mývatn is a population of marimo, a velvety green alga that forms extensive patches of fist-sized or larger balls on the

Marimo algal balls **Cladophora aegagrophila**.

lake bottom. Lake Akan in Japan and Lake Oisu in Estonia are the only other known places where such balls are found.

Over the centuries, Lake Mývatn has been a major source of food for the local farming community, which harvested duck eggs and arctic char in a sustainable way. Today, however, mining of lake sediments for the production of diatomite is also a major activity. Many of the local farmers claim that

Lake Mývatn, Iceland.

*Barrow's goldeneye **Bucephala islandica**.*

*Figure 57. The decline of Barrow's goldeneye **Bucephala islandica** at Lake Mývatn, Iceland.*

the mining caused a disruption of the ecosystem, as shown by substantial declines in fish and some waterfowl. The declines began soon after mining started in 1967, although a causal link to mining has not been found. Research has demonstrated, however, that the mining causes relocation of birds and the loss of the organic top layer of the sediment, which forms the basis of the food chain. While some waterfowl populations have declined, harlequin ducks are increasing – its colony by the Laxá River is the largest in the world.

In addition to concerns about the impacts of mining, rapidly increasing tourism and the tapping of geothermal energy pose imminent threats to the lake's ecosystem. With their rich and remarkable habitats, flora, and fauna, Lake Mývatn and the Laxá River constitute a unique ecosystem. In combination with the volcanic surroundings, the area is a true wonder of nature. If we cannot preserve places such as this, anything is at risk.

Árni Einarsson, Mývatn Lake Research Station, Institute of Biology, University of Iceland, Reykjavik, Iceland

Table 6. The number of male ducks in the Mývatn area in spring from 1975 to 1989.

	MEAN	MINIMUM	MAXIMUM	% OF TOTAL
Eurasian wigeon	1092	526	1887	12.3
Gadwall	53	25	99	0.6
Mallard	233	180	294	2.6
Pintail	44	20	77	0.5
Teal	122	50	196	1.4
Shoveler	2	0	8	0.03
Pochard	1	0	3	0.01
Tufted duck	3784	2466	6017	42.7
Greater scaup	1838	1383	2647	20.8
Common scoter	319	223	501	3.6
Long-tailed duck	164	100	300	1.9
Harlequin duck	127	32	265	1.4
Barrow's goldeneye	647	342	873	7.3
Red-breasted merganser	423	283	781	4.8
Goosander	6	1	19	0.1
All species	8855	6455	13097	100

rectly or with snow and rain. The large rivers of Siberia carry many contaminants to the Arctic, including heavy metals, hydrocarbons, and radionuclides. Some of these sink into river sediments, while others are carried all the way to the ocean. Many are taken up by the algae and other plants living in freshwater and become part of the food web. Concentrations depend on a variety of factors, including the amount of the contaminant that is introduced to the system and the characteristics of the food web through which it passes.

Freshwater areas can be harmed by ultraviolet radiation, which is increasing at certain times of the year as a result of the thinning of the stratospheric ozone layer. Clear water is especially susceptible, as ultraviolet rays pass through it more easily. Plankton in lakes, especially clear, shallow lakes, may suffer from increases in ultraviolet radiation. Elevated levels of ultraviolet radiation may also help break down dissolved organic carbon. This would increase the primary production of bacteria, increasing the emission of carbon dioxide from Arctic freshwater areas, which could affect global levels of greenhouse gases. Overall impacts, however, are hard to predict. Analyses of subarctic lake sediments have shown that past variations in climate caused major shifts in the amount of colored dissolved organic matter. It was estimated that this had a greater effect on underwater ultraviolet exposure than did moderate depletion of the ozone layer.

Climate change

Climate change is likely to cause a broad range of disturbances in freshwater areas. Inuit and Cree in Hudson Bay, for example, have noted lower water levels in rivers and thinner ice, impeding fishing and winter travel. Temperature changes might affect rates of evaporation, persistence of snow cover, and the timing of spring snowmelt. Shifts in climate are likely to change when and how much rain and snow fall over the Arctic. Recent data show that the water content of the upper soil layer at a research site at Barrow, Alaska decreased by about 20% over the last two decades of the 20th Century, concomitant with an increase in air temperatures. Such effects will have a strong influence on the availability of water for stream flow, the transfer of carbon and nutrients from land to water, and the flushing rate for lakes and ponds.

Some Arctic species are living near their upper temperature limit and may be severely affected by climatic warming. Such effects may occur directly. An ecosystem model for Toolik Lake, Alaska, for example, predicts that a 3°C rise in the surface water temperatures of the lake would cause the elimination of lake trout. A variety of indirect effects are also likely, such as the invasion of competing species from the south that are favored by warmer water conditions. Changes in the composition of insect species in streams is a likely response to climate warming that, in turn, may influence the fishes and birds that feed them.

Conservation

Several types of fishing take place in the Arctic, from traditional uses for personal or local consumption, to sport fishing, to commercial fishing. All of these are regulated to some degree by national, regional, or local governments. The regulations can include limits on the number of fish caught, the fishing gear used, the timing of the fishing season, and the size of the fish caught. They can also distinguish between different types of uses, giving preference to one or rotating among them to provide varied opportunities during the course of a fishing season. Fishing regulations can also be used to favor one species or to help shift the balance of populations in a river system.

Protected areas are an important part of freshwater conservation, both through large areas that include rivers and lakes and through specific designations for freshwater areas. When entire watersheds are protected, the ecological processes in a river system can function and change as a result of natural pressures. Even when only part of a river system is protected, important aspects of its ecology such as fish spawning or overwintering may be conserved effectively by limiting local disturbances. Protected areas and other local measures offer little protection, however, from global environmental pressures such as contaminants, ozone depletion, and climate change.

SOURCES/FURTHER READING

AMAP. 1998. *The AMAP Assessment Report: Arctic pollution issues.* Oslo: Arctic Monitoring and Assessment Program. xii +859p.

Bernes, C. 1996. *The Nordic arctic environment—unspoilt, exploited, polluted?* Copenhagen: The Nordic Council of Ministers. 240p.

Cauwet, G., and I. Sidorov. 1996. The biogeochemistry of Lena River: organic carbon and nutrients distribution. *Marine Chemistry* 53: 211-227.

Dynesius, M., and C. Nilsson 1994. Fragmentation and flow regulation of river systems in the northern third of the world. *Science* 266: 753-762.

Gibson, J.A.E., W.F. Vincent, B. Nieke, and R. Pienitz. 2000. Control of biological exposure to UV radiation in the Arctic Ocean: comparison of the roles of ozone and riverine dissolved organic matter. *Arctic* 53(4): 372-382.

Hobbie, J.E. 1980. *Limnology of tundra ponds, Barrow, Alaska.* Stroudsburg, Pennsylvania: Dowden, Hutchinson, and Ross. xiv + 514p.

Hobbie, J.E. 1984. *The ecology of tundra ponds of the Arctic Coastal Plain: a community profile.* Washington, DC: U.S. Department of the Interior, Fish and Wildlife Service. x + 52p.

Hobbie, J.E., B.J. Peterson, N. Bettez, L. Deegan, W.J. O'Brien, G.W. Kling, G.W. Kipphut, W.B. Bowden, and A.E. Hershey. 1999. Impact of global change on biogeochemistry and ecosystems of arctic Alaska freshwaters. *Polar Research* 18: 207-214.

Huntington, H.P., and the Communities of Buckland, Elim, Koyuk, Point Lay, and Shaktoolik. 1999. Traditional knowledge of the ecology of beluga whales (*Delphinapterus leucas*) in the eastern Chukchi and northern Bering seas, Alaska. *Arctic* 52(1): 49-61.

Jano, A.P., R.L. Jeffries, and R.F. Rockwell. 1998. The detection of vegetational change by multitemporal analysis of LANDSAT data: the effects of goose foraging. *Journal of Ecology* 86: 93-99.

Le Cren, E.D., and R.H. Lowe-McConnell. 1980. *The functioning of freshwater ecosystems.* Cambridge: Cambridge University Press. xxix + 588p.

Markager, S., W.F. Vincent, and E.P.Y. Tang. 1999. Carbon fixation in high Arctic lakes: implications of low temperature for photosynthesis. *Limnol. Oceanogr.* 44: 597-607.

McDonald, M, L. Arragutainaq, and Z. Novalinga. 1997. *Voices from the bay: traditional ecological knowledge of Inuit and Cree in the Hudson Bay bioregion.* Ottawa: Canadian Arctic Resources Committee and Environmental Committee of Municipality of Sanikiluaq. xiii + 90p.

Milner, A.M., and M.W. Oswood, eds. 1997. *Freshwaters of Alaska.* Ecological Studies 119. New York: Springer. xvii + 369p.

Oechel, W.C., A.C. Cook, S.J. Hastings, and G.L. Vourlitis. 1997. Effects of CO_2 and climate change on arctic ecosystems. In: S.J.,Woodin, and M. Marquiss, eds. *Ecology of arctic environments.* Oxford: Blackwell Science. p. 255-273.

Oswood, M.W. 1997. Streams and rivers of Alaska: a high latitude perspective on running waters. In: A.M. Milner, and M.W. Oswood, eds. *Freshwaters of Alaska.* Ecological Studies 119. New York: Springer. p. 331-356.

Pienitz, R., and W.F. Vincent. 2000. Effect of climate change relative to ozone depletion of UV exposure in subarctic lakes. *Nature* 404: 484-487.

Sand-Jensen, K., T. Riis, S. Markager, and W.F. Vincent. 1999. Slow growth and decomposition of mosses in Arctic lakes. *Canadian Journal of Fisheries and Aquatic Sciences* 456: 388-393.

Sedinger, J.S. 1997. Waterfowl and wetland ecology. In: A.M. Milner, and M.W. Oswood, eds. *Freshwaters of Alaska.* Ecological Studies 119. New York: Springer. p. 155-178.

Skinner, W.R., R.L. Jeffries, T.J. Carleton, R.F. Rockwell, and K.F. Abraham. 1998. Prediction of reproductive success and failure in lesser snow geese based on early season climatic variables. *Global Change Biology* 4: 3-16.

Tang, E.P.Y., W.F. Vincent, J. de la Noüe, P. Lessard, and D. Proulx. 1997. Polar cyanobacteria versus green algae for the tertiary treatment of waste-waters in cool climates. *Journal of Applied Phycology* 9: 371-381.

Vincent, W.F. 1997. Polar desert ecosystems in a changing climate: a North-South perspective. In: W. B. Lyons, C. Howard-Williams, and I. Hawes, eds. *Ecosystem processes in Antarctic ice-free landscapes.* Rotterdam: A.A. Balkema. p. 3-14.

Vincent, W.F., and J.E. Hobbie 2000. Ecology of Arctic lakes and rivers. In: M. Nuttall and T. Callaghan, eds. *The Arctic: environment, people, policies.* Reading, U.K.: Harwood Academic Publishers. p. 197-231.

Vincent, W.F., R. Pienitz, and I. Laurion. 1998. Arctic and Antarctic lakes as optical indicators of global change. In: W.F. Budd, ed. Antarctica and global change. *Annals of Glaciology* 27: 691-696.

Vincent, W.F., J.A.E. Gibson, R. Pienitz, V. Villeneuve, P.A. Broady, P.B. Hamilton, and C. Howard-Williams. 2000. Ice-shelf microbial ecosystems in the High Arctic and implications for life on snowball earth. *Naturwissenschaften* 87: 137-141.

8. The oceans and seas

The Arctic Ocean is dominated by sea ice, which provides important habitat for many species. Northern waters support coastal communities, as well as some of the world's richest fishing grounds. This chapter describes these features and the factors that affect them.

The Region

The central Arctic Basin, the deep waters beyond the continental shelves, is ice covered all year. The ice is typically several years old, moving with the currents that eventually push most of the ice out through Fram Strait into the North Atlantic. Beneath the ice are three water layers. The top 200 meters are Arctic surface water, which is usually less salty than the deeper waters. Below this is the Atlantic layer, from about 200 to 900 meters in depth, which flows to and from the Atlantic Ocean. Flows from the Pacific are limited to surface water by the shallow Bering Strait. The Atlantic layer is salty, above 0°C, and circulates counterclockwise, which is opposite to the circulation of the surface layer. Below the Atlantic layer is the colder Arctic deep-water layer, which is divided by the Lomonosov Ridge into the Eurasian and Canadian Basins.

Surrounding the Arctic Basin, except at Fram Strait, are the continental shelves of Eurasia and North America, also known as the marginal seas of the Arctic. The northward movement of Pacific water influences the Bering and Chukchi Seas. The Barents, Greenland, and Norwegian Seas are similarly influenced by the Atlantic Ocean, as well as, in the case of the Greenland Sea, the flow of sea ice from the Arctic Ocean through Fram Strait. The Bering, Chukchi, and Barents Seas are among the most productive marine ecosystems in the world. The Kara, Laptev, and East Siberian Seas across the north of Eurasia, and the Beaufort Sea above North America are influenced more by the inflow of freshwater from large rivers than by circulation from other oceans. Typically, they have more extensive ice cover that lasts longer than the other seas.

The shallow coastal waters covering the continental shelves are powerhouses in Arctic marine ecology. Where sunlight reaches the bottom, it boosts the productivity of bottom-dwelling plants and animals. Landfast ice in winter provides relatively stable conditions, although moving sea ice can scour the bottom. Rivers bring nutrients, and spring runoff can melt sea ice earlier than in areas without river inflows. Turbulence, from wind and from river flow, stirs up nutrients and minerals from the bottom, while also reducing the penetration of sunlight. Estuaries are productive and sensitive areas, vulnerable to changes in river flow and sedimentation. Soft shorelines are important habitat for some plants and invertebrates, and for birds that feed on them.

The ocean bottom – the benthic environment – is biologically important, particularly in shallower areas. The plants and animals of the seafloor are eaten by a variety of fish, birds, and mammals. Bacteria, invertebrates, and fish scavenge dead plants and animals that sink to the bottom, completing the circle. In the deeper parts of the Arctic Basin, the flora and fauna of the seafloor depend on organic material sinking from the upper layers or transported in from the

BOX 60. THE BIOICE RESEARCH PROGRAM MAPS BENTHIC BIODIVERSITY

▶ How much do we know about the sea floor? The BIO-ICE Research Program is designed to find out more about biodiversity and the distribution patterns of the species that live in and on the seabed in the waters surrounding Iceland. When the program began, about 2,000 species were known to live in the area. The program now estimates that it will double this total, and has already discovered and described 20 species that are entirely new to science. The program is a collaborative effort by several research institutes, government agencies, and the Sandgerdi town council.

Iceland is a part of the Greenland-Scotland Ridge, which marks the boundary between the biogeographic regions of the Arctic and the boreal North Atlantic. There are entirely different sets of species that live and thrive north and south of the Greenland-Scotland Ridge. BIOICE participants are gathering comprehensive information on which species of benthic invertebrates are thriving in Icelandic waters and on their physical environment, in order to promote international research on their ecology, taxonomy, systematics, and biogeography.

Samples are being collected from 600 stations on the sea floor. The samples have been taken by Icelandic, Norwegian, and Faroese research vessels. After sampling, the organisms are sorted to 150 major taxa at the Sandgerdi Marine Center, which is run by the program's participants. The sorted specimens are deposited in a scientific collection and the accompanying data are registered in a rela-

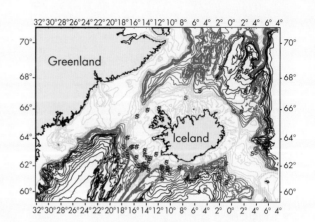

*Figure 58. The distribution of benthic marine species is usually restricted to limited areas of the sea floor. An example of this are four closely related benthic species of the genus **Pyrgo** (Foraminiferida); **P. sarsi** (red dots) is an endemic species of deep Arctic waters; **P. elongata** (blue-green dots) is distributed in shallow cold waters of the North Atlantic; **P. lucernula** (purple dots) thrives down to moderately deep and warm waters in most of the world oceans; **P. depressa** (blue dots) is distributed in the deep sea of the world oceans, except in the Arctic. (Source: Icelandic Institute of Natural History and Marine Research Institute, Iceland)*

Atlantic Ocean. Mud volcanoes and other geothermal spots on the seabed provide energy to support relatively high productivity over small areas.

Key characteristics

The dominant characteristic of the Arctic marine environment is sea ice, the extent of which varies greatly with the seasons. In September, it covers little more than the Arctic Basin and some marginal areas. At its maximum in spring, it covers the entire Arctic Ocean and much of the Bering and Barents seas, and may reach Iceland. Sea ice is a useful platform for some marine mammals. Seals haul out on the ice to bask and molt, and make their dens on or in the ice. Polar bears spend most of their lives on sea ice, where they hunt seals. The ice protects the water column from turbulence, providing a relatively stable habitat, but reduces sunlight reaching the water beneath. In coastal areas, sea ice disturbs the ocean floor as the keels of pressure ridges scour the seabed. Such scouring can destroy benthic communities.

Sea ice is habitat for an array of algae, which contribute substantially to the overall productivity of the Arctic Ocean. Ice algae start to grow within the sea ice in late winter and spring, especially in the bottom few centimeters of the ice. The bloom of ice algae is relatively brief, perhaps six weeks long. It

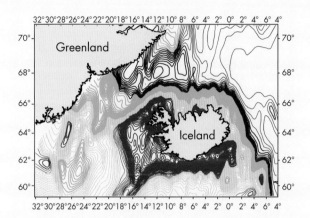

Figure 59. Distribution patterns of species are often strongly influenced by the annual mean temperature of water at the sea floor. The colored isotherms represent a thermal gradient from -1.2°C (blue lines) to +7.3°C (red lines).
(Source: Icelandic Institute of Natural History and Marine Research Institute, Iceland).

tional database, which is linked to a geographical information system. The resulting database holds information on:

- the sampling location
- how common the species are at different sampling locations
- taxonomical relationships among species

This information is combined with detailed information on bottom water temperature, salinity, and depth contour lines at the Marine Research Institute. This creates unique background information for analyzing species distributions, biological diversity, species communities, and evolutionary relationships.

After the initial sorting of the specimens, they are sent to about 80 collaborating scientists throughout the world. After completion of their research, the specimens are deposited in the taxonomic collection of the Icelandic Institute of Natural History and all additional data are registered in the database of the Marine Research Institute in Iceland. Each year several scientists and graduate students visit the Sandgerdi Marine Center to carry out their research. In addition, workshops have been held on the taxonomy of amphipods, polychaetes, and mollusks. Through this work, the BIOICE Program is making a major contribution to our understanding of biodiversity in a large part of the world.

Gudmundur Gudmundsson, Icelandic Institute of Natural History, Reykjavik

forms a band along the retreating ice edge, where meltwater provides a relatively fresh, stable layer in which a burst of phytoplankton growth takes place. Much of the ice algae and phytoplankton sinks to the bottom, but some is eaten by zooplankton. These, in turn, are consumed by arctic cod, capelin, and young herring. Arctic cod, though small, are found in great numbers in dense schools. Two schools alone in the Canadian High Arctic numbered at least 900 million fish in an area of 59 hectares. These fish are a key food item for marine mammals, seabirds, and other fish. In the Barents Sea, the arctic cod, capelin, and herring support some 1.3 million seals and 14 million seabirds.

Other productive sites include polynyas – areas of open water surrounded by ice. The North Water between Ellesmere Island and Greenland, the Great Siberian Polynya in the Laptev Sea, and systems of nearshore leads are among the regular and well-known areas. Other polynyas, which may not be as consistent from year to year, are nonetheless important. Many seabirds and marine mammals congregate at polynyas. Some remain in the Arctic all winter, whereas others use the open water in spring until the ice melts along the coast.

The circulation and mixing of water layers influences the productivity and structure of Arctic marine ecosystems. In the relatively shallow Bering, Chuk-

BOX 61. VENT AND SEEP COMMUNITIES ON THE ARCTIC SEAFLOOR

▶ Deep under perpetual ice cover, with reduced photosynthesis and thus little organic matter coming from above, the sea floor of the central Arctic Ocean is a marine desert, its life more sparse than in other ocean basins. But photosynthesis is not the only basis for life: locally, methane and hydrogen sulfide seep from the ocean floor, supporting dense oases of organisms that depend on bacteria able to consume these chemicals. Such "chemosynthetic" life does not directly depend on sunlight and can thrive even at great ocean depths.

Two basic kinds of oases are now known from the oceans. Hot vents, first discovered in 1977, are now known from several places along the Mid-Oceanic Ridge. These seafloor geysers belch out seawater heated to 400°C and laden with nutritious chemicals. Dramatic colonies of large clams, giant tubeworms, and other strange life-forms have been discovered at hot vents.

The second broad category of chemosynthetic oasis is the "cold seep," which usually involves the upward seepage of methane dissolved in water or as small bubbles. Mud volcanoes and related cold-seep features form over great sediment accumulations in which bacteria digest buried organic matter, producing methane as a waste product. Specially evolved bacteria oxidize the methane, forming the foundation of a food chain. Different bacteria have evolved to oxidize the foul-smelling hydrogen sulfide, itself the waste product of yet other bacteria living below the ocean floor, which oxidize sulfate ions of seawater origin. The conspicuous and, by bacteria standards, large sulfur bacteria (*Beggiatoa* spp.) form thin, snow-like mats on the seafloor where seepage takes place. Bacterial mats form at hot vents also, but *Beggiatoa* is common at cool oceanic seeps and some non-seep environments where hydrogen sulfide rises close to the seafloor and oxygen is present in the water.

Neither hot nor cold vents had been found in the Arctic until, in 1995 and 1996, a team of American, Norwegian, and Russian scientists discovered an active cold-seep mud volcano on the continental margin west of the Barents Sea, at 72°N. The Haakon Mosby mud volcano is about a kilometer in diameter, and stands only a few meters high in water 1250 meters deep. The mud volcano has a patchy white cover of sulfur bacteria mats (which few creatures eat), and is home to a diverse community of creatures whose food web is based on methane-consuming bacteria.

In the summer of 1998, scientists inspected the Haakon Mosby mud volcano first hand, aboard the Russian *Mir* submersibles that had been used to explore the *Titanic*. The submersible sampled bottom sediments and used a "slurp gun" to suck in unsuspecting bottom fish in order to study them and augment museum collections. Two species of small tubeworms were found on the mud volcano, one of them

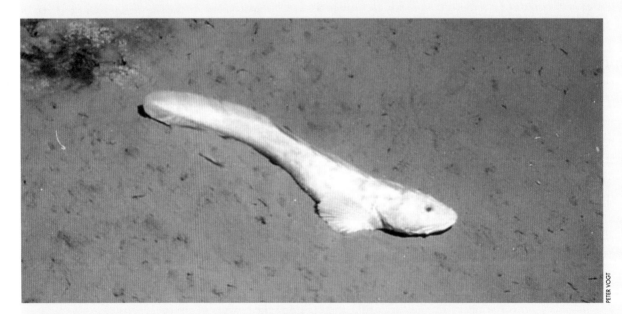

Scalebelly eelpout **Lycoides squamiventer**: *a common fish at the Haakon Mosby mud volcano.*

previously known only from the Antarctic. A number of small, sand-sized animal species sifted from the seafloor mud of the mud volcano also depend on chemosynthesis.

The most common fish found were scalebelly eelpout. Stomach contents of these fish included tubeworms and other chemosynthesis-dependent creatures, confirming that the fish are indeed part of the local ecosystem. The eelpout was several hundred times more abundant on the mud volcano than elsewhere on the bottom of the surrounding Greenland-Norwegian Sea, and is apparently attracted to the mud volcano by its abundant food supplies.

A variety of other fish were photographed, but not captured, on the mud volcano surface. Conspicuous are several species of skate, which probably form the top of the local food chain by consuming smaller bottom fish like eelpout. The habit eelpout have of lying inside small bottom depressions may be their way of escaping detection and consumption by skates searching for prey. The largest predator fish in these cold waters is the Greenland shark, but scientists aboard the *Mir* submersibles were not lucky enough to glimpse this rare animal.

The search for hot vents along the Arctic extension of the Mid-Oceanic Ridge, which separates the North American and Eurasian plates, has just begun. To date, the northernmost known hot vent is just north of Iceland, in a shallow-water environment atypical of hot vents studied elsewhere. There is every reason to expect such vents to exist farther north, probably with some organisms new to science, or at least distinct from those found farther south.

Peter Vogt, Naval Research Laboratory, Washington, DC, USA

SOURCES/FURTHER READING

Herman, Y. 1989. *The Arctic Seas – climatology, oceanography, geology, and biology*. New York: Van Nostrand Reinhold Co. 888p.

Humphris, S.L., R.A. Zierenberg, and I. Mullineaux 1995. *Seafloor hydrothermal systems*. Geophysical Monograph 91. Washington: American Geophysical Union. 466p.

Milkov, A., P. Vogt, G. Cherkashev, G. Ginsburg, N. Chernova, and A. Andriashev. 1999. Sea-floor terrains of Haakon Mosby mud volcano as surveyed by deep-tow video and still photography. *Geomarine Letters* 19: 38-47.

Pimenov, N., A. Savvichev, I. Rusanov, A. Lein, A. Egorov, A. Gebruk, L. Moskalev, and P. Vogt. 1999. Microbial processes of carbon cycle as the base of food chain of Haakon Mosby mud volcano benthic community. *Geomarine Letters* 19: 89-96.

Vogt, P.R., J. Gardner, and K. Crane. 1999. The Norwegian-Barents-Svalbard (NBS) continental margin: introducing a natural laboratory of mass wasting, hydrates, and ascent of sediment, pore water, and methane. *Geomarine Letters* 19: 2-21.

chi, and Barents seas, the inflow of warmer water from the Pacific or Atlantic Ocean adds to the nutrients stirred up from the continental shelf. Where these waters meet the colder Arctic waters, for example at the Barents Sea's polar front, productivity is high. These zones feed the food webs of their respective seas, as well as the Arctic Basin and deeper waters, where currents carry some of the material northward and downward.

River flow into the marginal seas brings nutrient-rich freshwater to the ocean. The Lena River, for example, spreads nutrients far offshore, feeding a plankton bloom in the Laptev Sea. Other large rivers have similar effects, although the variation in salinity caused by the periodic influx of freshwater limits the number of species that can survive in such an environment. The sediment load of the river also affects the degree to which nutrients can be used. Muddy waters allow little light to penetrate, reducing growth.

The extremes of climate and sunlight in the Arctic drive seasonal variations in production. Nutrients that build up during the winter support the spring algal and phytoplankton blooms. Most primary production occurs in a brief pulse. Grazing by herbivores and predation by carnivores extend the cycling of nutrients throughout the year. Some zooplankton, for example, are able to graze intensively, then descend to deeper waters to await the next period of high productivity. Fish, birds, and marine mammals may migrate to other habitats or build up fat reserves to live off during the winter.

Functionality

The burst of primary production in spring is greatest in the marginal seas and along the ice edge, though under-ice productivity in the Arctic Basin is greater than was once thought. Ice algae and phytoplankton convert nutrients and energy into living matter, supporting the food web. Once in the Arctic Ocean, nutrients are typically used and recycled in a highly efficient manner. Dead plants and animals are scavenged effectively on the ocean floor. Animals that feed on the bottom, such as walrus, bearded seals, and some

Drift ice at 89°N in April.

birds, bring the organic matter back up into the water column. The ability of the marine ecosystem to re-use organic matter is significant for many reasons, among them the fact that long-lasting contaminants are recycled along with the nutrients.

In the water column, microalgae are eaten by small, floating, grazing animals known as herbivorous zooplankton. These in turn are food for predatory zooplankton, the numerous small invertebrates such as copepods and euphausiids. Zooplankton feed many fish, seabirds, and marine mammals. The fish are also eaten by marine mammals and birds, as well as by larger fish. Marine mammals may be eaten by other marine mammals, such as polar bears and killer whales. Some walruses eat seals. Seabirds are eaten by large fish like halibut and Pacific cod and by seagulls.

Dead algae and other organic matter sinks to the sea floor, where it may be eaten by a variety of bottom-dwelling animals or reduced to nutrients by bacteria. Some of these live on the bottom and others, such as mollusks, live in the sediments of the sea floor. The shallower seas and areas where currents bring organic matter downward are the most productive areas of the ocean bottom. Deeper areas also support benthic flora and fauna, but in far lower densities. These are eaten by bottom-feeders such as halibut, eiders, bearded seals, walruses, and gray whales.

The migrations of marine mammals and birds reflect seasonal patterns of production and feeding. Walruses and many species of seals move north in summer with the pack ice, feeding in the productive zone of the ice edge, which also provides favor-

Drift ice in the Denmark Strait.

able habitat for large numbers of seabirds. Some walruses and seals establish summer haul-outs on land, where they rest between feeding trips. Others stay with the ice throughout the year. Polar bears typically travel with the ice, though in some areas, such as southern Hudson Bay, they spend time on land after the ice melts, awaiting new ice for their return to the ocean.

Many Arctic marine species are not found in temperate waters, and are thus specifically adapted to life in the Arctic. The bowhead whale, found only in waters that have sea ice, is able to make breathing holes by breaking sea ice up to 20 centimeters thick. Of those species that are also found outside the Arctic, some have Arctic populations that display special adaptations. For example, the Baltic clam is found in Western European waters as well as in the White and Pechora Seas. The growth of this clam decreases in more northerly populations, reflecting the decrease in overall quality of habitat. Its sub-population in the Pechora Sea, however, grows as quickly as do its more southerly populations, likely reflecting specific genetic adaptations to the Arctic.

Changes in sea ice and other patterns are likely to have substantial impacts in Arctic marine systems. The severity of these changes can be seen in the effects of extreme events in the past. In 1978, for example, a polynya in Lancaster Sound in Canada did not form. This caused a breeding failure in northern fulmars, thick-billed murres, and black guillemots. Freezing of polynyas also have caused catastrophic die-offs of eiders in Hudson Bay. Similarly, the timing and location of the plankton blooms in spring influence the species that depend on them. If the bloom occurs early, less

BOX 62. LIFE WITHIN SEA ICE

▶ Pioneering polar biologists reported life in Arctic sea ice as early as the mid-1800s, as its strong brownish coloration and odor caught their attention. Today we know that both the color and the "Arctic scent" are produced by masses of diatoms, a group of microalgae that grow attached to the bottom of the ice.

The ice interior is inhabited by a large variety of organisms called protists. In sea ice, these are mainly microscopic, unicellular algae such as prasinophytes, dinoflagel-

*Microalgae from Greenland Sea: a Chrysophyte algae **Dictyoha speculum**; a phytoplankton species, which is sometimes found in sea ice (below); a heterotrophic euglenid flagellate, a typical sea ice species.*

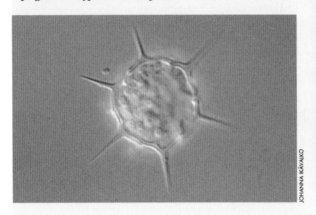
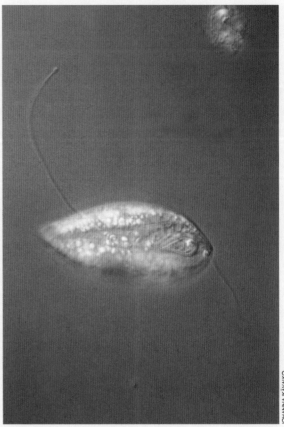

food will remain when migratory species arrive. If the bloom is late, the progression from algae to zooplankton may be delayed, changing the pattern of the food web.

Interactions with other regions

In the Arctic, the coast is primarily in the region of tundra and polar desert, though the forest-tundra does reach the shore in some areas such as northwestern Russia, Fennoscandia, Iceland, southern Greenland, and parts of Alaska. Along this land-water boundary, there can be some interaction of species. Seabirds nest on land but feed in the water. Arctic foxes scavenge on sea ice, but breed on land. Polar bears, usually considered marine mammals, often den in snowbanks on land. The landfast ice blurs the distinction between land and sea, as caribou and other land animals may walk considerable distances across the ice. Peary caribou in the Canadian High Arctic move across the sea ice, often in search of better foraging.

The ocean receives nutrients that wash off the land. Rivers carry great quantities of organic matter, though the patterns of river flow determine whether those materials are carried far into the ocean or deposited in sediments close to river mouths. Because rivers help lates, and chrysophytes, in addition to diatoms. Photosynthesizing algae are the primary producers of the ice-associated food web in the Arctic. The size of ice algae is typically 5-40 μm. Heterotrophs (first level consumers) feed on ice algae as well as bacteria and dissolved organic matter within sea ice. Common heterotrophic protists in Arctic sea ice are ciliates (ca. 50-100 μm), choanoflagellates, non-photosynthesizing dinoflagellates, and euglenids (ca. 20-50 μm). The sea ice thus contains a complete foodweb, from primary producers to consumers and, finally, decomposers (bacteria). Some of these organisms live their entire life cycle within sea ice, whereas others spend only part of their lives in the ice.

How is it possible for life to exist in such a dark, cold, and harsh environment? The answer lies in the structure of sea ice and in the adaptations of the organisms. Sea ice is composed of two main elements: non-saline ice crystals, and brine channels filled with brine (a highly saline solution). The organisms live not in the ice crystals but in the brine channels. As the air temperature drops, leading to ice crystal formation and growth, the salinity of the brine increases. In practice, this means that sea ice organisms must be highly tolerant of extremes in their environment. Protists within sea ice are also capable of adapting to very low light levels. Even during the Arctic winter, diatoms and other algae attached to the bottom of the ice sheet in some areas can photosynthesize using the little light available.

Sea ice organisms come from the water column, bottom sediments, or from rivers. During ice formation in fall, these algae and other microscopic organisms become attached to the ice crystals that later form the solid ice cover. Species diversity and biomass are usually fairly low during the early stages of ice development. In spring, when the ice cover is significantly thicker, a drastic increase in algal production is spurred as air and ice temperatures increase and more light is available. This is followed by an increase in the diversity and abundance of consumers. Multi-year ice that has gone through several such cycles thus has internal bands of protist communities from different years.

Johanna Ikävalko, Department of Ecology and Systematics, Division of Hydrobiology, University of Helsinki, Helsinki, Finland

SOURCES/FURTHER READING

Dickie, G. 1852. Notes on the algae. In: P.C. Sutherland, ed. *Journal of a voyage in Baffin's Bay and Barrow Straits, in the years 1850-51, performed by H. M. ships "Lady Franklin" and "Sophia", under the command of Mr. William Penny, in search of the missing crews of H. M. ships "Erebus" and "Terror."* Vol. 2. London: Longman, Brown, Green and Longmans.

Ehrenberg, C.G. 1841. *Einen Nachtrag zu dem Vortrage über Verbreitung und Einfluss des mikroskopischen Lebens in Süd- und Nord-Amerika.* Akad. Wiss. Berlin Monatsber. 1841:202-207. (In German)

Maykut, G.A. 1985. The ice environment. In: R. Horner, ed. *Sea ice biota.* Boca Raton, Florida: CRC Press. p. 21-82.

Monti, D., L. Legendre, J.-C. Therriault, and S. Demers. 1996. Horizontal distribution of sea-ice microalgae: environmental control and spatial processes (south-eastern Hudson Bay, Canada). *Mar. Ecol. Prog. Ser.* 133:229-240.

---

determine the temperature, salinity, and nutrient content of the seas into which they flow, changes in the timing and volume of river flows may cause physical changes to their estuaries and the surrounding sea water. The biota and food webs of those areas would change as a result.

Moving in the opposite direction, some organic matter is transferred from the ocean into freshwater or onto land. Some seabirds nest inland, where predators such as arctic foxes, snowy owls, and falcons may eat them. At seabird colonies, droppings fertilize terrestrial plants, creating locally lush growth. Anadromous fish may do most of their growing in the ocean, but return to freshwater to spawn and often to die. Highly abundant species such as salmon provide much food for bears, eagles, and other scavengers. Their carcasses increase nutrient concentrations in lakes and rivers.

In addition to the circulation of water and nutrients, the Arctic marine environment interacts biologically with the rest of the world, most obviously through migrations. Large numbers of seabirds and marine mammals follow seasonal patterns to and from northern regions. They feed extensively on Arctic flora and fauna in summer, and return south to winter in warmer regions. Gray whales migrate thousands of

BOX 63. LIVING WATER IN THE FROZEN ARCTIC: THE GREAT SIBERIAN POLYNYA

▶ The system of flaw polynyas on the Siberian shelf of the Laptev Sea is known as the Great Siberian Polynya. The prevailing southerly winds in winter and heat transport from the underlying water layer support these extensive stretches of open water, which are up to 200 kilometers wide even at extremely low air temperatures. It is the most stable system of flaw polynyas in the Russian Arctic, lasting from early October until late June each year.

An early indication that the region was inhabited by birds was the observation by the Zarya expedition of early spring migrations of ducks over the fast ice of the Chukchi Sea. The existence of the non-migrating population of Laptev walrus was suggested in connection with the existence of open water throughout the winter. The first scientific investigations of the Polynya, however, did not take place until the winter and spring of 1996 and 1999 on the Russian-German TRANSDRIFT expeditions.

Although the benthic biomass beneath the Polynya is not higher than on adjacent bottom areas, the open water allows Laptev walruses to forage throughout the winter. Our estimate of 1,430 walruses in the Polynya in May 1996 equates to a density of 0.25-0.28 individuals per kilometer of fast ice edge. These animals occur individually and in groups of up to six, mainly on the fast ice edge and rarely on drifting ice floes. Males predominated in the wintering population of Laptev walrus, suggesting that females winter in other waters.

Ringed seal and bearded seal also winter in the Polynya. The ringed seal was widely distributed on the fast ice, whereas the bearded seal was found only at the edge. The density of bearded seals was 0.05-0.07 individuals per kilometer of fast ice edge. Habitat use by polar bears and arctic foxes was recorded as well. Foxes prefer the fast ice edge, whereas polar bear footprints were observed as far as 35 kilometers inshore of the open water.

The stock of plankton and benthic fauna also provides food for the large numbers of birds arriving in spring. The majority of seabirds do not stay in the Polynya, but only follow the open water during their migration from the Bering Sea to Siberian breeding grounds. Tens of thousands of oldsquaw (or long-tailed ducks) and king eiders pass through. The king eiders migrate in flocks of up to 300 birds, flying across drift ice fields from one lead to another. Black-legged kittiwakes and pomarine jaegers (or pomarine skuas) migrate individually or in loose flocks of up to 6 birds, their paths following the configurations of the leads. In addition to the seabirds, a songbird, the snow bunting, was occasionally recorded on its migration from the mainland to the New Siberian Islands.

The Great Siberian Polynya.

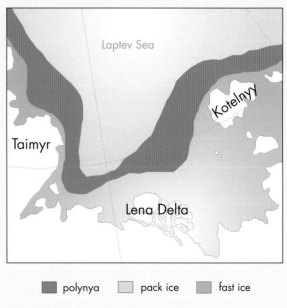

Figure 60. The Great Siberian Polynya.

The intensive bird migration passes to the west by the end of May. Although early spring surveys from mid-April to early May 1999 did not indicate any birds other than individual thick-billed murres (or Brünnich's guillemots), it is highly likely that some duck and gull flocks arrive at the Polynya early in spring and stay there until summer. Ducks were observed off the New Siberian Islands as early as April. Stomachs of king eiders taken at the Polynya contained the remains of Gastropoda and Bivalvia shells with *Leonucula bellotii* and *Cryptonatica clausa* among them. Biocenoses (species communities) dominated by *Leonucula bellotii* were found beneath the Polynya, with an average biomass of 111-188 g/m². The impact on benthic fauna of the entire 20,000-bird eider population, even over the month of spring staging, does not exceed 0.17% of the Polynya's benthic biomass.

On 8 May 1999, our helicopter took off from the drift ice for the last time. The strip of water beneath an extremely dark-grey "water sky" remained behind. Our short excursions had allowed us to raise a little the fog that covers the undisturbed and little-known wildlife of the Great Siberian Polynya.

Diana Solovieva, Lena-Delta State Reserve, Tiksi, Sakha Republic, Russia

SOURCES/FURTHER READING

Abramova, E.N. 1998. Composition, abundance and population structure of spring-time zooplankton in the shelf-zone of Laptev Sea. In: H. Kassens, H.A. Bauch, I. Dmitrenko, H. Eicken, H.-W. Hubberten, M. Melles, J. Thiede, and L. Timochov, eds. *Land-ocean system in the Siberian Arctic: dynamics and history*. Berlin: Springer-Verlag. p. 161-168.

Birulya, A.A. 1907. Notes on the life of birds along the polar coast of Siberia. *Zapiski Imperatorskoy Academii Nauk po fiz.-mat. sekcii* 18(2):1-157. (In Russian)

Sirenko, B.I., V.V. Petriashov, E. Rachor, and K. Hinz. 1995. Bottom biocoenoses of the Laptev Sea and adjacent areas. In: H. Kassens, D. Piepenburg, J. Thiede, L. Timokhov, H.W. Hubberten, and S. Priamikov, eds. *Berichte zur Polarforshchung*, 176. Bremenhaven: Alfred Wegener Institute for Polar and Marine Research. p. 211-221.

Solovieva, D.V. 1998. Spring stopover of birds on the Laptev Sea Polynya. In: H. Kassens, H.A. Bauch, I. Dmitrenko, H. Eicken, H.-W. Hubberten, M. Melles, J. Thiede, and L. Timochov, eds. *Land-ocean system in the Siberian Arctic: dynamics and history*. Berlin: Springer-Verlag. p. 189-195.

Zakharov, N.P. 1996. *Sea ice*. St.Petersburg: Gydrometeoizdat. 178p. (In Russian)

kilometers to the subtropical regions where they breed, having done most of their yearly feeding in the richer northern waters of the Arctic and subarctic. Salmon spawning in the Arctic grow and mature in temperate oceans before returning to their natal streams. The northern seas even attract shearwaters, which migrate from breeding areas in the southern hemisphere to "winter" in the Arctic summer.

Natural variability

Over long periods, the Arctic marine environment varies tremendously. During the last Ice Age, lower sea levels exposed a thousand-kilometer-wide plain between Asia and North America, eliminating the connection between the Pacific and Arctic Oceans. Only for the past 10,000 years or so have ocean water and marine animals been able to move through the Bering Strait. In the past few thousand years, pack ice in the Bering Sea has at times extended to the Aleutian Islands, far to the south of its usual limits today. These changes have shifted the distribution of marine fish, mammals, and birds, with resulting effects on the human communities that used them.

Large shifts in environmental conditions also occur over years and decades. The extent of sea ice varies considerably from year to year. In the early 20th Century, the Barents Sea was largely ice-free, although now it is typically ice-covered in winter. Walrus hunting records from the Bering Strait area from the 1920s to the present indicate substantial environmental variability. Winds and sea ice affect the movements and availability of walrus in the Bering and Chukchi Seas, causing wide fluctuations in annual harvests.

The flow of water to and from the Arctic Ocean varies from year to year. In some years, high inflows of water can raise water temperatures across large areas. In the Chukchi Sea in 1998, the ice retreated hundreds of kilometers offshore, much farther than usual. In the Barents Sea, changes in the flow of Atlantic water help determine the productivity of the region, both through temperature and through the

BOX 64. THE ANADYR CURRENT AND OCEAN PRODUCTIVITY IN THE BERING STRAIT AREA

▶ Locally, water flowing through the Bering Strait has two principal sources. The Anadyr Current on the western side originates at a depth of about 200 meters along the edge of the continental shelf as a branch of the Bering Slope Current. Owing to its source in the deep waters of the Bering Sea, this current is cold, saline, and nutrient-rich. It contrasts with the Alaskan Coastal Current on the eastern side, which originates on the shallow continental shelf of the eastern Bering Sea and can be traced southward into the coastal zone of the Gulf of Alaska. In summer, the Alaskan Coastal Current is comparatively warm, fresh, and nutrient-poor.

The northward flow of the Anadyr Current water through the Bering Strait completes the pole-to-pole circulation of ocean water that began with deep water formation in the North Atlantic, returning water to the western Arctic that was originally lost from the eastern Arctic. Sequestered beneath the euphotic zone (the depth to which sunlight penetrates sea water) during their long journey, the deep waters of the North Pacific and Bering Sea have been enriched for centuries with nutrients from the detritus that rains down from surface layers. Apart from eutrophic coastal zones receiving runoff from agricultural areas, the water welling upward in the northern Bering Sea contains the highest concentrations of nutrients – nitrate, phosphate, and silicate – in the world ocean.

As the Anadyr Current flows northward from the shelf edge, it brings an essentially endless and tremendously rich supply of nutrients into the euphotic zone of the shallow Bering-Chukchi continental shelf, creating one of the world's most highly productive regions. In contrast to most areas of the Arctic, nutrients are not exhausted by the spring bloom

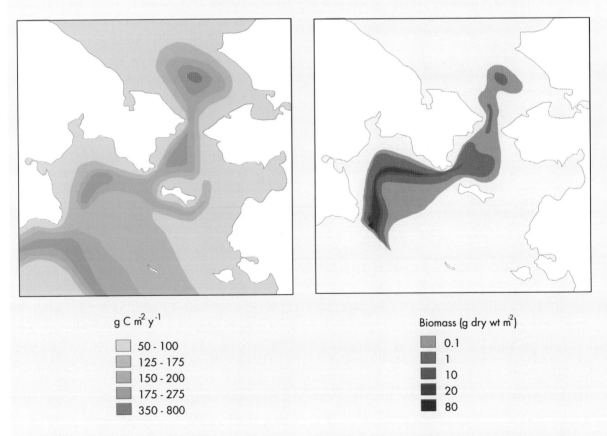

$g\ C\ m^{-2}\ y^{-1}$

- 50 - 100
- 125 - 175
- 150 - 200
- 175 - 275
- 350 - 800

Biomass (g dry wt m^{-2})

- 0.1
- 1
- 10
- 20
- 80

Figure 61. Generalized pattern of primary production in the Bering Sea.

Figure 62. Distribution of oceanic zooplankton on the northern Bering-Chukchi continental shelf.

and the phytoplankton grow continually from the retreat of seasonal sea ice in spring until fall, when the light level falls below that necessary for photosynthesis. The prolific primary production supports rich benthic communities that feed a huge biomass of species at higher trophic levels. Gray whales, for example, migrate annually from Mexico to the Bering Strait area to feed on abundant crustaceans, and clams and other bottom fauna support the large population of Pacific walruses that depend on them.

In addition to nutrients, the Anadyr Current also carries with it a great quantity of large zooplankton, common in the basin of the Bering Sea but absent from the shallow continental shelf. These zooplankton are the primary prey of several species of small auklets, whose populations on islands in the Bering Strait region – St. Lawrence Island and the Diomede Islands – number in the millions and are hardly rivaled anywhere else within their range. The zooplankton also provide the bulk of the year's food for bowhead whales in the region.

By so enriching the waters of the region, the Anadyr Current is the defining feature of the northern Bering Sea and southern Chukchi Sea. Without it, we would lack the great abundance of species that are key components of the ecosystem and critical to the way of life of local people.

Alan M. Springer, University of Alaska Fairbanks, Fairbanks, Alaska, USA

SOURCES/FURTHER READING
Coachman, L.K., K. Aagaard, and R.B. Tripp. 1975. *Bering Strait: regional physical oceanography*. Seattle: University of Washington Press.
Grebmeier, J.M., C.P. McRoy, and H.M. Feder. 1988. Pelagic-benthic coupling on the shelf of the northern Bering and Chukchi Seas. I. Food supply source and benthic biomass. *Mar. Ecol. Prog. Ser.* 48: 57-67.
Pickard, G.L., and W.J. Emery. 1982. *Descriptive physical oceanography*. New York: Pergamon Press.
Sambrotto, R.N., J.J. Goering, and C.P. McRoy. 1984. Large yearly phytoplankton production in western Bering Strait. *Science* 225: 1147-1150.
Springer, A.M., and C.P. McRoy. 1993. The paradox of pelagic food webs in the northern Bering Sea-III. Patterns of primary production. *Cont. Shelf Res.* 13: 575-599.
Springer, A.M., C.P. McRoy, and K.R. Turco. 1989. The paradox of pelagic food webs in the northern Bering Sea – II. Zooplankton communities. *Cont. Shelf Res.* 9: 359-386.
Springer, A.M., D.G. and Roseneau. 1985. Copepod-based food webs: auklets and oceanography in the Bering Sea. *Mar. Ecol. Progr. Ser.* 21: 229-237.
Walsh, J.J., C.P. McRoy, L.K. Coachman, J.J. Goering, J.J. Nihoul, T.E. Whitledge, T.H. Blackburn, P.L. Parker, C.D. Wirick, P.G. Shuert, J.M. Grebmeier, A.M. Springer, R.D. Tripp, D.A. Hansell, S. Djenidi, E. Deleersnijder, K. Henriksen, B.A. Lund, P. Andersen, F.E. Muller-Karger, and K. Dean 1989. Carbon and nitrogen cycling within the Bering/Chukchi seas: source regions for organic matter effecting AOU demands of the Arctic Ocean. *Prog. Oceanog.* 22: 277-359.

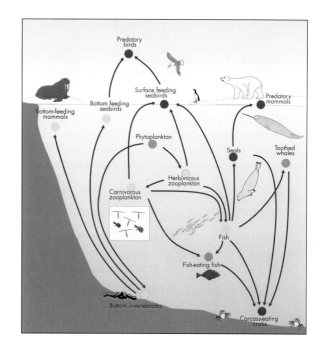

● Third and higher level predators
● Second level predators
◐ First level predators
○ Grazers and filter feeders
● Primary producers

Figure 63. Simplified Arctic marine food web.

transport of nutrients and plankton. Large-scale ocean processes have substantial impacts as well. The North Atlantic Oscillation has an influence from East Greenland to Fennoscandia and beyond. El Niño in the tropical Pacific affects the Bering Sea by altering winds and ocean temperatures across the entire ocean.

Other changes mark shifts in the structure of the environment and the ecological functions it supports. In the Bering Sea in 1976, the size and abundance of fish, predation patterns by marine mammals, and several other physical and biological factors appear to have changed at the same time. Combined with human activities, these changes may have led to a cascading effect throughout much of the Bering Sea ecosystem. Whether ecosystems undergo such regime shifts regularly, or whether they are extraordinary events, is not known.

A seal breathing hole.

*Figure 64. Distribution of bearded seal **Erignathus barbatus**.*

Bearded seal **Erignathus barbatus**.

Seabird colony on Svalbard.

Figure 65. Distribution of narwhal **Monodon monoceros**.

Narwhals **Monodon monoceros**.

THE OCEANS AND SEAS 197

The fishing industry accounts for two-thirds of Iceland's export.

Human uses and impacts

Fishing and marine mammal hunting are substantial around the Arctic. These activities support the indigenous cultures that have thrived in the region for millennia, and contribute to local economies today. Whalers were among the first to exploit Arctic marine resources for commercial purposes, followed by fishers and sealers.

Commercial hunting of whales, walrus, and seals caused the demise of some species and stocks during the past few centuries. The Steller's sea cow was driven to extinction in the North Pacific in the 18th Century. The right whale and some stocks of bowhead whale remain at the brink of extinction, with populations numbering in the tens or hundreds. Others, including some Atlantic populations of walrus and some beluga whale stocks, remain low and are possibly overexploited in some areas. Nonetheless, most marine mammal populations in the Arctic appear capable of sustaining current harvests.

Commercial fishing is a large-scale and lucrative industry, especially in the Norwegian, Barents, and Bering seas. The Bering Sea, for example, accounts for 2-5% of the world's fishery production, and over half of the U.S. fish catch. About four-fifths of Iceland's export income is from fish.

Commercial fishing removes a considerable amount of fish biomass, both target fish and bycatch. It also affects habitat, particularly as a result of bottom trawling. Some fish stocks collapsed in the 20th Century. The herring population in the northeast Atlantic, which had been one of the largest fish stocks in the world, was reduced by a factor of one thousand between 1950 and 1969. The fisheries in the Barents Sea and Iceland then switched from herring to capelin. Capelin has remained an important fishery species in Iceland, but collapsed in the Bering Sea in the mid-1980s. As a result, many seabirds died from starvation. In the Bering Sea, great changes in the relative abundance of fishes are probably the result of a combination of fishing pressure and environmental change. The removal or alteration of a major component of the marine ecosystem can have great impacts on the entire food web. Prey species may flourish and predators may suffer. Increases in some fish species to replace others, however, can make impacts more difficult to identify and assess.

Shipping has been a large-scale activity since the 1930s, with the development of the Northern Sea Route in Russia. Barges and cargo vessels around the Arctic take goods to coastal communities and transport ore and metal from mines. The likelihood that climate change will lead to longer shipping seasons has spurred considerable interest in further use of the Northern Sea Route as the shortest passage from Europe to East Asia.

Shipping affects the surrounding environment in several ways. The noise of ships may make some animals move away from a busy area, though many marine mammals and birds appear capable of adapting to noise in some circumstances. Icebreakers are loud and leave tracks of broken ice and open water. Though these often re-freeze, they can affect the movements of some animals, as well as people traveling over the sea ice, in addition to occasionally trapping marine mammals. Ports and harbors in the Arctic, supporting both mineral development and shipping, lead to physical disturbance and pollution.

*Remains of two gray whales **Eschrichtius robustus** in Sireniki, Chukotka, Russia.*

The extraction of oil and gas from the continental shelf has also attracted attention and activity. Offshore development is currently underway in Russia and Alaska. Oil and gas exploration and development produce several types of impacts. Seismic testing of the seafloor uses air guns, which can kill larval fish within a few meters of the gun and displace larger fish. The noise may disturb marine mammals. The installation and operation of drilling platforms, from human-built islands to mobile rigs, can disturb the ocean floor and alter coastal currents and fish movements. The migratory paths of marine mammals may shift to avoid some structures, possibly affecting their availability to hunters. Pipelines for transporting oil to land or to tankers may disrupt the seafloor. Oil spills have the potential

At Svalbard, early whalers who died from scurvy, botulism, gangrene, and other causes were often buried on the shoreline.

BOX 65. BELUGA WHALE HUNTING IN RUSSIA

▶ The beluga whale is a small, toothed whale, and is white as an adult. It is found throughout the Arctic and into the subarctic, including four of the coastal regions of the Russian Federation. The annual beluga harvest in the former Soviet Union in the 1930s and 1940s is estimated at 5,000 animals. By the 1950s and 1960s, this had declined to between 1,500 and 3,000 whales per year. For the last three decades, there has been no commercial harvest of belugas, although a few individuals have been taken for food in indigenous communities, for scientific purposes, and for use in aquaria. Beluga hunting in Russia is regulated by conservation and hunting rules established in 1986, which authorize the capture only of individuals older than one year using nets, seines, or togglehead harpoons. The use of firearms and the capture of lactating females and calves are forbidden.

Due in part to the extreme economic conditions in the region, the Russian Federation issued permits in August 1999 for a commercial harvest of up to 200 belugas from the Sea of Okhotsk, the meat and blubber to be exported to Japan. More than 13.2 tons of meat and blubber from 36 belugas were actually exported. In September 1999, the Russian Federation decided to withdraw the permits, limiting hunting once again to traditional and scientific purposes only. The pressures for such harvests remain a conservation concern in the region.

The degree to which Russian beluga stocks are shared with other countries is unclear. On the Alaska side of the Bering and Chukchi Seas, four stocks of belugas have been identified, but the details of their migratory patterns to Canada and Russia are not known. A recent study documenting traditional ecological knowledge in Chukotka helped identify migratory patterns and other indications of connections between stocks in Russia and those in Alaska and Canada. Recent satellite tagging of belugas in Canada and Alaska corroborates these findings, and points the way to future collaborative research to discover more about the connections across the Bering Strait, including shared conservation concerns.

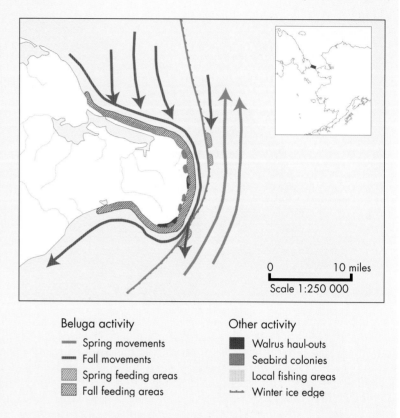

*Figure 66. Traditional ecological knowledge of the movements of beluga whales **Delphinapterus leucas**, Uelen, Chukotka, Russia.*

Arkady Tishkov, Institute of Geography, Russian Academy of Sciences, Moscow, Russia
Henry Huntington, Huntington Consulting, Eagle River, Alaska, USA
Valentin Iljashenko, Ministry of Nature Resources of Russian Federation, Moscow, Russia

SOURCES/FURTHER READING
Mymrin, N. I., The Communities of Novoe Chaplino, Sireniki, Uelen, and Yanrakinnot, and H. P. Huntington. 1999. Traditional knowledge of the ecology of beluga whales (*Delphinapterus leucas*) in the northern Bering Sea, Chukotka, Russia. *Arctic* 52(1):62-70.
O'Corry-Crowe, G. M., R. S. Suydam, A. Rosenberg, K. J. Frost, and A. E. Dizon. 1997. Phylogeography, population structure and dispersal patterns of the beluga whale *Delphinapterus leucas* in the western Nearctic revealed by mitochondral DNA. *Molecular Ecology* 6:955-970.

A walrus **Odobenus rosmarus** colony, Bering Strait, Russia.

for catastrophic impacts over large areas. Clean-up would be difficult, if not impossible, in ice-covered waters.

As on land, diffuse disturbances are present in the Arctic marine environment. Ocean currents, winds, and rivers all bring pollution to the Arctic Ocean, including radionuclides, heavy metals, persistent organic pollutants (POPs), and hydrocarbons. Some of these, such as radionuclides, tend not to be taken up in the marine food web. Others, such as some heavy metals and most organic pollutants, concentrate in living things, especially in the fatty tissues that are common in Arctic animals. Because the marine environment is efficient in recycling organic matter, pollutants can move through the food web over and over. Although some heavy metals are naturally occurring and have been found in tissues hun-

Figure 67. Distribution of walrus **Odobenus rosmarus**.

THE OCEANS AND SEAS 201

BOX 66. THE INTERACTION OF CLIMATE AND FISH STOCKS IN THE BARENTS SEA

▶ The waters off northern Norway are some of the most profitable fishing grounds in the world. The catches, however, have always been variable, making fishing a risky business. This variability is caused by a combination of the physical characteristics of the area and the biological characteristics of the cod, herring, and capelin stocks.

In this region, the interface of the warm Atlantic and cold Arctic currents produces two zones of high plankton productivity: the Polar Front in the middle of the Norwegian Sea, and the marginal ice zone of the Barents Sea. The largest stocks of plankton feeders are herring in the Norwegian Sea and capelin in the Barents Sea, both of which migrate to the Norwegian coast to spawn. The herring and capelin in turn are prey to a large number of species, of which the Atlantic cod is the most important. The cod feed in the Barents Sea and spawn on the central Norwegian coast. As a result, this coast has supported large fisheries since ancient times.

Productivity in the Barents Sea is affected by variations in the inflow of Atlantic water. Warm periods have occurred with a frequency of eight to ten years since the 1950s, and

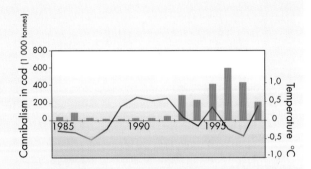

Figure 69. Cannibalism of cod (bars) in relation to deviations from long-term average sea temperature (solid line) in the southwestern Barents Sea.

coincide with good recruitment for the fish stocks. The increased flow from the Atlantic, however, also transports large numbers of juvenile herring into the Barents Sea, where they stay for three to four years as they mature. The herring have a devastating impact, as a strong herring year class will devour most of the capelin fry, blocking capelin recruitment for several years, and greatly reducing the spawning stock of that fish. When the adult herring leave the Barents Sea, the cod in turn suffer from the lack of food, since neither herring nor capelin are present in sufficient numbers.

When neither herring nor capelin are present in sufficient numbers, older cod eat younger ones. In the early 1990s, which was a warm period, herring and cod produced strong year classes, and the capelin were greatly depleted. In 1994-95, the stock of juvenile cod increased considerably, just as the herring began to leave the area. The cod turned to cannibalism, eating an estimated 2 million tonnes of smaller cod between 1993 and 1998. By these processes, the interactions between climate and fish stocks in the Northeast Atlantic are of fundamental importance to fish production and, in turn, the success of the region's fisheries.

Johannes Hamre, Institute of Marine Research, Bergen, Norway

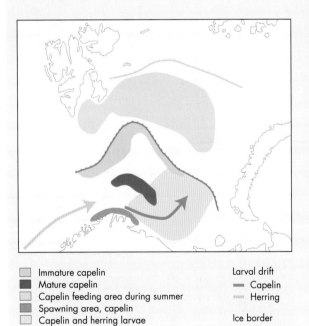

Immature capelin
Mature capelin
Capelin feeding area during summer
Spawning area, capelin
Capelin and herring larvae

Larval drift
— Capelin
····· Herring

Ice border
— September
— January

Figure 68. Distribution and migration of Barents Sea capelin **Mallotus villotus** and Atlantic herring **Clupea harengus** stocks.

SOURCES/FURTHER READING

Hamre, J. 1994. Biodiversity and exploitation of the main fish stocks in the Norwegian Sea – Barents Sea ecosystem. *Biodiversity and Conservation* 3: 273-292.

Hamre, J. 1999. Capelin and herring as key species for the yield of cod. Contribution to the Russian – Norwegian Symposium "Management strategies for fish stocks in the Barents Sea." Bergen, 15-16 June 1999.

Hamre, J. and E. Hatlebakk. 1998. System model (Systmod) for the Norwegian Sea and the Barents Sea. In: T. Rødseth, ed. *Models for multispecies management*. Heidelberg: Physica-Verlag. p. 93-115.

BOX 67. BYCATCH: A THREAT TO SEABIRDS IN ARCTIC WATERS

▶ Foraging seabirds frequently come in contact with fishing gear and are accidentally killed. Some seabirds take baited hooks at the surface during hook-and-line (longline) operations and are dragged underwater. Others get entangled and drown in gillnets when diving for food. Seabirds are regularly drowned in both types of fishing gear in Arctic fisheries. Such incidental mortality or bycatch, however, is not currently monitored, except in some Alaska fisheries. Information on the magnitude of the problem, therefore, is patchy and incomplete.

In longline fisheries, the northern fulmar accounts for the greatest number of birds killed, with annual mortality estimated at 9,000 in the Bering Sea and 50,000-100,000 in the North Atlantic. The global population of northern fulmar is 10-12 million, however, so the bycatch does not threaten the species as a whole. Bycatch of the three albatross species in the North Pacific longline fisheries, although on a much smaller scale, is considered a serious conservation issue because the short-tailed albatross is endangered and the other two species are declining. At present, only two Arctic countries (USA and Norway) have conducted mitigation research for their longline fisheries. Few seabird bycatch mitigation measures are in place.

A variety of diving seabird species drown in gillnet fisheries in Arctic countries. The common murre (or common guillemot), thick-billed murre (or Brünnich's guillemot), black guillemot, common eider, and razorbill are the species most frequently recorded. There are historical as well as recent records of over 100,000 individuals being taken in one season from Greenland, Norway, and eastern Russia. Some fisheries that historically have taken many seabirds, such as for Atlantic salmon along the coast of western Greenland, have ceased. However, seabird bycatch in cod nets is probably underestimated.

The major impediments to managing seabird bycatch in Arctic waters have been fragmentary information and insufficient attention to the problem. The Circumpolar Seabird Working Group of CAFF is actively addressing this issue in collaboration with fishers and fishery managers in the Arctic countries. Proposed mitigation measures include outreach and education, routine monitoring and assessment, improvements in fishing gear, and avoiding areas of high seabird concentrations. In areas where such measures have been tried, bycatch has been greatly reduced without hampering the fishing industry.

Snorri Baldursson, CAFF International Secretariat, Akureyri, Iceland

SOURCES/FURTHER READING
CAFF (Conservation of Arctic Flora and Fauna). 1998. *Technical Report No 1: Incidental Take of Seabirds in Commercial Fisheries in the Arctic Countries* (Bakken, V. and J. Chardine, eds.). Akureyri: CAFF. 50p.
CAFF (Conservation of Arctic Flora and Fauna). 2000. *Technical Report No 7: Workshop on Seabird Incidental Catch in the Waters of Arctic Countries – Report and Recommendations* (Chardine, J., Porter, J.M. and K. Wohl, eds.). Akureyri: CAFF. 73p.

---

dreds of years old, the levels of many have increased in recent decades. Organic pollutants, on the other hand, are largely or exclusively the product of industrial processes. Although some, such as DDT and PCBs, have been banned or greatly reduced, they are still abundant in the Arctic because of their persistence in the environment and the length of time that they take to travel northwards in air and water.

The position of an organism in the food web is the largest factor in determining the concentrations of contaminants it will have, though age, sex, and geographic location are important factors as well. The polar bear, feeding at the top of the food web, typically has the highest concentrations of organic pollutants. Other predators have similarly high levels. By contrast, baleen whales that feed lower in the food web tend to have much lower levels. The effects of these contaminant burdens are not known. Humans, too, are at the top of the food web in the Arctic, and high levels of many contaminants have been found in consumers of marine mammals.

Increases in ultraviolet radiation (UV), through depletion of ozone in the upper atmosphere, may have significant effects on primary production in the Arctic Ocean. The plankton in the upper water layers may be susceptible to changes in the amount and type

BOX 68. IMPACTS OF TRAWLING ON SEABED ANIMALS AND HABITATS

▶ Trawling for fish and shrimp is one of the most common fishing methods in Arctic and subarctic waters. However, it can kill animals and damage benthic habitats on the seabed.

The most widely used gear for bottom fishing in the North Atlantic is the otter trawl, which consists of a large net bag rigged with ground rope and headline to keep the net open vertically. Otter boards are connected to the ground rope with heavy wire to keep the net spread horizontally and to hold the trawl down as it is towed through the water. In general, the effects of otter trawling in shallow areas with a soft seabed are relatively minor for most species. The otter boards penetrate the seabed and make furrows up to one meter wide, stirring large amounts of sediment and many smaller benthic animals into the water. In shallow waters, however, storms stir up far more sediment than trawling. It is likely that the animals found on these bottoms are resilient to such disturbances, settling back unharmed when the storm or trawl passes. An exception to this pattern are the larger benthic animals, such as sea urchins, mollusks and

The reef forming deep-sea coral, **Lophelia pertusa** *(white coral in upper left-hand corner), is common on the continental shelf and shelf break off Norway. The red gorgonian,* **Paragorgia arborea**, *is common on the reefs. The brittle star,* **Gorgonocephalus caputmedusae** *(yellow in the middle), is frequently found sitting on top of the gorgonians to take advantage of stronger currents.*

*Fragments and larger pieces of dead corals, **Lophelia pertusa**, from a trawling ground on the Norwegian continental shelf at a depth of about 190 meters. The bottom has been severely disturbed.*

large polychaetes. These animals are more likely to be damaged or killed when hit by trawling gear, unlike the smaller animals, which are stirred up into the water column.

With depth, the disturbance from storms decreases. Large structural biota such as cold-water corals and sponges provide habitat for many other species. Otter trawls passing over these habitats can easily break down fragile structures that rise above the seabed. Due to the slow recovery of such biota, effects of otter trawling in deeper waters are considered more serious than in shallow waters. The coral Lophelia pertusa, for example, is widely distributed in subarctic waters. The diversity of fauna found among intact corals is very high, dominated mainly by fragile forms such as sponges and soft corals. Studies in Norwegian waters have found that otter-trawls destroyed large areas that had been dominated by this coral. Although the distribution of corals several decades ago is not known, it is highly likely that they and other sensitive habitats have become more fragmented as a result of trawling.

Only in the last decade has active research begun on the effects of fishing methods on benthic communities. More research is needed, particularly in Arctic and subarctic waters. Priorities include mapping the distribution of sensitive habitats and studying the species found there to establish a baseline for measuring impacts.

Stefán Áki Ragnarsson, Marine Research Institute, Reykjavík, Iceland

SOURCES/FURTHER READING

Auster, P.J., R.J. Malatesta, R.W. Langton, L. Watling, P.C. Valentine, C.L.S. Donaldson, E.W. Langton, A.N. Shepard, and I.G. Babb. 1996. The impacts of mobile fishing gear on seafloor habitats in the Gulf of Maine (Northwest Atlantic): Implications for conservation of fish populations. *Reviews in Fishery Science* 4:185-202.

Bergman, M.J.N., and M. Hup. 1992. Direct effects of beamtrawling on macrofauna in a sandy sediment in the southern North Sea. *ICES Journal of Marine Science* 49:5-11.

Churchill, J.H. 1989. The effects of commercial trawling on sediment resuspension and transport over the Middle Atlantic Bight continental shelf. *Continental Shelf Research* 9:841-864.

Collie, J.S., G.A. Escanero, and P.C. Valentine. 1997. Effects of bottom fishing on the benthic megafauna of Georges Bank. *Marine Ecology – Progress Series* 155:159-172.

Fosså, J.H., P.B. Mortensen, and D.M. Furevik. 2000. Lophelia-korallrev langs Norskekysten forekomst og tilstand. In: *Fisken og Havet*, Vol. 2: Havforskningsinstitutet. (In Norwegian)

Gilkinson, K., M. Paulin, S. Hurley, and P. Schwinghamer. 1998. Impacts of trawl door scouring on infaunal bivalves: results of a physical trawl door model/dense sand interaction. *Journal of Experimental Marine Biology and Ecology* 224:291-312.

Hall, S.J. 1999. The effects of fishing on marine ecosystems and communities. In: *Fish Biology and Aquatic Resources Series 1*, p. 274. London: Blackwell Science.

Lindegarth, M., D. Valentinsson, M. Hansson, M., and M. Ulmestrand. 2000. Interpreting large-scale experiments on effects of trawling on benthic fauna: an empirical test of the spatial confounding in experiments without replicated control and trawled areas. *Journal of Experimental Marine Biology and Ecology* 245:155-169.

Oliver, J.S., P.N. Slattery, L.W. Hulberg, and J.W. Nybakken. 1980. Relationships between wave disturbance and zonation of benthic invertebrate communities along a subtidal high-energy beach in Monterey Bay, California. *Fishery Bulletin* 78:437-454.

Prena, J., P. Schwinghamer, T.W. Rowell, D.C. Gordon Jr., K.D. Gilkinson, W.P. Wass, and D.L. McKeown. 1999. Experimental otter trawling on a sandy bottom ecosystem of the Grand Banks of Newfoundland: analysis of trawl bycatch and effects on epifauna. *Marine Ecology – progress series* 181:107-124.

BOX 69. SEABIRDS, FOXES, AND RATS: CONSERVATION CHALLENGES ON BERING SEA ISLANDS

▶ The islands of the Bering Sea – the Aleutians, the Pribilofs, St. Matthew, and St. Lawrence – are superb habitat for many species of sea birds. Nesting on islands is particularly attractive, as the birds are close to rich marine feeding areas and are safe from some of the predators that would be present on mainland areas. Many of the Aleutian Islands, for example, had no terrestrial mammals, so seabirds could safely nest on the ground or in earthen burrows. Unfortunately, many of these islands were also ideal locations for introducing foxes to be raised for the fur trade, a practice that began shortly after Vitus Bering sailed from Russia to Alaska in 1741 and reached its apex between 1900 and 1930. The foxes, mostly left to fend for themselves, either eliminated or severely reduced the numbers of birds on the islands they shared, pushing the Aleutian Canada goose to the brink of extinction.

In 1949, the U.S. Fish and Wildlife Service began a series of efforts to remove the foxes to allow the bird populations to recover. The removal of introduced species is often difficult if not impossible, but 36 islands have been made fox-free again. A careful study monitoring 12 species of birds on two of the islands has documented an impressive increase for all species, the total population climbing from about 5,000 individuals in the mid-1970s to about 14,000 by 1990. Also, the Aleutian Canada goose has recovered from endangered status, in part due to fox removals.

The removal of one alien species has greatly helped avian conservation on the islands, but another alien threat looms large: the rat. Large fleets of fishing boats, trawlers, processors, fuel barges, and other ships work in nearshore areas in the Bering Sea, and often run aground. Cargo ships traveling from the west coast of the U.S. to East Asia follow a great circle route that brings them through the Aleutian Chain, raising the risk of a wreck and the escape of rats to the islands. Such an incident occurred in 1990, when

The arctic fox **Alopex lagopus** *is a common predator of seabirds.*

the Korean ship Chil Bo San No. 6 ran aground of Unalaska Island, releasing swarms of Norwegian rats. The only reason the incident was not an ecological disaster was that rats had already inhabited Unalaska, which has a major port, for a century. Most islands have no rats, and are highly susceptible to a "rat spill".

In response, the U.S. Fish and Wildlife Service has established a shipwreck response program, with emergency kits and trained personnel, to stop rats before they spread from a grounded ship. The problem is especially acute at the Pribilof Islands, home to millions of sea birds and hundreds of thousands of northern fur seals. New harbors and increasing ship traffic open an invasion route for rats. To combat this threat, residents of the island participate in rat patrols, and the harbors are fortified with rat traps and rat poison. Prevention efforts have killed several rats to date and there is no evidence they have gotten beyond the defenses.

G. Vernon Byrd, U.S. Fish and Wildlife Service, Homer, Alaska
Arthur L. Sowls, U.S. Fish and Wildlife Service, Homer, Alaska

SOURCES/FURTHER READING

Byrd, G.V. 1999. Aleutian Canada goose rebounds. *Arctic Bulletin* 3.99: 20.

Byrd, G.V., J.L. Trapp, and C.F. Zeillemaker. 1994. Removal of introduced foxes: a case study in restoration of native birds. *Trans. 59th No. Am. Wildl. & Natur. Resour. Conf.* p. 317-321.

O'Harra, D. 1995. The rat zone. *Anchorage Daily News*, April 9, 1995.

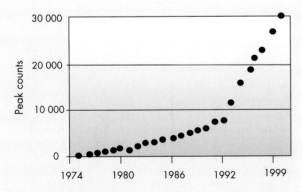

Figure 70. Recovery of the Aleutian Canada goose **Branta canadensis** *population.*

of solar radiation they receive. Many zooplankton may be killed in their early life stages if exposed to high UV levels. Fish eggs in shallow areas or floating in the water may also be susceptible to increased UV. Mammals and birds may suffer eye damage, in addition to the effects of changes to the food web.

Climate change

Climate change has the potential for the most dramatic impacts of any environmental shift in Arctic marine ecosystems. Water temperature largely determines fish and shellfish distribution. In particular, some fish migrations depend on thresholds in water temperature. Pacific salmon have reached new areas in the western Canadian Arctic, likely an indication of warmer ocean temperatures. How the salmon will affect the areas into which they are spreading remains to be seen. More dramatic changes may also occur. Changes to the extent of sea ice – and perhaps the loss of permanent sea ice altogether, as predicted in some models – would clearly have tremendous impacts on algae, plankton, fish, and marine mammals that use sea ice.

There is evidence of change already. The average thickness of sea ice over the Arctic Basin has decreased from 3.1 meters in the 1960s to 1.8 meters in the 1990s, a decline of over 40 percent. The extent of sea ice has decreased in the Bering Sea and the Arctic Ocean as a whole, though it has increased in the North Atlantic. In Canada, caribou cross from Victoria Island to the mainland in fall. But freeze-up is now 10-14 days later than it used to be. Some animals break through the ice, and Inuit hunters have recorded hundreds of animals dying. In Hudson Bay, polar bears spend more time ashore because the sea ice melts earlier. Deprived of their main prey, female bears have less fat reserves as a result.

Changes in the seasonal patterns of sea ice would also affect ocean currents and weather patterns that are determined to some degree by ice. Changes in precipitation would change the inflow of freshwater from rivers. Precipitation also affects the top of sea ice and thus the degree to which light can penetrate it. Less sea ice in the fall could lead to increased erosion, as coastal areas would be exposed to fall storms. While the details of such impacts are hard to predict, there is little doubt that they are likely to be substantial.

Conservation

Conservation of the Arctic marine environment is a complex undertaking. Many species migrate over large distances, perhaps across two or more national jurisdictions, as well as into international waters. Quantitative information on population size and status is often poor or missing for many species. The intensity of the use of many species, particularly fish of commercial interest, complicates matters as different interests and nations compete for them. The complexity of ecological interactions between species confounds predictions of the impacts of fishing and hunting. The possible impacts caused by climate change add to the complexity and uncertainty.

A number of national and international conservation measures attempt to address this situation. Several fisheries organizations set catch limits for shared stocks and high seas fishing, but enforcement is difficult. Seabirds are protected under a variety of treaties on migratory species. The International Whaling Commission sets harvest limits for large whales worldwide. Regional and bilateral organizations facilitate cooperative management, especially for small whales and seals, which are not covered by other fora.

At the national level, Arctic countries address marine conservation in several ways. Fishing and hunting regulations control harvest levels. Coastal zone management and environmental impact assessment cover broader threats and impacts. Marine protected areas, though few in the Arctic, are receiving more attention as a means of identifying areas of special significance and attempting to conserve regions and systems in addition to species and uses. Ecosystem approaches to management are being promoted in several areas, although the actual practice of predicting interactions and balancing uses remains problematic.

SOURCES/FURTHER READING

AMAP. 1998. *The AMAP Assessment Report: Arctic pollution issues.* Oslo: Arctic Monitoring and Assessment Program. xii +859p.

Babaluk, J.A., J.D. Reist, J.D. Johnson, and L. Johnson. 2000. First records of sockeye (*Oncorhynchus nerka*) and Pink salmon (*O. gorbuscha*) from Banks Island and other records of Pacific salmon in Northwest Territories, Canada. *Arctic* 53:161-164.

Barnes, P.W., D.M. Schell, and E. Reimnitz. 1984. *The Alaskan Beaufort Sea: ecosystems and environments.* Orlando, Florida: Academic Press. xvi + 466p.

Bernes, C. 1996. *The Nordic Arctic environment—unspoilt, exploited, polluted?* Copenhagen: The Nordic Council of Ministers. 240p.

Beverton, R.H.J. 1990. Small marine pelagic fish and the threat of fishing: are they endangered? *Journal of Fish Biology* 37 (Supplement A):5-16.

Born, E.W., I. Gjertz, and R.R. Reeves. 1995. *Population assessment of Atlantic walrus.* Oslo: Norsk Polarinstitutt. 100p.

Burns, J.J., J.J. Montague, and C.J. Cowles, eds. 1993. *The bowhead whale.* Lawrence, Kansas: The Society for Marine Mammology. xxxvi + 787p.

CAFF (Conservation of Arctic Flora and Fauna). 1998. *Technical Report No 1: Incidental Take of Seabirds in Commercial Fisheries in the Arctic Countries* (V. Bakken, and J. Chardine, eds.) Akureyri: CAFF. 50p.

CAFF (Conservation of Arctic Flora and Fauna). 2000. *Technical Report No 7: Workshop on Seabird Incidental Catch in the Waters of Arctic Countries – Report and Recommendations* (J. Chardine, J.M. Porter, and K. Wohl, eds.) Akureyri: CAFF. 73p.

Crawford, R.E., and J.K. Jorgenson. 1996. Quantitative studies of Arctic cod (*Boreogadus saida*) schools: important energy stores in the arctic food web. *Arctic* 49:181-193.

Ebbesmeyer, C.C., D.R. Cayan, D.R. McLain, F.H. Nichols, D.H. Peterson, and K.T. Redmond. 1991. 1976 step in the Pacific climate: forty environmental changes between 1968-1975 and 1977-1984. In: J.L. Betancourt and V.L Tharp, eds. 1991. *Proceedings of the Seventh Annual Pacific Climate (PACLIM) Workshop, April 1990.* California Department of Water Resources, Interagency Ecological Studies Program Technical Report 26, pp. 115-126.

Fay, F.H. 1974. The role of ice in the ecology of marine mammals of the Bering Sea. In: D.W. Hood and E.J. Kelley, eds. *Oceanography of the Bering Sea.* University of Alaska Institute of Marine Science. Occasional Publication No 2. p. 383-399.

Gjøsæter, H. 1995. Pelagic fish and the ecological impact of the modern fishing industry in the Barents Sea. *Arctic* 48:267-278.

Gregor, D.J., H. Loeng, and L. Barrie, eds. 1998. The influence of physical and chemical processes on contaminant transport into and within the Arctic. In: Arctic Monitoring and Assessment Program. *AMAP assessment report: Arctic pollution issues.* Oslo: AMAP. p. 25-116.

Hansen, J.C., A. Gilman, V. Klopov, and J.Ø. Odland, eds. 1998. Pollution and human health. In: Arctic Monitoring and Assessment Program. *AMAP assessment report: Arctic pollution issues.* Oslo: AMAP. p. 775-844.

Hummel, H., A.J. Löhr, P.P. Strelkov, A.A. Sukhotin, and Alexander B. Tzetlin. 1998a. *An integrated assessment of environmental and socio-economic aspects of the coastal zone of the sub-arctic White Sea, the southern Barents Sea, and the Arctic Pechora Sea.* Yerseke, Netherlands: Netherlands Institute of Ecology. vi + 83p.

Hummel, H., R. Bogaards, T. Bek, L. Polishchuk, K. Sokolov, C. Amiard-Triquet, G. Bachelet, M. Desprez, A. Naumov, Peter Strelkov, S. Dahle, S. Denisenko, M. Gantsevich, and L. de Wolf. 1998b. Growth in the bivalve *Macoma balthica* from its northern to its southern distribution limit: a discontinuity in North Europe because of genetic adaptations in Arctic populations? *Comparative Biochemistry and Physiology Part A*

120: 133-141.

Huntington, H.P., ed. 2000. *Impacts of changes in sea ice and other environmental parameters in the Arctic.* Final Report of the Marine Mammal Commission Workshop, Girdwood, Alaska, 15-17 February 2000. Bethesda, Maryland: Marine Mammal Commission.

Krupnik, I.I., and L.S. Bogoslovskaya. 1999. Old records, new stories: ecosystem variability and subsistence hunting in the Bering Strait area. *Arctic Research of the United States* 13 (Spring-Summer): 15-24.

Lavigne, D.M. 1996. Ecological interactions between marine mammals, commercial fisheries, and their prey: unravelling the tangled web. In: Studies of high-latitude seabirds. 4. Trophic relationships and energetics of endotherms in cold ocean systems. Canadian Wildlife Service. Occasional paper 91: 59-71.

Lankester, K. 1999. *Whaling in the Arctic and beyond: elements for sustainability.* An advice to the WWF International Arctic Programme. Amsterdam: Scomber Consultancy. iii + 87p.

Marine Mammal Commission. 1998. *Annual report to Congress, 1997.* Bethesda, Maryland: Marine Mammal Commission. xvi + 236p.

McDonald, M., L. Arragutainaq, and Z. Novalinga. 1997. Voices from the Bay: traditional ecological knowledge of Inuit and Cree in the Hudson Bay bioregion. Ottawa, Canadian Arctic Resources Committee.

National Research Council. 1996. *The Bering Sea ecosystem.* Washington, DC: National Research Council, Polar Research Board. x + 307p.

Rothrock, D., Y. Yu, and G. Maykut. 1999. The thinning of the Arctic ice cover. *Geophysical Research Letters*, December 1999.

Stirling, I. 1997. The importance of polynyas, ice edges, and leads to marine mammals and birds. *Journal of Marine Systems* 10: 9-21.

Stirling, I., N.J. Lunn, and J. Iacozza. 1999. Long-term trends in the population ecology of polar bears in western Hudson Bay in relation to climate change. *Arctic* 52:294-306.

Tynan, Cynthia T., and Douglas P. DeMaster. 1997. Observations and predictions of arctic climate change: potential effects and marine mammals. *Arctic* 50: 308-322.

9. Status and trends in species and populations

Quantitative data are available on the status and trends of various species and groups of species. This chapter presents some of the existing data, both to help assess the state of the Arctic environment today, and as a baseline for evaluating changes in the future. It is important to note that available data often reflect economic or other human interests. Marine fishes, for example, are relatively well documented, as are geese. By contrast, little or no information is available for most plants. The degree to which these species indicate overall ecosystem health, however, is not clear.

The information in this chapter is presented in two ways. First, as data for current populations and trends of various species, supplemented by notes where additional information is available. Second, as a list of globally threatened species, a designation that interprets population levels and trends, as well as threats. Interpreting the information, however, requires caution for several reasons:

- *Population fluctuations.* The populations of many Arctic species fluctuate naturally, sometimes widely over a few years or decades. Increasing or declining trends may be part of a natural fluctuation, or they may be a response to human activities. Distinguishing between the two or their interactions is difficult at best.

- *Data uncertainty.* Estimating the abundance of wild species is rarely straightforward. Population counts

Little auks **Alle alle** over a fjord in Svalbard.

typically include some uncertainty, especially for marine species. Evaluating trends for a population or stock of one species requires several years of data, particularly for Arctic species that exhibit wide annual fluctuations in abundance and productivity. Estimating abundance on a circumpolar basis is even more complicated. Several population counts must be compared, each having its own degree of uncertainty.

- *Data quality and comparability.* For some species there are population estimates, for others only data on total biomass, and finally for some species only indices of abundance from harvest levels. Comparing these figures is difficult or impossible. Ideally, both accurate estimates of population size and time series data are needed to assess trends. There are also a variety of methods for obtaining these data, many of which have changed over time. Thus, for example, to compare counts from several decades ago with counts from today requires considering the changes in methods, as well as environmental changes. Comparing counts from different parts of the Arctic involves similar concerns, as well as making sure that the same years and seasons are being compared. In many cases, data on the same species but from different times or regions simply cannot be compared.

- *Definitions.* Identifying a trend as increasing, declining, or stable requires defining those terms and the length of time that will be considered. Statistically reliable trends involve an assessment of the

BOX 70. BLACK GUILLEMOTS IN ICELAND: A CASE-HISTORY OF POPULATION CHANGES

▶ Black guillemots have been studied and monitored intensively on the island of Flatey, Breiðafjörður Bay, in northwest Iceland, for nearly 30 years. The rapid increase of this species in the 1970s and early 1980s was due largely to immigration of birds from other areas. The rise came to a halt in 1986, and the population remained stable for several years, at which time it was probably the largest black guillemot colony in Iceland. After 1989, the population declined drastically and by 2000 was less than half its former size.

Studies in recent years have tried to explain the decline. Various factors have been suggested as explanations:

Human disturbance. Despite the differences in disburbance levels in different parts of the Bay, corresponding population changes took place in all areas. Thus, disturbance by humans was not regarded as a significant factor.

Incidental take. Large numbers of black guillemots drown in fishing nets, but no change could be shown in annual adult survival over a period of two decades. Deaths of adult birds in nets cannot be the cause of the decline, although net mortality of immature birds has not yet been evaluated.

Competition with puffins. During the study period, the number of puffins on Flatey increased. Their main colony happened to be the principal breeding area of black guillemots on the island, which has experienced the largest decline. The guillemot population decline, however, occurred over a wider area, and therefore cannot have been due solely to competition for nest sites.

Tick parasitism. In the early years of the study, ticks were rarely found on guillemot chicks. The frequency of infested chicks increased steadily from fewer than 10% in 1990 to about 35% by 1995. Heavy infestations are rare, however, and the infected chicks usually host only a few ticks. Whether viruses or bacteria carried by the ticks could be factors is not yet known.

Black guillemot **Cepphus grylle**.

Changes in the food resources. From about 1996, the breeding success of most seabirds in the Flatey region has been very poor. Since birds with different feeding tactics and foods have been affected, their breeding success is most likely influenced by a common factor. In recent years, the Breiðafjörður region has seen unusually large occurrences of Atlantic cod, which are notorious for the diversity of their prey. The greatly increased biomass of cod may have reduced the food available for most seabirds, lowering their breeding success.

The trends on Flatey may be the result of a combination of several of these factors. Although the changes appear simple on the surface, they are likely to reflect complex environmental interactions. Long-term studies are essential to our understanding of such phenomena.

Aevar Petersen, Icelandic Institute of Natural History, Reykjavik, Iceland

SOURCES/FURTHER READING

Frederiksen, M. and A. Petersen 1999. Adult survival of the Black Guillemot *Cepphus grylle* in Iceland. *Condor* 101(4):589-597.

Petersen, A. 1979. [The breeding birds of Flatey and some adjoining islets, in Breidafjördur, NW. Iceland.] *Náttúrufr.* 49(2-3):229-256. (In Icelandic, English summ.).

Petersen, A. 1981. *Breeding biology and feeding ecology of black guillemots.* University of Oxford. D.Phil. thesis. 378 p.

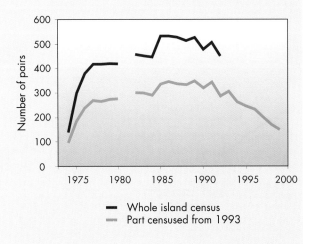

Figure 71. Changes in the black guillemot **Cepphus grylle** *population on Flatey island, Breiðafjörður, from 1974 to 1999.*

variance in the data and then determining the probability of detecting a specific degree of change over a stated interval. The quality of available data and the circumstances of its use affect the standards used in trend analysis. One definition may require a 50% change in population before declaring that a trend is increasing or declining, whereas another definition may only require a 20% change or even less. A population that has increased for four decades and then declined slightly in the fifth decade may be regarded as increasing overall, or as declining if only the final period is considered.

- *Scarcity of data.* The data that are available today are by no means complete or comprehensive. An increase in the number of endangered species may reflect deteriorating environmental conditions. On the other hand, it may mean that better information is available about these species, providing a more reliable way of determining whether they are at risk. Furthermore, in most cases the best data are available for large animals, which are only a minute fraction of the total biomass of the Arctic. It is not clear how the status of large vertebrates relates to the overall status of ecosystems.

With these cautions in mind, the available data do offer a useful starting point for considering and comparing the state of the Arctic across regions and over time. The CAFF Program has begun establishing biodiversity monitoring networks for a variety of species groups. The data produced by these networks will provide a more extensive and reliable quantitative basis for future assessments. Further studies are needed to examine the relationships of various trends, to determine the links between them, and to identify the causes of environmental shifts and cycles.

Population Levels and Trends

■ The monitoring of Arctic species has been increasing in recent decades, providing information on trends for more and more species. Nonetheless, increases or declines are often hard to detect. Identifying the causes of those changes is even harder. Some of the available data are presented as tables for marine mammals, seabirds, marine fishes, terrestrial mammals, and terrestrial birds.

The data are at times inconsistent, imprecise, and uneven. For some species, there is current, reliable information not only on population size, but also on other biological and ecological parameters such as habitat, hunting, and stock interactions. For others, even population figures are many years out of date. The vast majority of Arctic species, including all flora and invertebrates, do not appear in the tables at all because there is little or no information about their abundance. In addition to the tables and accompanying notes, the boxes in this chapter provide more information about selected species for which circumpolar information is available, plus two examples of the challenges and results of monitoring specific populations.

*Figure 72. Distribution of Steller's sea lion **Eumetopias jubatus** in the Arctic.*

Globally, there have been several recent attempts to create an overall species index, a collective index of trends in abundance for many species. In the World Wildlife Fund's (WWF) Living Planet Report 2000, the United Nations Environment Programme's World Conservation Monitoring Centre (UNEP-WCMC)

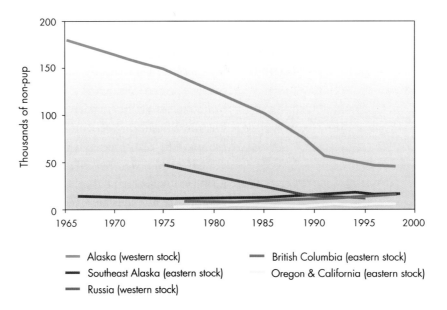

*Figure 73. Estimated numbers of non-pup (adults and juvenile) Steller's sea lions **Eumetopias jubatus** in different areas of the Arctic.*

created an overall Living Planet Index (LPI). More than 730 species have been included in the index, which has been compiled for forest, marine, and freshwater ecosystems. The index shows a 33% loss of the Earth's natural ecosystems over the last 30 years, with all three categories declining.

Currently there are not adequate data to compile a similar index for the Arctic. A compilation of what is available across the region and across various species groups is attempted below. It would be highly desirable to create an overall species index for the Arctic region. This way the Arctic could be included in future global assessments and compared with other regions.

Marine mammals

Careful population estimates have been made for some stocks of marine mammals. For other stocks, the estimates are more conjectural. Table 7 shows estimates for various stocks of several species, with an assessment of the population trend. The sources of data, their reliability, and the time span covered by the trend analysis vary. Question marks by the population figures indicate great uncertainty. Species and stocks for which no information was available have been omitted.

The world population of polar bears *Ursus maritimus* is currently estimated at 22,000-28,000 animals. They occur in about 20 populations, and genetic analysis indicates that interbreeding is minimal. Although the species appears to be secure at the moment, knowledge of population trends and the accuracy of estimates vary geographically (see Figure 74). Population research is both labor-intensive and expensive, which often limits the extent of the studies undertaken. Principal threats include increasing levels of pollutants, climatic warming, and the risk of over-harvesting.

Seabirds

Eiders

Eiders are considered seabirds due to their preference for marine habitats. While breeding in the Arctic, however, some species – particularly the king eider, *Somateria spectabilis* – travel up to several hundred kilometers inland. With some exceptions, there is insufficient data to estimate either population sizes or to determine trends for world eider populations (Table 8). Most estimates of population sizes for king and common eiders, *Somateria mollissima*, are simply best guesses. Also, there are estimates for some areas and not others. For example, there is an estimate for king eiders in western Canada but not eastern Canada. Some estimates are from wintering grounds and others are from breeding grounds. Wintering areas usually contain birds from several breeding areas, so estimates are not always comparable.

An estimated 600,000 to 750,000 common eiders winter off North America. Arctic North American breeding populations have probably declined by about 50% since 1970. An estimated 900,000 to 1,000,000 common eiders overwinter in Iceland, including birds from northeast Greenland. In western Greenland, some formerly large colonies have almost disappeared, but quantitative information is too scarce to estimate the magnitude of an overall decline. Common eider populations in Russia are believed to be stable or declining. In the Baltic area, south of the Arctic, common eider populations expanded rapidly during 1950-1985, but since 1985 have been stable or declining.

An estimated 370,000 king eiders nest in western North America and winter in the Bering Sea and North Pacific. An additional 100,000 or more from the same wintering area nest in Russia. An estimated 280,000 winter off southwestern Greenland, the majority nesting in the eastern Canadian Arctic and an unknown portion in Greenland. King eiders in western North America have declined by about 50% since the mid-1970s. In eastern North America and western Greenland, they have declined by 50% since the 1950s. King eiders in Russia seem to be stable, but no accurate numbers are available.

An overall population trend for the spectacled eider, *Somateria fischeri*, is not yet available. A common wintering site for the world population was discovered in 1994 and harbors over 363,000 birds. This species has declined by 95% in western Alaska in the past three decades. Trends for the North Slope of Alaska are unknown. The species is thought to have de-

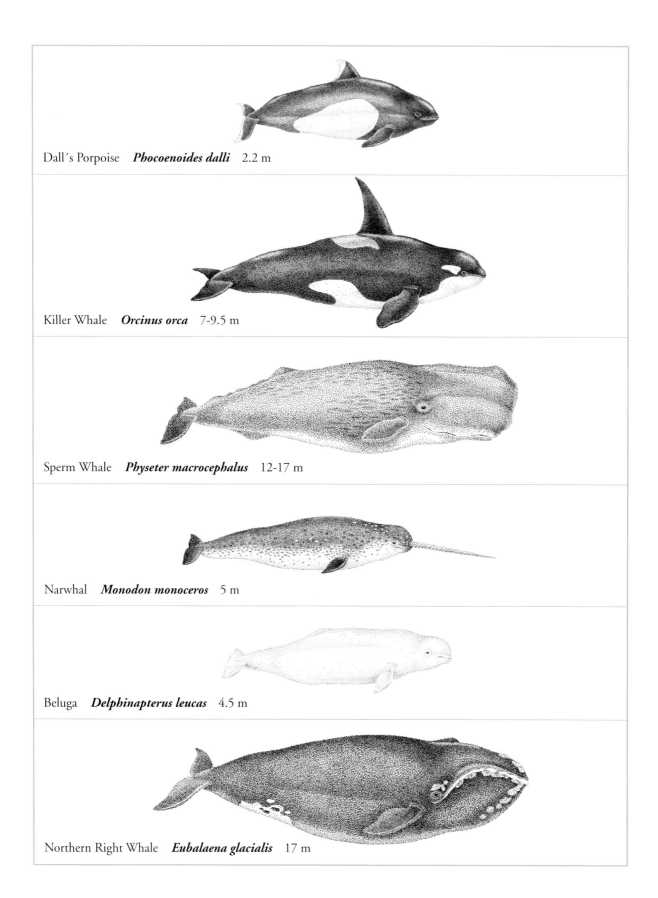

Dall's Porpoise *Phocoenoides dalli* 2.2 m

Killer Whale *Orcinus orca* 7-9.5 m

Sperm Whale *Physeter macrocephalus* 12-17 m

Narwhal *Monodon monoceros* 5 m

Beluga *Delphinapterus leucas* 4.5 m

Northern Right Whale *Eubalaena glacialis* 17 m

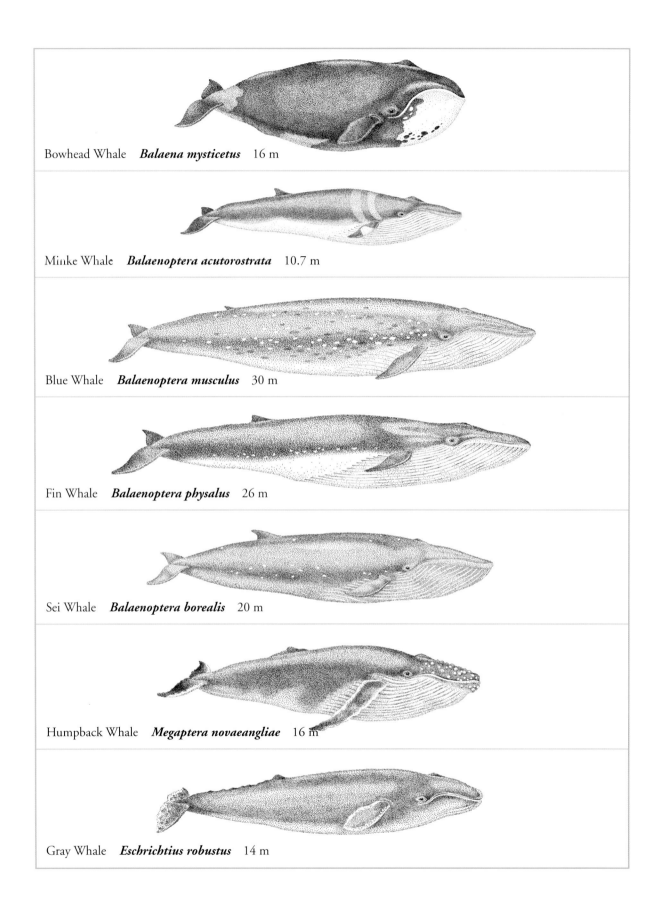

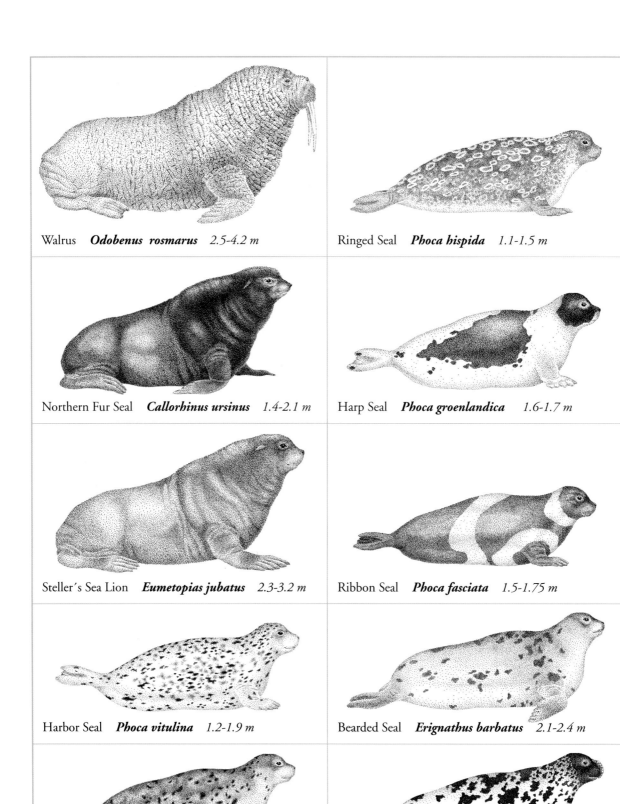

Table 7. Estimated population size and trends for marine mammal stocks in the Arctic.

SPECIES	STOCK	POPULATION	TREND
Gray whale *Eschrichtius robustus*	Eastern Bering Sea	21,600	Stable
Beluga whale *Delphinapterus leucas*	Beaufort Sea	39,000	Stable
	Eastern Chukchi Sea	3,700	Stable
	Eastern Bering Sea	20,000	Stable
	Bristol Bay, Alaska	1,500	Stable
	White Sea - Dvinskiy Bay	Few thousands	Stable
	White Sea - Solovetsky Island	<1,000	Stable
	Barents Sea - Pormor Strait off Kolgyej	2,000-3,000	Stable
	South Greenland: Qaqortoq to Maniitsoq	disappeared	Endangered
	Southwest Greenland: Maniitsoq to Disko	3,500-13,000	Declining
	Northwest Greenland: Avanersuaq and Upernavik	Few thousands	Declining
	Canadian High Arctic	14,000-60,000	Declining
	Southeast Baffin - Pangnirtung	2,000	Increasing
	Ungava Bay	~20	Increasing?
	Eastern Hudson Bay (coastal)	~1,000	Declining?
	Western Hudson Bay	>23,000	Unknown
	Southern Hudson Bay	>1,000	Unknown
	James Bay	>3,000	Unknown
Narwhal *Monodon monoceros*	Franz Josef Land	~60 (1996)	Stable
	Svalbard	~100 (1998)	Stable
	East Greenland - Scorebysund	~100 (1984)	Stable
	West Greenland - Disko Bay	1,300-21,000	Unknown
	West Greenland - Uummannaq	~400 (1996)	Unknown
	West Greenland - Avanersuaq	2,000	Unknown
	Baffin Bay	34,000	Stable
	Hudson Bay	3,500	Unknown
Killer whale *Orcinus orca*	Eastern North Pacific	700	Unknown
Sperm whale *Physeter macrocephalus*	North Pacific	141,000	Unknown
Dall's porpoise *Phocoenoides dalli*	Alaska Stock	830,000	Unknown
Humpback whale *Megaptera novaeangliae*	Western North Pacific	400?	Unknown
Fin whale *Balaenoptera physalus*	Northeast Pacific	15,000	Unknown
	West Greenland	520-2,100	Unknown
	North Atlantic	17,000	Declining

SPECIES	STOCK	POPULATION	TREND
Minke whale *Balaenoptera acutorostrata*	Eastern Bering Sea	3,000	Unknown
	Central Atlantic	25,000	Unknown
	West Greenland	2,400-16,900	Unknown
Northern right whale *Eubalaena glacialis*	North Pacific	50	Unknown
Bowhead whale *Balaena mysticetus*	Western Arctic (Bering-Chukchi-Beaufort Stock)	8,200	Increasing
	Davis Strait-Baffin Bay	315-435	Increasing
	Spitsbergen	10	Unknown
	Hudson Bay-Fox Basin	345	Increasing
Walrus *Odobenus rosmarus*	Pacific	300,000	Increasing
	Atlantic	Unknown	Declining
	Laptev Sea	Unknown	Declining
Northern fur seal *Callorhinus ursinus*	Pribilof Islands	700,000	Declining
Bearded seal *Erignathus barbatus*	Bering / Chukchi	250,000-300,000	Unknown
	Canadian Arctic	190,000	Unknown
Steller's sea lion *Eumetopias jubatus*	Gulf of Alaska	9,782	Declining
	Bering Sea/Aleutians	12,434	Declining
Harp seal *Phoca groenlandica*	Northwest Atlantic	5,200,000	Unknown
	Greenland Sea	~500,000	Unknown
	White Sea and Barents Sea	2,000,000	Unknown
Hooded seal *Cystophora cristata*	Greenland Sea	~135,000	Unknown
	Atlantic	400,000-450,000	Unknown
Harbor seal *Phoca vitulina*	Southeast Alaska	30,000	Unknown
	Bering Sea	30,000	Declining
	Iceland	20,000	Declining
Spotted seal *Phoca largha*	Global	400,000	Unknown
	Bering Sea	200,000-250,000	Stable
Ribbon seal *Phoca fasciata*	Global	240,000	Unknown
	Bering Sea	55,000	Stable
Ringed seal *Phoca hispida*	Global	2,300,000-7,000,000	Unknown
	Alaska	1,000,000-1,500,000	Unknown

Sources: Zeh et al. 1993, Cleator, H.J. 1996, Fisheries and Oceans Canada 1998 a, b, 1999, 2000, Hill and DeMaster 1998, Trites et al. 1999, NAMMCO Annual Report 2000.

Polar bear **Ursus maritimus**.

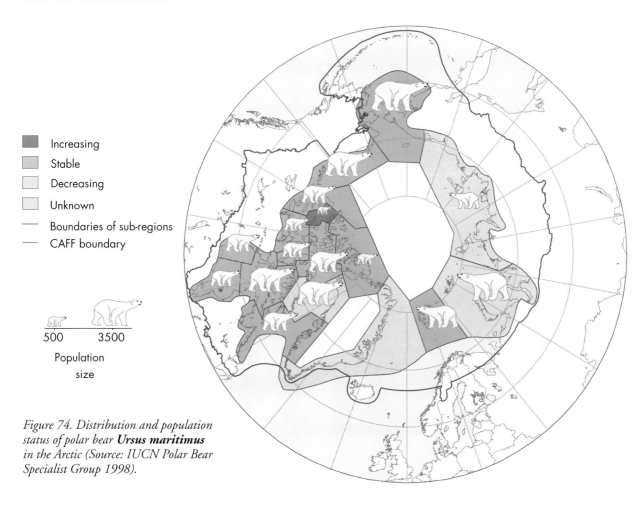

Figure 74. Distribution and population status of polar bear **Ursus maritimus** in the Arctic (Source: IUCN Polar Bear Specialist Group 1998).

Table 8. Estimated population size and trends for eiders in the Arctic region.

SPECIES	REGION	POPULATION (individuals)	TREND
Common eider *Somateria mollissima*	Canada west	70,000	Declining
	Canada east	350,000-450,000	Declining
	Finland	150,000-200,000	Declining/Stable?
	Greenland		Declining
	Iceland	300,000	Stable
	Norway & Russia	300,000-550,000	Stable
	Svalbard/Franz Josef Land	40,000-80,000	Stable
	Sweden	300,000-400,000	Increasing
	NW North America/Russian Far East	130,000-200,000	Stable?
King eider *Somateria spectabilis*	Western Canadian Arctic	360,000	Declining
	Eastern Canadian Arctic	Unknown	Unknown
	Greenland, NE Europe, Siberia	300,000	Stable?
	Svalbard	1,000	Unknown
	Eastern Siberia	100,000	Stable?
	Alaska	12,000	Unknown
Spectacled eider *Somateria fischeri*	Eastern Asia/Alaska	180,000	Unknown
Steller's eider *Polysticta stelleri*	Western Siberia	40,000	Increasing
	Eastern Siberia	180,000	Declining?
	Alaska	>1,000	Declining

Sources: Asbirk et al. 1997, Anker-Nielsen et al. 2000, CAFF Circumpolar Seabird Working Group 1998, Durinck and Falk 1996, Goudie et al. 1994, Mosbech and Boertmann 1999, Pihl 1997, Solovieva 1997, Suydam et al. 2000, Wetlands International 1999.

Common eiders **Somateria mollissima** with one King eider **S. spectabilis**.

Spectacled eider **Somateria fisheri**.

Steller's eider **Polysticta stelleri**.

clined in Russia, though documentation is lacking.

The Steller's eider, *Polysticta stelleri*, has declined severely in Alaska, leaving only a remnant breeding population in western Alaska. It continues to breed sporadically in small numbers near Barrow and perhaps elsewhere in Alaska. Population trends in Russia are unknown, but its range appears to be spreading to the west. Increasing numbers have been seen in wintering areas in northern Norway and in the Baltic, indicating a positive population trend for the western Eurasian population. The total population has recently been estimated at about 220,000 birds.

Murres/guillemots

The world population of the common murre/common guillemot, *Uria aalge*, is estimated in the *CAFF International Murre Strategy and Action Plan* to be 12-15 million breeding birds, of which 8-11 million breed in the Arctic. Of these, 260,000-300,000 individuals breed in the Barents Sea Region, two-thirds of them on Bear Island. Populations in many parts of the North Atlantic declined in the 1800s due to hunting and human activity, but started to recover as soon as harvest-

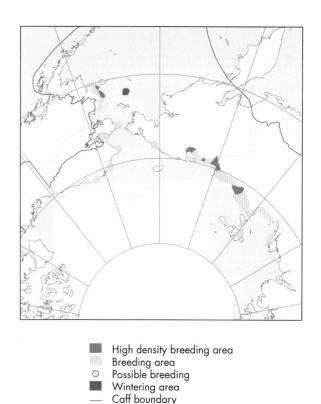

■ High density breeding area
▦ Breeding area
○ Possible breeding
■ Wintering area
— Caff boundary

Figure 75. Breeding and wintering distribution of spectacled eider **Somateria fischeri**.

STATUS AND TRENDS IN SPECIES AND POPULATIONS

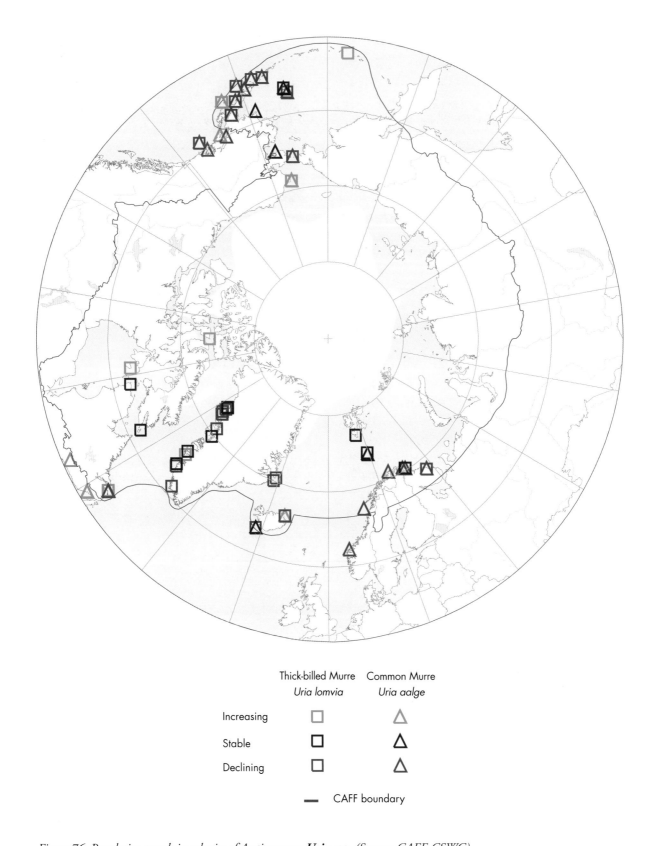

*Figure 76. Population trends in colonies of Arctic murres **Uria** spp. (Source: CAFF-CSWG).*

224 STATUS AND TRENDS IN SPECIES AND POPULATIONS

ing and disturbance were reduced. Recent declines have been recorded in the Faeroes, northwestern Norway, whereas the Icelandic population is stable or increasing. The rest of the European population has increased within most of its range, despite some local declines.

The Arctic is home to the entire population of thick-billed murre/Brünnich's guillemot, *Uria lomvia*, which is currently estimated to be 14 million birds. The largest colony on Novaya Zemlya, in Bezymyannaya Bay on the west coast of the southern island, may once have been the largest colony of seabirds in the Northern Hemisphere. Researchers estimated that 1,600,000 birds were breeding there in the 1930s. Today, there are far fewer birds. A large decline occurred in the 1940s due to intensive harvesting of eggs and birds, and in 1948 the population was estimated at 200,000 individuals. A count in 1992 recorded only 81,000 birds. In 1994, however, another census put the number at 141,000. The difference in the two recent estimates is considered to be the result of different counting methods.

Counts of colonies in Svalbard in 1981, 1985, and 1989 demonstrate that trends differ from area to area. The Kongsfjorden-Krossfjorden area saw a decline of 31% between 1981 and 1985, whereas northwestern Spitsbergen had a 21% increase in the same period. In Hornsund, the increase between 1985 and 1989 was almost 600%. In Svalbard overall, murres increased by 17% (from 145,000 to 171,000 individuals) between 1981 and 1989. At the two Icelandic thick-billed murre colonies presently monitored, a 2-4% annual decline has been detected over the past 15 years. Over-hunting in Greenland and Canada in winter is suspected as the cause of the decline.

Thick-billed murres are the most abundant seabird in the Canadian Arctic in summer. They are subjected to significant harvests on their wintering grounds in both Newfoundland/Labrador and southwestern Greenland. They nest at 12 colonies, 11 of which are in the eastern Arctic. The colonies range from 1,000 to 800,000 nesting birds, with a combined total of approximately 2,952,000 individuals. Complete colony counts of the largest colonies are problematic and rarely conducted. Instead, population trend information is derived from counts of small-er study plots. Most colonies appear to be stable or to have increased slightly. Prince Leopold Island has increased by 1.7% per year since the late 1970s, Digges Sound by 0.6%, and Coats Island by 1.8%. It should be noted that these rates are averaged over the whole monitoring period, which obscures changes in magnitude and direction from year to year.

There are 696 seabird colonies with murres/guillemots that are known and have been surveyed around the Arctic (see Figure 32). Some of these have been surveyed twice or more. Trend information exists for only a small proportion of these, however, due to the lack of long-term monitoring at what are often very remote sites. Data on trends have been included in the map only if they cover at least a ten-year period and the most recent survey was between 1990 and 2000. Only 57 colonies meet these criteria. The seabird population at a colony is considered stable as long as it has not changed by more than 20% over a ten-year period (see Figure 76).

Gulls

In general, gull populations have increased in the circumpolar Arctic since the 1960s. These changes may reflect the development of fisheries in the Barents Sea and North Atlantic, which supplied gull populations with a reliable food source in the form of waste and offal. Great black-backed gull, *Larus marinus*, populations, for example, increased and expanded their range in eastern North America. More recently, collapses in commercial fisheries have brought further changes. The over-harvesting of fish in the Barents Sea in the late 1970s, for example, deprived large gulls in Russian colonies of their main food source. In subsequent years, many gulls failed to breed. Of the few chicks that hatched, many died of predation or starvation. A steep decline in great black-backed gull has also been observed in Iceland in recent years, after a steep previous increase over many decades.

The ivory gull, *Pagophila eburnea*, is often considered the most Arctic of all the gulls, breeding only in high latitudes and rarely venturing farther south. In Canada, the ivory gull is rare, with less than 4,000 birds breeding at 14 small colonies. These are thought to be about one quarter of the global population. Unfortu-

*Ivory gull **Pagophila eburnea**.*

nately, no population surveys or detailed research has taken place since 1985. Consequently, population data is extremely out of date or non-existent. Inuit from Grise Fiord, Nanisivik, Arctic Bay, and Resolute Bay in Canada have recently reported severe declines in the sightings of ivory gulls, suggesting that the Canadian population may be in decline. This is supported by a recent aerial survey on the Brodeur Penninsula, Baffin Island, where known colonies were found empty.

Another gull that spends its life cycle almost entirely in the Arctic is the Ross's gull, *Rhodostethia rosea*, famous for its pink breeding plumage. According to recent surveys in its main breeding range in northern Yakutia, the gull is more widespread than had been assumed. The current population estimate of some 100,000 birds may be low.

Other seabirds

There is no evidence for any decline in seabirds before 1976 in the Bering Sea, but scientists noticed declining stocks in the North Atlantic as early as the 1940s. Major die-offs occurred between 1976 and 1984 in the Eastern Bering Sea and also in some colonies in the North Atlantic. In the Barents Sea, the populations of four seabird species, among them the gannet, *Sula bassana*, and razorbill, *Alca torda*, have clearly increased; nine appear to be stable; and six have declined, including the shag, *Phalacrocorax aristoteles*, four species of gull, and the arctic tern, *Sterna paradisaea*.

Marine fishes

Northern waters, particularly the North Atlantic and the Bering Sea, are some of the world's most important marine fisheries. Nonetheless, there are many gaps in our knowledge of stock status and trends. Recent data continue to indicate problems with overfishing in many areas, but current management efforts, such as those for cod around Iceland, offer hope of improvements.

*Ross's gull **Rhodostethia rosea**.*

Atlantic cod *(Gadus morhua)*
The Norwegian and Barents Sea stock is currently considered to be below safe biological limits. Recent fishing mortality has been high, and the spawning stock biomass has substantially declined. Major reductions in fishing mortality are now needed to rebuild it.

The Icelandic spawning stock biomass has steadily increased since the 1995 when harvests were limited to 25% of the "available biomass". While recruitment into the Icelandic stock was generally poor between 1985 and 1996, subsequent year classes are estimated to be stronger.

The Greenlandic stock is reported to be below safe biological limits, with the offshore component severely depleted since 1990. The dramatic decline in stock abundance was associated with changes in environmental conditions, emigration, and high fishing mortality. The inshore fishery is unregulated, and there are indications that this stock is declining, too.

The biomass of Atlantic cod in the eastern Davis Strait and Labrador Sea remains very low with weak year class recruitment during the 1990s. Offshore, there are no signs of recovery.

Pacific cod *(Gadus macrocephalus)*
While the Eastern Bering Sea stock is currently reported to be relatively stable, it is projected to decline as a result of recent below-average year classes. The stock is not considered to be overfished.

Little substantial material is available on the Western Bering Sea stock. Overall abundance in the region seems to be lower than in the Eastern Pacific. The highest regional abundance appears to be in the Navarin area of the northeastern Bering Sea.

Navaga *(Eleginus navaga)*
The navaga is commercially fished primarily by the Russian fleet in the White, Barents, and Kara Seas. There is little information available on the status of

Table 9. Trends for commercial fish stocks in the Arctic.

SPECIES	STOCK	TREND
Atlantic cod *Gadus morhua*	Norwegian and Barents Seas	Declining
	Iceland	Recovering
	Greenland	Declining
	Labrador Sea and Davis Strait	Declining
Pacific cod *Gadus macrocephalus*		
	Eastern Bering Sea	Stable / Declining
	Western Bering Sea	Unknown
Navaga *Eleginus navaga*	White, Barents, Kara Seas	Declining?
Saffron cod *Eleginus gracilis*	Western North Pacific	Unknown
Polar/Arctic cod *Boreogadus saida*	Northwest Atlantic, Barents and White Seas	Declining?
Greenland cod *Gadus ogac*	Greenland	Unknown
Walleye pollock *Theragra chalcogramma*	Eastern Bering Sea	Increasing
Yellowfin sole *Limanda aspera*	Bering Sea and Aleutian Islands	Stable
Greenland halibut *Reinhardtius hippoglossoides*	Bering Sea and Aleutian Islands	Declining
	North East Arctic (east of Iceland to Nova Zemlya)	Declining
	Iceland and eastern Greenland	Stable ?
	Western Greenland/Davis Strait	Increasing
Deepwater redfish *Sebastes mentella*	North East Arctic (east of Iceland to Nova Zemlya)	Declining
Capelin *Mallotus villosus*	Barents Sea	Declining
	Iceland and eastern Greenland	Increasing
Atlantic herring *Clupea harengus*	North East Arctic (east of Iceland to Nova Zemlya)	Increasing
	Iceland	Increasing
Wolffishes *Anarhichas* spp.	Western Greenland/Davis Strait	Declining

Source: COSEWIC 2000; FAO 1990, 2000; ICES 1999, 2000; NAFO 1999, 2000; Witherell 2000.

this species, although the catch of navaga reported to the FAO has declined substantially since the early 1990s.

Saffron cod *(Eleginus gracilis)*
There is little information on the status of this Western North Pacific species. Catches reported to the FAO increased during the mid-1990s.

Polar/Arctic cod *(Boreogadus saida)*
Catches reported to the FAO have declined greatly in the Northwest Atlantic, Barents and White Seas since

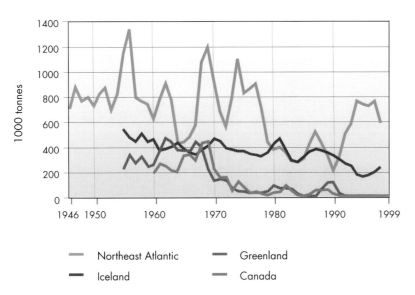

*Figure 77. Total catch of Atlantic cod **Gadus morhua** in the Northeast Atlantic (including Barents Sea), Iceland, Davis Strait (Greenland), and Labrador Sea (Canada) (Sources: ICES and NAFO reports 2000).*

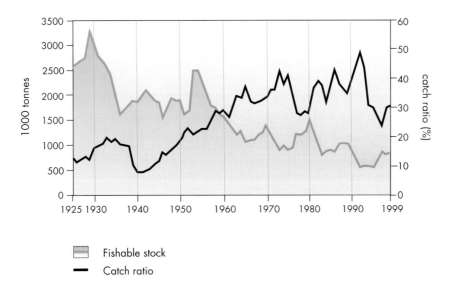

Figure 78. The fishable stock and proportion caught (catch ratio) of Icelandic cod from 1925 to 1999 (Source: Marine Research Institute, Reykjavik, Iceland).

the mid 1980s. The status of polar cod stocks is unclear.

Greenland cod *(Gadus ogac)*
This local Greenlandic species appears to be only of minor, local importance.

Walleye pollock *(Theragra chalcogramma)*
The most abundant groundfish in the Eastern Bering Sea, the pollock is also one of the world's most commercially important fishes. The stock is currently not overfished, has an exploitable biomass above management targets, and is increasing.

STATUS AND TRENDS IN SPECIES AND POPULATIONS 229

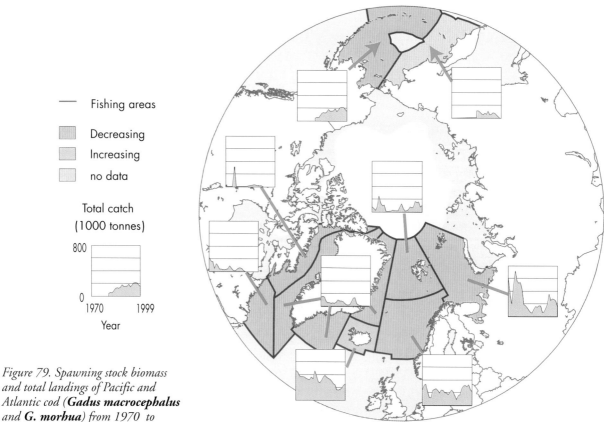

*Figure 79. Spawning stock biomass and total landings of Pacific and Atlantic cod (**Gadus macrocephalus** and **G. morhua**) from 1970 to 1999. (Source: UNEP-WCMC).*

Yellowfin sole *(Limanda aspera)*
This relatively slow-growing and long-lived species is distributed throughout much of the North Pacific. The Eastern Bering Sea population currently has an exploitable biomass above the management target and the stock size is stable. The stock is not overfished.

Greenland Halibut *(Reinhardtius hippoglossoides)*
The species, generally referred to as Greenland turbot in the northern Pacific fisheries, is not currently considered to be overfished in the Eastern Bering Sea and Aleutian Islands. The catch and exploitable biomass, however, have declined substantially since 1980.

The North-East Arctic stock is currently considered to be outside safe biological limits. A steady decline has been observed since the 1970s, with recent spawning stock biomass among the lowest recorded. Recruitment has declined over the last three decades. The fishery is limited to small, coastal longline and gillnet vessels.

The status of the halibut stock surrounding Iceland and eastern Greenland is difficult to determine. While there are indications that stock size has been relatively stable in recent years, recruitment appears to have been poor. Because of the long period needed by juveniles to reach maturity, recruitment failure will only be detected several years after it occurs.

Inshore stocks in the fjords of the Western Greenland and Davis Strait area depend for recruitment on the offshore spawning stock of the Davis Strait. Little spawning occurs inshore, and inshore fish do not appear to contribute to the offshore stock. The biomass of the offshore stock appears to have declined from the mid-1980s to the mid-1990s. Since then, it has recovered to levels approaching those of the late 1980s.

Deepwater Redfish *(Sebastes mentella)*
The deepwater redfish or ocean perch is mainly found in the deeper areas of the North Atlantic. It is generally slow growing and long-lived. The Northeast Arctic stock is currently considered to be outside of safe biological limits. The spawning stock is close to its historical low, and recruitment was poor throughout the 1980s.

Capelin *(Mallotus villosus)*
This small (~20 cm) circumpolar species is caught for food and for use as fishmeal. It is also an important prey species for cod and herring. The Barents Sea stock is currently considered to be within safe biological limits. The fishery was reopened in the spring of 1999 after the stock was estimated to have recovered to exploitable levels.

Deepwater redfish **Sebastes mentella** *with a flatfish in the middle*

The Iceland and Eastern Greenland stock is considered to be within safe biological limits. Although the spawning stock fell at the start of the 1990s, good recruitment in the following years allowed a quick recovery, and currently it appears fairly strong.

Atlantic Herring *(Clupea harengus)*
Herring is the most important Atlantic pelagic fish, occurring throughout northern areas. It plays an important role in the food web, linking plankton to larger fish species, seabirds, and marine mammals. The distinct Norwegian spring-spawning stock is one of the populations that collapsed during the 1960s and 1970s. The stock has recovered significantly during the 1980s and 1990s. Its biomass is currently considered to be within safe biological limits.

The overexploitation of Icelandic herring resulted in a stock collapse during the 1960s. A fishing ban was enforced between 1972 and 1975, and the stock has gradually recovered. The stock is considered to be within safe biological limits, with the spawning stock biomass at its observed maximum of 583,000 tonnes.

Wolffishes *(Anarhichas* spp.*)*
Two species are commercially fished in the Western Greenland/Davis Strait area: Atlantic wolffish, *A. lupus*, and spotted wolffish, *A. minor*. The species are often combined in catch data. Both species occur throughout much of the North Atlantic. In the early 1980s, the spawning stock biomass of both species declined drastically. Current stocks remain severely depleted despite a steady increase in recruitment since the early 1980s (recorded only for *A. lupus*). In Canada, the Atlantic wolffish has been classed as a Species of Special Concern.

Terrestrial mammals

Information on status and trends in terrestrial mammals is far scarcer than for birds and fish. However, there are data, at least at a regional level, for reindeer/caribou, muskoxen, and some predators.

Caribou and reindeer *(Rangifer tarandus)*
Caribou and wild reindeer populations tend to fluc-

BOX 71. POPULATION FLUCTUATIONS IN SVALBARD REINDEER

▶ The Svalbard reindeer is one of seven extant subspecies of reindeer/caribou. The High Arctic archipelago of Svalbard, where the animals are found, was discovered by Willem Barentz in 1596 at which time the islands supported a large population of the reindeer. Between 1860 and 1920, the animals were severely persecuted by hunters, fishermen and tourists: official records show that at least 20,000 carcasses were imported to Norway. Probably an equal number of reindeer were consumed locally and not recorded. At the end of this period there were believed to be only about 1,000 reindeer left, probably around 10% of the original total, and the animals were given legal protection from hunting.

Today there are approximately 8,000 Svalbard reindeer distributed in four main regions around the islands. The animals are neither gregarious nor migratory, and there appears to be little exchange between the different subpopulations. Svalbard reindeer have no natural competitors and are not subject to significant levels of predation, hunting, husbandry or active management. Instead, local populations are allowed to settle at levels dictated by the prevailing abundance of resources.

The population of reindeer in Adventdalen, an area of approximately 150 km² on the main island of Spitsbergen, has been carefully monitored annually since 1979. Adventdalen supports a population of 400-800 reindeer. The mean rate of increase of this population is zero, but its numbers undergo large fluctuations from year to year. Strong evidence is now available that the size of this population is determined by a combination of density-dependent regulation resulting from intra-specific competition (competition within species) for food, and perturbations resulting from random variation in climate. Climatic effects operate in part through changes in the availability of forage in winter and, most likely, in the quality of forage in summer.

Dag Vongraven, Norwegian Polar Institute, Tromsø, Norway

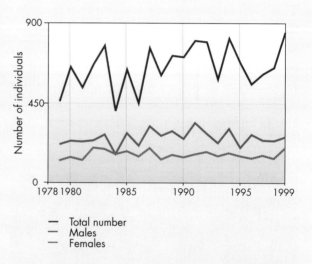

Figure 80. Population fluctuations of Svalbard reindeer **Rangifer tarandus platyrhynchus** *from 1978 to 1999. (Courtesy of Nicholas Tyler, University of Tromsø, Norway).*

---

tuate. These natural patterns were affected by the introduction of firearms, commercial hunting and increases in the domestic reindeer industry during the first half of the 20th Century. Currently, there are about 4.5 million caribou and wild reindeer in the Arctic. Most herds are increasing (Table 10), although both Peary caribou *Rangifer tarandus pearyi* and Arctic-island caribou declined in the Canadian High Arctic during the 1980s and 1990s.

In North America and Russia, the numbers of grizzly or brown bears, *Ursus arctos*, wolves, *Canis lupus*, and wolverines, *Gulo gulo*, partially track the fluctuations in caribou and wild reindeer. Human harvesting can affect local abundance. In these countries overall, the three large carnivores are relatively abundant, their populations stable or increasing. In the more intensively populated Fennoscandia, where there is greater conflict with domestic reindeer and sheep, large carnivores are scarcer.

Terrestrial birds

Little information exists on the overall population trends of Arctic terrestrial birds, with the exception of geese and some shorebirds.

Svalbard reindeer **Rangifer tarandus platyrhynchus**.

Geese

Twelve of the world's 15 goose species breed in the Arctic and their trends are relatively well known (Table 11). Most goose populations are increasing, their populations at the highest levels ever counted. Only the lesser white-fronted goose, *Anser erythropus* is declining sharply and is threatened with extinction. Of the 43 goose populations in the Arctic, 17 are increasing (40%), 12 are stable (28%), 8 are declining (18%), and for 6 the trend is unknown (14%) (See Table 11 and Figure 84).

Greater snow goose *(Chen caerulescens atlantica)*
The greater snow goose provides one of the most extreme examples of an extraordinary population increase. The population rose from less than 50,000 in 1965 to over 800,000 in 1999. The overall population of snow geese, including other subspecies, increased from 462,000 in 1969 to almost 4 million (with some estimates up to 10 million) birds in 2000. Although hunting restrictions have played a role, changes in agricultural practices in the species' wintering range provided much better food and are believed to be the main cause of the population explosion.

BOX 72. WOLVES

▶ Gray wolves currently number about 150,000 worldwide. Some 80% live in the circumpolar countries, although the number actually living in the Arctic is unknown. Wolf management practices reflect a long history of exploitation and predator control by humans. The high reproductive potential and dispersal behavior of wolves, however, contribute to their persistence in many areas and to their repopulation of historical range in others.

Canada
Canada has the most wolves of any arctic country, with some 50,000 to 63,000 animals inhabiting about 86% of their historical range. About 16,000 of the wolves live north of 60°N. Most of these Arctic populations are considered stable and not at risk. Wolves in the High Arctic region, however, were classified as indeterminate in 1999, due to insufficient information.

Alaska
Wolves remain widely distributed over their historical range in Alaska, with a stable or slightly increasing population currently estimated at 7,500 wolves statewide. They may be hunted and trapped. Proposed wolf control measures, largely intended to benefit moose and caribou populations, continue to be highly controversial and contentious.

Greenland
There are about 75 wolves in Greenland today. They are fully protected throughout all of the island, except for an area in the southeast. Some live on the northern tip of Greenland, about 800 kilometers from the North Pole, almost certainly the northernmost wolf population in the world. Wolves were probably never abundant in Greenland, and were absent for much of the 20th Century.

Fennoscandia
The total wolf population in Finland is estimated to be 100 animals, about 30 of which are shared with Russia. They are protected south of the reindeer areas, and subject to a five-month harvest season within the reindeer herding areas. The total kill ranges from 5-15 wolves per year. In winter 2000-2001, 29 animals were killed, which was about 25% of the total population. Sweden currently has about 50-70

Wolf **Canis lupus**.

wolves. Wolves have been protected throughout the country since 1966, but negative attitudes still persist. Norway has approximately 22-24 wolves. Wolves have been protected in Norway since 1973, although conflicts with livestock led to culling of nine wolves in early 2001. The wolf populations in the northern parts of all three countries are unstable, and consist mostly of animals wandering from other regions. Conflicts with reindeer herding complicate the protection of wolves in the CAFF region of Fennoscandia. A recently proposed policy would facilitate the dispersal of wolves from Finland and Russia to both Norway and Sweden.

Russia

Russia has between 40,000 and 50,000 wolves, found across three-quarters of the country. With the dissolution of the Soviet Union, however, wolf research and monitoring declined. Russia has no legal protection for wolves except in nature reserves. Hunting can occur throughout the year. An estimated 15,000 wolves are killed annually. Although wolf populations in most areas are thought to be stable or increasing, recent estimates in northwestern Russia (Kola and Karelia regions) show declines in the late 1990s.

While wolves are abundant in Alaska, northern Canada, and Russia, local overharvests may occur. Habitat loss continues to be a concern for wolf conservation, especially in areas with recovering wolf populations. Wolves are also regarded by many as a nuisance species, hampering management and recovery plans. The challenge continues to be the development and public acceptance of a flexible conservation plan that accommodates wolves in wilderness, but allows for local conflict management.

H. Dean Cluff, Northwest Territories Department of Resources, Wildlife and Economic Development, Yellowknife, Northwest Territories, Canada.

REFERENCES

Aronson, Å., P. Wabakken, H. Sand, O.K. Steinset, and I. Kojola. 2000. The wolf in Scandinavia. Status report of the 1999-2000 winter. Hedmark, Norway: Høgskolen i Hedmark.

Bibikov, D. 1993. Wolves in Russia. *International Wolf* 3(2):18-19.

Carbyn, L.N., S.H. Fritts, and D.R. Seip, eds. 1995. *Ecology and conservation of wolves in a changing world.* Canadian Circumpolar Institute, Occasional Publication No. 35. 642 pp.

Carbyn, L.N., and C.G. van Zyll de Jong. 1999. COSEWIC status report on the gray wolf (*Canis lupus*) in Canada.

Dawes, P.R., M. Elander, and M. Ericson. 1986. The wolf (*Canis lupus*) in Greenland: A historical review and present status. *Arctic* 39(2):119-132.

Wolf Specialist Group. 2000. Minutes of the 2000 Wolf Specialist Group (IUCN/SSC) meeting. Duluth, MN (23 February). http://www.wolf.org

*Figure 81. Distribution of wolf **Canis lupus** in the Northern Hemisphere.*

BOX 73. WOLVERINE

▶ The wolverine is the largest terrestrial mustelid and is found throughout the Arctic. Its status and ecology remain poorly understood, largely because it is elusive and occupies remote habitats. Wolverine occur in small numbers throughout their range and their productivity is often closely tied to the fate of caribou and wolf populations. Globally, the World Conservation Union (IUCN) has listed the wolverine as "vulnerable."

Canada

Wolverines are believed to be relatively more abundant in the west and central portions of the Northwest Territories and Nunavut mainland, with a sparser distribution to the east. They are designated as vulnerable west of Hudson Bay. Although sightings are sporadic, wolverine also inhabit the Arctic archipelago. The richest wolverine habitat in northern Canada is probably the southwest Yukon Territory. Currently, there are insufficient data to assess wolverine population abundance or trends. It is believed that the wolverine population in the Northwest Territories and Nunavut combined is above 1,000 and stable, and that the same is true for the population in the Yukon Territory.

Alaska

Most of Alaska supports a healthy wolverine population, with the highest densities probably found in terrain that includes both mountain and lowland areas. Wolverines

Wolverine *Gulo gulo*.

are regularly harvested in Alaska, and are susceptible to population declines if harvest levels are too high. Maintaining sustainable wolverine populations requires sufficiently large tracts of proper habitat to be protected. Although harvest levels of wolverine are regularly monitored in Alaska, there are insufficient data to assess wolverine population abundance or trends. Current estimates show more than 4,500 wolverine ranging throughout Alaska.

Eurasia

Wolverine are currently found in parts of Estonia, Finland, Norway, and Sweden, and across the taiga and forest-tundra zones of northern Russia. It is believed that there are more than 26,000 wolverine in Eurasia. The population in northwestern Russia has decreased in the past few years, but in Russia overall wolverines appear to be stable or increasing. In Finland, there were about 120 individuals in 1999, and the wolverine is considered threatened. Some 30-40% of these animals are shared with other neighboring countries. In Norway, the wolverine population is about 200, and in Sweden between 250 and 300. The populations in Nordic countries are stable or increasing, but conflicts with reindeer herders and sheep herders complicate conservation efforts.

Wolverine have naturally low densities, relatively low reproductive potential, and are dependent on large tracts of isolated wilderness. These characteristics make them vulnerable to the pressures of over-harvest and human disturbance. The push to develop non-renewable resources in all northern regions is likely to contribute to habitat fragmentation and increased hunting pressure, possibly leading to a decline in wolverine abundance.

Robert J. Gau and Robert Mulders, Northwest Territories Department of Resources, Wildlife and Economic Development, Yellowknife, Northwest Territories, Canada.

REFERENCES

Banci, V. 1999. Updated (Draft) Status Report on the Wolverine in Canada in 1999. Prepared for the Committee on the Status of Endangered Wildlife in Canada, Ottawa, Ontario, Canada. 37p.

Holbrow, W. C. 1976. *The biology, mythology, distribution and management of the wolverine in western Canada*. Winnipeg: MNRM Practicum, Natural Resources Institute, University of Manitoba. 214p.

Kojola, I. 2000. Suurpetojen lukumäärä ja lisääntyminen vuonna 1999 (In Finnish). *Riistantutkimuksen tiedote* 165. (*Report of Finnish Game and Fisheries Institute* 165). 8p.

Smit, C. J., and A. Van Wijngaarden. 1981. *Threatened mammals in Europe*. Wiesbaden, Germany: Akademische Verlgsgesellschaft. 259p.

Figure 82. Distribution of wolverine **Gulo gulo** *in the Northern Hemisphere.*

BOX 74. BROWN BEAR

▶ The brown bear has the widest distribution of any bear. The species and its subspecies are known by many names, including grizzly, Kodiak, and Gobi bear. Its specific status and ecology in many northern environments is not well-known.

CANADA
It is currently estimated that the mountains and tundra of the Northwest Territories and the mainland of Nunavut are home to 5,050 brown bears. An additional 6,300 live in the mountainous Yukon Territory. Although the northern Canadian population of brown bears is considered stable, the Canadian population as a whole has been designated vulnerable because of the potential for habitat degradation and loss.

ALASKA
Alaska has the largest population of brown bears of any state or province in North America, and bears currently occupy all of their historical range. The brown bear population is estimated at 31,700 and is considered stable throughout most of the state. The highest densities occur along the southern coast.

FENNOSCANDIA
Only 45-50 brown bears are found at any given time in Norway, which shares most of its population with Sweden. The Swedish population has grown from an estimated 294 animals in 1942 to about 1,000 today, and its range has increased as well. Finland, too, has been successful in restoring its brown bear population. Since the late 1970s, despite hunting pressures, the population has grown to more than 800 bears.

RUSSIA
Russia has the largest brown bear population in the world, with an estimated population of at least 125,000 animals. It is a common game species in most areas, and European Russia experienced a considerable increase in numbers and range in the late 1970s and 1980s. In the late 1990s, however, the populations in Kola and Karelia decreased. The range of brown bears is expected to remain stable in Asian Russia, although extensive poaching in the Far East will likely cause regional declines.

Humans are the most significant source of mortality for adult brown bears. Hunting, human-bear conflicts, and poaching can become significant problems. This is especially true in areas where the human presence is increasing and bears begin to associate humans with garbage and other sources of food. Habitat fragmentation and loss from industrial and other development contributes substantially to overall brown bear mortality. In a few areas, brown bear populations have been reduced to increase the populations of other animals, such as moose. Although the global population of brown bears is in good shape today, the combination of these and other impacts such as climate change are threats to the species in many areas.

Robert J. Gau, Northwest Territories Department of Resources, Wildlife and Economic Development, Yellowknife, Northwest Territories, Canada.

REFERENCES
Pasitschniak-Arts, M. 1993. *Ursus arctos. Mammalian Species* 439:1-10.
Servheen, C., S. Herrero, and B. Peyton. 1999. *Bears: status survey and conservation action plan.* IUCN/SSC Bear and Polar Bear Specialist Groups. Cambridge, U.K.: IUCN. 309p.

Brown bear **Ursus arctos.**

Pacific or black brant *(Branta bernicla nigricans)*
The breeding range of Pacific brant includes the Arctic regions of western Canada, Alaska, and eastern Russia. Most of the Arctic population nests in large colonies on the Yukon-Kuskokwim Delta of western Alaska. The entire population, including the birds now breeding as far west as the Olenek Delta in Russia, gathers at Izembek Lagoon on the Alaska Peninsula (a Ramsar site) to prepare for the trans-Pacific migration in early November. From there a large population migrates along the Pacific coast of North America, wintering in California and Mexico. A smaller population migrates to China and Korea and is believed to be the remnant of a larger population that migrated to Southeast Asia. It is believed that this population was hunted heavily in China, leading to a severe decline. Today, only 7,000 Pacific brant winter in Asia and their breeding range has gradually been taken over by the North American population. The Pacific brant is an important resource throughout its range. This population has been stable or increasing under a conservative harvest regime during fall and winter. Since 1984, the Pacific brant has been cooperatively managed in the U.S. through the Yukon-Kuskokwim Delta Goose Management Plan. This co-management process involves state and federal wildlife agencies, as well as hunters throughout the range.

Barnacle goose *(Branta leucopsis)*
The Svalbard population of barnacle goose numbered only 300 birds in 1948. The main causes of the decline were overharvesting and disturbances in winter habitats in the Solway Firth region of Scotland. The species became totally protected in Great Britain in 1954 and in Svalbard in 1955. The establishment of Caerlaverock National Nature Reserve in Scotland in 1957 was of great importance. By the mid-1960s, the population had recovered, reaching 3,000-4,000 birds. Additional reserves have been added in Scotland and Svalbard. By 1973, most nesting sites were protected. In early 1997, the total population was estimated at 23,000. Similar increases have occurred in Greenland and Russia (Table 11).

Shorebirds

Of the 57 species of shorebirds breeding in the Arctic, most are found above the treeline. Of the world's 24 species of sandpipers, 17 breed exclusively in the Arctic. Table 12 summarizes the scarce information available on trends in shorebirds, which are usually monitored at their wintering grounds outside the Arctic. Virtually all of Canada's Arctic-nesting shorebirds are declining. No obvious explanation emerges for these widespread declines, and it is likely that several factors are involved.

Long-term monitoring data are available from the wintering sites of Eurasian populations. For most Eurasian species, no trend is apparent. Some sandpiper species, however, seem to be declining. Monitoring of shorebirds at breeding sites in the Arctic started only recently in a few selected sites. At Zackenberg, Greenland, the sanderling, *Calidris alba*, the ringed plover, *Charadrius hiaticula*, and the turnstone, *Arenaria interpres*, have been stable over five consecutive years of monitoring. According to preliminary data, the knot, *Calidris canutus*, appears to be declining and the dunlin, *Calidris alpina*, increasing (See also Table 12).

Passerines

There is a low diversity of passerines in the Arctic, and most of those present have a widespread breeding range. However, little is known about their life history and status. Passerine monitoring programs that are conducted in southern Canada cannot monitor northern boreal and Arctic breeders. The Canadian Landbird Monitoring Strategy has established a priority ranking system for Canadian landbirds, based on a species' vulnerability to disturbance, its population trends, and the proportion of its range that falls within Canada. The Strategy identifies six Arctic-breeding species, including four passerines, that are a high priority for research or conservation action.

BOX 75. MUSKOXEN

▶ The circumpolar status of muskoxen is a conservation success story. The recovery of populations in Canada and Greenland and their re-colonization of former ranges is partly due to protection from unregulated hunting. Re-introductions have been successful in Alaska, Russia, west Greenland, and Quebec, though less so in Norway.

Canada
Of the world's 150,000 muskoxen, 130,000 are found in Canada's Northwest and Nunavut Territories. Complete protection was legislated in 1917 after unregulated commercial hunting had reduced muskoxen on the Canadian mainland to perhaps only 500 by the 1930s. Recovery was slow, but by the late 1990s muskoxen had recolonized much of their former range on the mainland and numbered 24,000. Although muskoxen continue to spread on the periphery of their range, some more central areas have experienced declines, especially along the central mainland coast. The reasons for the declines are uncertain.

On the large islands of Banks and Victoria, there was no evidence that commercial hunting was a factor, but unusually severe ice storms may have caused population declines by the early 1900s. By 1994, however, there were 64,000 muskoxen on Banks Island and 50,000 on Victoria Island, although the Banks Island population subsequently declined to 46,000 by 1998. Elsewhere in the High Arctic, muskoxen persist despite local fluctuations in numbers, due in large part to periodic die-offs and low reproduction during winters with extreme snow depth and ice storms.

Alaska
Muskoxen disappeared from Alaska by the late 1800s. In 1930, re-introductions began when muskoxen from Greenland were released on Nunivak Island in western Alaska. Subsequently, muskoxen from Nunivak Island were released on Nelson Island, in the Arctic National Wildlife Refuge, in northwestern Alaska, and on the Seward Peninsula. Those populations now total over 3,300 animals. Populations are now stabilized by hunting on Nunivak Island, hunting and emigration on Nelson Island, and decreasing production and survival and emigration in the Arctic National Wildlife Refuge. On the Seward Peninsula, numbers and range have been increasing for over 20 years.

Muskox *Ovibos moschatus.*

Greenland
The muskox population in northeast Greenland in 1991 was between 9,500 and 12,500, most of which were in the Northeast Greenland National Park. Muskoxen from northeast Greenland were translocated to four sites in west Greenland. By 1991 their numbers had increased to 2,670.

Russia
Muskoxen were re-introduced to the Wrangel Island in 1975 from Canada's Banks Island and to the Taimyr Peninsula in the early 1980s from Alaska's Nunivak Island. The Taimyr herd now numbers 2,000 animals, and the Wrangel herd 600. In the 1990s, muskoxen were re-introduced near the Lena Delta and to the northern Ural Mountains. The Lena herd now has about 100 animals, most of which live on the edge of the Lena Delta Nature Reserve.

Threats

The prospects for muskox populations depends in part on the value that people place on them, which varies widely. In some communities, muskoxen are seen as a threat to caribou or reindeer and are thus regarded as a nuisance. In other areas, they are seen as a valuable resource. Immediate threats to muskoxen are relatively few and localized. Most hunting is done under a quota system. Oil and gas development is identified as a potential threat in areas such as Alaska's North Slope and Greenland's Jameson Land. Although muskoxen have existed for thousands of years in an extreme and variable environment, some pervasive threats may accumulate over time. Muskoxen have extremely low genetic variability, and thus may be particularly susceptible to diseases and parasites. It is unclear if and how global climate change added to air-borne pollution may affect muskoxen.

Anne Gunn, Northwest Territories Department of Resources, Wildlife and Economic Development, Yellowknife, Northwest Territories, Canada

REFERENCES

Boertmann, D., M. Forchhammer, C.R. Olesen, P. Aastrup, and H. Thing. 1991. The Greenland muskox population status 1990. *Rangifer* 12:5-12.

Coady, J.W., P.E. Reynolds, J.R. Dau, G.M. Carroll, R.J. Seavoy, and P.J. Bente. 1999. *Muskoxen in Alaska*. Poster presented at the 10th Arctic Ungulate Conference, 9-12 August 1999, Tromso, Norway.

Fournier, B., and A. Gunn. 1998. *Muskox numbers and distribution in the Northwest Territories, 1997*. Yellowknife: Northwest Territories Department of Resources, Wildlife and Economic Development. File Rep. No. 121. 55p.

Van Coeverden de Groot, P.J. 2000. *Microsatellite variation in the muskox* Ovibos moschatus. Unpublished Ph.D. Thesis, Queens University, Kingston, Ontario.130pp.

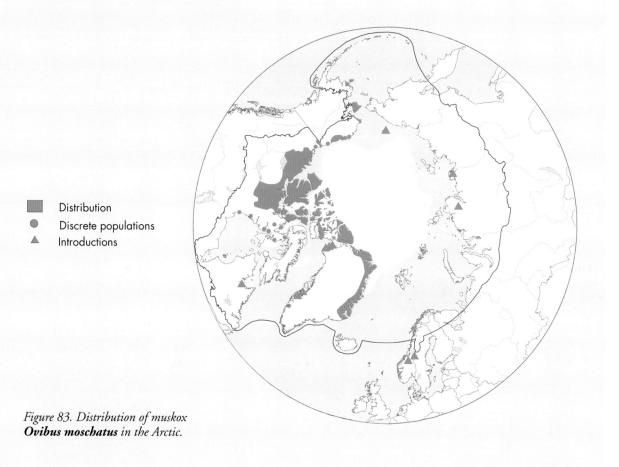

■ Distribution
● Discrete populations
▲ Introductions

Figure 83. Distribution of muskox **Ovibus moschatus** *in the Arctic.*

Table 10. Estimated population size and trends of Arctic terrestrial mammals.

SPECIES	REGION	POPULATION	TREND
Caribou/reindeer *Rangifer tarandus*	Mulchatna	200,000	Increasing
	Nelchina	50,000	Increasing
	Fortymile	22,000	Stable
	Western Arctic	450,000	Increasing
	Teshekpuk	25,000	Increasing
	Central Arctic	18,000	Increasing
	Porcupine	129,000	Increasing
	Bluenose (3 herds)	122,000	Increasing
	Bathurst	350,000	Stable
	Queen Maud Gulf	200,000	Increasing
	Beverly	275,000	Increasing
	Qamanirjuaq	460,000	Increasing
	Northeastern Mainland	120,000	Declining
	Northern Hudson Bay	30,000	Unknown
	Baffin Island	140,000	Unknown
	George River	780,000	Increasing
	Leaf River	250,000	Unknown
	Arctic-island caribou (Banks, Victoria, Prince of Wales, Somerset)	30,000	Stable/ Declining
	Peary caribou (western High Arctic)	1,100	Declining
	Peary caribou (eastern High Arctic)	Unknown	Unknown
	West Greenland	30,000	Increasing
	Svalbard	8,000	Stable
	Iceland	3,000	Stable
	Archangelsk	17,000	Increasing?
	Komi	3,500	Declining?
	Taimyr	600,000	Increasing
	Yakutia	240,000	Stable
	Chukotka (+ Magadan area)	35,000	Increasing?
Muskox *Ovibos moschatus*	Alaska	3,300	Increasing
	Canada	130,000	Stable?
	Greenland	15,000	Increasing
	Russia	1,000	Increasing
Brown bear *Ursus arctos*	Alaska	31,700	Stable
	Yukon Territory	6,300	Stable
	Northwest Territories and Nunavut	5,050	Stable
	Fennoscandia	1,800	Increasing
	Russia	125,000	Declining

SPECIES	REGION	POPULATION	TREND
Wolverine *Gulo gulo*	Alaska	4,500	Unknown
	Yukon Territory	1,000	Stable
	Northwest Territories and Nunavut	1,000	Stable
	Fennoscandia	550-600	Stable
	Eurasia	26,000	Increasing
Wolf *Canis lupus*	Alaska	7,500	Stable/ Increasing
	Canada north of 60°N	6,000	Stable
	Greenland	75	Stable?
	Fennoscandia	160	Declining
	Russia	40,000-50,000	Stable/ Increasing

Sources: Born 1998, Kojola 1999, Pavlov et al. 1996, A. Gunn, pers. comm., A. Tishkov, pers. comm.

The lesser white-fronted goose **Anser erythropus** *breeds in the forest-tundra and southern tundra of Eurasia. The population has declined sharply throughout its range and is now globally threatened. In Fennoscandia, only about fifty nesting pairs remain. WWF and the European Union have concentrated efforts to protect this species.*

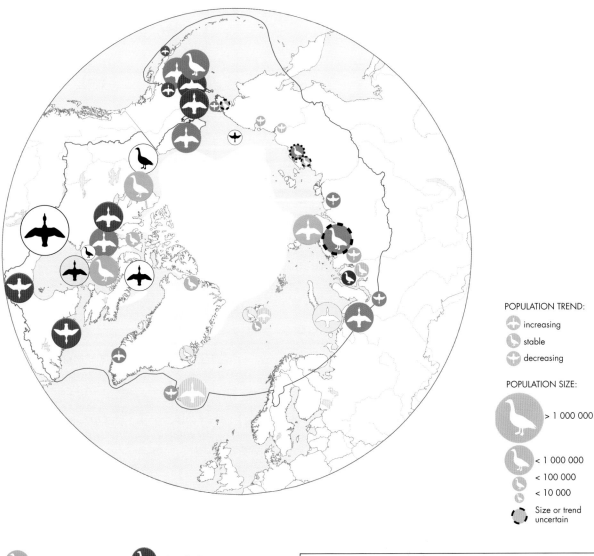

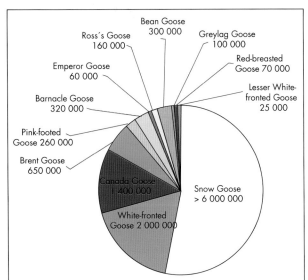

Figure 84. Trends of Arctic goose populations (Source: Madsen et al. 1996).

244 STATUS AND TRENDS IN SPECIES AND POPULATIONS

Table 11. Estimated population size and trends of geese in the Arctic region.

SPECIES	REGION	POPULATION (individuals)	TREND
Greater snow goose *Chen caerulescens atlantica*	Northern Canada	650,000+	Increasing
Lesser snow goose *Chen caerulescens*	Northern Canada	6,000,000+	Stable/Increasing
Pacific or Black brant *Branta bernicla nigricans*	W Canada, Alaska, Eastern Siberia (Pacific)	130,000	Stable
	Atlantic Population	160,000	Stable
Dark-bellied brent goose *Branta bernicla bernicla*	Western Palaearctic	300,000	Stable
Light-bellied brent goose *Branta bernicla hrota*	Canada	20,000	Stable?
	Svalbard	5,000	Increasing
Greater white-fronted goose *Anser albifrons frontalis*	Mid-continent Population	950,000	Increasing
	Pacific Population	420,000	Increasing
	Northeastern Siberia	100,000-150,000	Declining?
Greater white-fronted goose *Anser albifrons albifrons*	Western Palaearctic	1,000,000-1,500,000	Increasing?
Greenland white-fronted goose *Anser albifrons flavirostris*	West Greenland	33,000	Increasing
Lesser white-fronted goose *Anser erythropus*	Eurasia (Norway to Chukotka)	20,000-25,000	Declining
Cackling Canada goose *Branta canadensis minima*	Alaska	210,000	Increasing
Aleutian Canada goose *Branta canadensis leucopareia*	Alaska	30,000	Increasing
Taverner's Canada goose *Branta canadensis taverneri*	Alaska	40,000-50,000	Unknown
Emperor goose *Chen canagica*	Alaska	60,000	Stable
Tundra bean goose *Anser fabalis rossicus*	Russia	300,000 - 600,000	Stable
Pink-footed goose *Anser brachyrhynchus*	Greenland/Iceland	230,000	Increasing
	Svalbard	37,000	Increasing
Graylag goose *Anser anser*	Iceland	80,000	Declining
Barnacle goose *Branta leucopsis*	Greenland	40,000 - 45,000	Increasing
	Svalbard	23,500	Increasing
	Russia	260,000	Increasing
Red-breasted goose *Branta ruficollis*	Russia	90,000 to 100,000	Increasing

Sources: Madsen et al. 1996, Madsen et al. 1999, Filion and Dickson 1999, Miyabayashi and Mundkur 1999, U.S. Fish and Wildlife Service 2000.

Table 12. Estimated population sizes and trends of shorebirds breeding in the Arctic.

SPECIES **North American Populations**	POPULATION (individuals)	TREND
Black-bellied plover *Pluvialis squatarola*	150,000	Declining
American golden plover *Pluvialis dominica*	150,000+	Declining/Stable?
Semi-palmated plover *Charadrius semipalmatus*	150,000	Declining/Stable?
Whimbrel *Numenius phaeopus*	50,000	Declining/Stable?
Marbled godwit *Limosa fedoa*	95,000	Declining
Turnstone *Arenaria interpres*	235,000	Stable?
Knot *Calidris canutus*	251,000	Declining?
Sanderling *Calidris alba*	300,000	Declining
Semi-palmated sandpiper *Calidris pusilla*	3,500,000	Declining
Least sandpiper *Calidris minutilla*	600,000	Declining
White-rumped sandpiper *Calidris fuscicollis*	500,000	Stable?
Baird's sandpiper *Calidris bairdii*	500,000	Stable?
Pectoral sandpiper *Calidris melanotos*	450,000	Declining/Stable?
Purple sandpiper *Calidris maritima*	10,000	Declining/Stable?
Dunlin *Calidris alpina*	1,000,000	Declining
Stilt sandpiper *Micropalama himantopus*	200,000	Declining
Buff-breasted sandpiper *Tryngites subruficollis*	25,000	Declining
Short-billed dowitcher *Limnodromus griseus*	400,000	Declining
Common snipe *Gallinago gallinago*	1,000,000	Declining
Red-necked phalarope *Phalaropus lobatus*	2,500,000	Declining
Red phalarope *Phalaropus fulicarius*	900,000	Declining

SPECIES **Eurasian Populations**	POPULATION	TREND
Grey plover *Pluvialis squatarola*	168,000 (Atlantic)	Increasing
Snipe *Gallinago gallinago*	750,000 (Iceland)	Stable
Black-tailed godwit *Limosa limosa*	65,000 (Iceland)	Increasing
Bar-tailed godwit *Limosa lapponica*	115,000 (West palaearctic)	Stable/declining?
Turnstone *Arenaria interpres*	80,000 (West palaearctic)	Increasing
Knot *Calidris canutus*	260,000 (Siberia)	Declining
Knot *Calidris canutus*	400,000 (Eastern Canada/Greenland)	Stable?
Sanderling *Calidris alba*	123,000 (East Atlantic)	Increasing?
Little stint *Calidris minuta*	211,000 (Europe)	Stable
Purple sandpiper *Calidris maritima*	50,500 (Eastern Atlantic)	Stable
Dunlin *Calidris alpina*	1,370,000 (Europe, Central Siberia)	Declining
Dunlin *Calidris alpina*	800,000 (Iceland/Greenland)	Stable
Curlew sandpiper *Calidris ferruginea*	436,000 (Siberia)	Stable
Broad-billed sandpiper *Limicola falcinellus*	40,000 - 60,000 (Europe/Siberia)	Declining?
Ruff *Philomachus pugnax*	3,300,000 (Siberia)	Declining
Spoon-billed sandpiper *Eurynorhynchus pygmaeus*	2,000 (Chukotka)	Declining

Sources: Hagermeijer and Blair 1997, Morrison et al. 2000, Tomkovich et al. 2000, Wetlands International 1999.

Globally Threatened Species

■ The concept of threatened species has become increasingly important for biological and legal reasons in recent decades. In addition to domestic lists made in accordance with national laws in all Arctic countries, there have been various attempts to create international or global lists of threatened species. More and more information is being collected in support of adding species to such lists, making the lists increasingly useful. Because many of the newly listed species had not been studied before, however, the rising number of listed species does not necessarily indicate an overall environmental trend. The fact that very few species have been removed from such lists demonstrates that threatened species do not recover easily. One use for such lists is to identify key areas, species, and species groups that require focused conservation efforts.

Table 13: Globally threatened species and populations within the Arctic Region. The Xs indicate the country or countries in which the species or population is found. Under "Category," VU= vulnerable; EN = endangered; CR = critically endangered. Species classified as "low risk" have been omitted. Note: Several rare plant species do not have an English common name.

SCIENTIFIC NAME	COMMON NAME	RUS	CAN	USA	GRL	ISL	NOR	SWE	FIN	CATEGORY
Mammals										
Balaena mysticetus	Bowhead whale (Baffin Bay, Davis Strait stock)		X		X					EN
Balaena mysticetus	Bowhead whale (Svalbard, Barents Sea Stock)	X			X		X			CR
Balaenoptera borealis	Sei whale	X	X	X		X	X			EN
Balaenoptera musculus	Blue whale	X	X	X	X	X	X			EN
Balaenoptera physalus	Fin whale	X	X	X	X	X	X			EN
Eubalaena glacialis	Northern right whale	X	X	X	X	X				EN
Megaptera novaeangliae	Humpback whale	X	X	X	X	X	X			VU
Physeter catodon	Sperm whale	X	X	X	X	X	X			VU
Delphinapterus leucas	Beluga	X	X	X	X		X	X		VU
Phocoena phocoena	Harbour porpoise	X	X	X	X	X	X	X	X	VU
Eumetopias jubatus	Steller's sea lion	X	X	X						EN
Callorhinus ursinus	Northern fur seal	X	X	X						VU
Halichoerus grypus	Grey seal					X	X	X	X	EN
Enhydra lutris	Sea otter	X		X						VU
Lutra lutra	Eurasian otter	X					X	X	X	VU
Dicrostonyx vinogradovi	Vinogradov's lemming	X								CR
Barbastella barbastellus	Western Barbastelle	X								VU
Gulo gulo	Wolverine	X	X				X	X	X	VU
Moschus moschiferus	Siberian musk deer	X								VU
Rangifer tarandus pearyi	Peary caribou		X							EN
Ovis nivicola borealis	Putorean snow sheep	X								VU
Birds										
Anas formosa	Baikal teal	X								VU
Anser erythropus	Lesser white-fronted goose	X					X	X	X	VU
Brachyramphus marmoratus	Marbled murrelet	X								VU
Branta ruficollis	Red-breasted goose	X								VU
Eurynorhynchus pygmaeus	Spoonbill sandpiper	X								VU
Grus americana	Whooping crane		X	X						EN
Grus leucogeranus	Snow crane	X								CR
Haliaeetus pelagicus	Steller's sea-eagle	X								VU
Numenius borealis	Eskimo curlew		X	X						CR
Numenius tahitiensis	Bristle-thighed curlew			X						VU
Phoebastria albatrus	Short-tailed albatross	X		X						VU
Rissa brevirostris	Red-legged kittiwake	X								VU

SCIENTIFIC NAME	COMMON NAME	RUS	CAN	USA	GRL	ISL	NOR	SWE	FIN	CATEGORY
Fishes										
Acipenser baerii	Siberian sturgeon	X								VU
Acipenser baerii baerii	Siberian sturgeon	X								EN
Acipenser baerii stenorrhynchus	Lena River sturgeon	X								VU
Acipenser fulvescens	Lake sturgeon		X							VU
Coregonus zenithicus	Shortjaw cisco		X							VU
Pleuronectes ferrugineus	Yellowtail flounder		X							VU
Salvelinus confluentus	Bull trout			X						VU
Salvethymus svetovidovi	Svetovidov's long-finned char	X								VU
Sebastolobus alascanus	Shortspine thornyhead	X	X							EN
Raja (Dipturus) laevis	Barndoor skate		X							VU
Plants										
Artemisia aleutica				X						VU
Astragalus gorodkovii		X								VU
Beckwithia glacialis alaskensis				X						VU
Cardamine sphenophylla		X								VU
Claytoniella vassilievii petrovskyi		X								VU
Douglasia beringensis		X								VU
Hedinia czukotica		X								VU
Jamesoniella undulifolia					X					VU
Polystichum aleuticum				X						EN
Potentilla murrayi				X						VU
Puccinellia jenissejensis		X								VU
Puccinellia rosenkrantzii					X					VU
Puccinellia svalbardensis							X			VU
Pucciphippsia czukczorum		X								VU
Ranunculus wilanderi							X			VU
Salicornia borealis			X							VU
Salix stolonifera carbonicola		X								VU
Taraxacum leucocarpum		X								VU
Taraxacum nanaunii		X								VU
Taraxacum tolmaczevii		X								VU

Source: Hilton-Taylor 2000.

The IUCN Red List of Threatened Species 2000 has four categories for threatened species: low risk, vulnerable, endangered, and critical. According to the list, 1806 mammal species are globally threatened, not taking into account subspecies and stocks. Of those, 43 species (including subspecies) and stocks are threatened within the Arctic region as defined by CAFF. Of the 43 species, most occur in the Russian Federation (36 species) and in Canada (28 species), which of course also have the most territory in the Arctic.

There are 1913 globally threatened bird species, 16 of which are within the Arctic. Most are within the borders of the Russian Federation (9 species). Two species listed in 1996, the Steller's eider, *Polysticta stelleri*, and the spectacled eider, *Somateria fischeri*, are no longer considered globally threatened, although they remain on the United States list.

There are 817 globally threatened fish species, of which 12 are found in the Arctic. Two species of skates are included, the big skate, *Dipturus binoculata* (listed as low-risk), and the barndoor skate, *Dipturus laevis* (listed as vulnerable), both in Canada. One species of lamprey, *Lampetra fluviatilis*, is listed as low-risk in Greenland.

There are 7,299 globally threatened plant species (80 Bryophyta species and 7,219 Tracheophyta species), 73 of which (including subspecies) are within the Arctic.

REFERENCES

Anker-Nilssen, T., and Barrett, R.T. 1991. Status of seabirds in northern Norway. *British Birds* 84: 329-341.

Anker-Nilssen, T., V. Bakken, H. Strøm, A.N. Golovkin, V.V. Bianki, and I.P. Tatarinkova. 2000. The status of marine birds breeding in the Barents Sea Region. Norsk Polarinstitutt Rapport Nr. 113. Tromsø. 213p.

Aronson, Å., P. Wabakken, H. Sand, O.K. Steinset, and I. Kojola. 2000. The wolf in Scandinavia. Status report of the 1999-2000 winter. Hedmark, Norway: Høgskolen i Hedmark.

Asbirk, S.A., L. Berg, G. Hardeng, P. Koskimies, and A. Petersen. 1997. Population sizes and trends of birds in the Nordic Countries 1978-1994. TemaNord 1997:614. 88p.

Banci, V. 1999. Updated (Draft) Status Report on the Wolverine in Canada in 1999. Prepared for the Committee on the Status of Endangered Wildlife in Canada, Ottawa, Ontario, Canada. 37p.

Bateson, P.P.G., and R.C. Plowright. 1959. The breeding biology of the Ivory Gull in Spitsbergen. *British Birds* 54: 272-277.

Bibikov, D. 1993. Wolves in Russia. *International Wolf* 3(2:18-19).

Birkhead, T. R., and D. N. Nettleship. 1980. Census methods for murres, *Uria species*: an integrated approach. Canadian Wildlife Service Occasional Paper.

Boertmann, D., M. Forchhammer, C.R. Olesen, P. Aastrup, and H. Thing. 1991. The Greenland muskox population status 1990. *Rangifer* 12:5-12.

Born, E.W., ed. 1998. Grønlandske fugle, havpattedyr og landpattedyr – en status over vigtige ressourcer (In Danish). Pinngortitaleriffik Teknik rapport Nr.16.

CAFF Circumpolar Seabird Working Group. 1996. *International Murre Conservation Strategy and Action Plan*. Ottawa, Canada: CAFF. 16 pp.

CAFF Circumpolar Seabird Working Group. 1998. *The Circumpolar Eider Conservation Strategy and Action Plan*. Akureyri, Iceland: CAFF.

Carbyn, L.N., and C.G. van Zyll de Jong. 1999. COSEWIC status report on the gray wolf (*Canis lupus*) in Canada. Ottawa: Committee on the Status of Endangered Wildlife in Canada (COSEWIC) Secretariat, Canadian Wildlife Service, Department of the Environment.

Carbyn, L.N., S.H. Fritts, and D.R. Seip, eds. 1995. *Ecology and conservation of wolves in a changing world*. Canadian Circumpolar Institute, Occasional Publication No. 35. 642 pp.

Cleator, H.J. 1996. The status of bearded seal, *Erignathus barbatus*, in Canada. *Can. Field. Nat.* 110: 501-510.

Coady, J.W., P.E. Reynolds, J.R. Dau, G.M. Carroll, R.J. Seavoy, and P.J. Bente. 1999. *Muskoxen in Alaska*. Poster presented at the 10th Arctic Ungulate Conference, 9-12 August 1999, Tromso, Norway.

COSEWIC (Committee on the Status of Endangered Wildlife In Canada). 2000. *COSEWIC Species Assessments – November 2000*. Ottawa: Committee on the Status of Endangered Wildlife in Canada (COSEWIC) Secretariat, Canadian Wildlife Service, Department of the Environment.

Dawes, P.R., M. Elander, and M. Ericson. 1986. The wolf (*Canis lupus*) in Greenland: A historical review and present status. *Arctic* 39(2:119-132).

Dragoo, D. E., G. V. Byrd, and D. B. Irons. 2000. Breeding status and population trends of seabirds in Alaska in 1999. U.S. Fish and Wildlife Service Report AMNWR 2000/02.

Durinck, J. and Falk K. 1996. The distribution and abundance of seabirds off southwestern Greenland in autumn and winter 1988-1989. *Polar Research* 15(1), 23-42.

Falk, K., and K. Kampp. 1997. A manual for monitoring Thick-billed Murre populations in Greenland. Pinngortitaleriffik Technical Report 7. 90 pages.

FAO. 1990. *FAO species catalogue: vol. 10. gadiform fishes of the world (Order Gadiformes)*.

FAO. 2000. *FISHSTAT Plus*. Rome: Food and Agriculture Organization. (October, 2000). Fishery statistics database available to download from: http://www.fao.org/fi/statist/FISOFT/FISHPLUS.asp

Filion, A. and K.M. Dickson. 1999. Population status of migratory game birds in Canada: November 2000. CWS Migratory Birds Regulatory Report Series No. 1. Canadian Wildlife Service. Ottawa, Canada. 91 pp.

Fisheries and Oceans Canada 1998a. Baffin Bay Narwhal. *Stock Status Report* E5-43.

Fisheries and Oceans Canada 1998b. Hudson Bay Narwhal. *Stock Status Report* E5-44

Fisheries and Oceans Canada 1999. Hudson Bay – Foxe Basin Bowhead Whales. *Stock Status Report* E5-52.

Fisheries and Oceans Canada (DFO). 1999. Hornady River Arctic Char. DFO Science Stock Status Report DF5-68.

Fisheries and Oceans Canada 2000. Northwest Atlantic Harp Seal. *Stock Status Report* E1-01.

Fournier, B., and A. Gunn. 1998. *Muskox numbers and distribution in the Northwest Territories, 1997*. Yellowknife: Northwest Territories Department of Resources, Wildlife and Economic Development. File Rep. No. 121. 55p.

Gärdenfors, U., J.P. Rodriguez, C. Hilton-Taylor, C. Hyslop G. Mace, S. Molur, and S. Poss. 1999. Draft guidelines for the Application of IUCN red list criteria at national and regional levels. *Species* 31/32: 58-70.

Gaston, A.J. 1999. Trends in thick-billed murre populations in eastern Canadian Arctic. *Bird Trends* 7: 7-10.

Goudie, R.I., S. Brault, B. Conant, A.V. Kondratyev, M.R. Petersen, and K. Vermeer. 1994. The status of sea ducks in the North Pacific Rim: towards their conservation and management. *Transactions of the North*

American Wildlife and Natural Resources Conference 59:27-49.

Hagermeijer, W.J.H., and M. Blair. 1997. *The EBCC Atlas of European Breeding birds: their distribution and abundance.* London: T & AD Poyser. 903p.

Hill, P.S, and D.P. DeMaster. 1999. *Alaska Marine Mammal Stock Assessment.* National Report. Seattle: National Marine Fisheries Service. 149p.

Hilton-Taylor, C. (Compiler) 2000: 2000 IUCN Red List of Threatened Species. IUCN, Gland, Switzerland and Cambridge, UK. Xviii + 61p.

Holbrow, W.C. 1976. *The biology, mythology, distribution and management of the wolverine in western Canada.* Winnipeg: MNRM Practicum, Natural Resources Institute, University of Manitoba. 214p.

Ianelli, J.N., and V. Wespestad. 1998. *Trends in North Pacific cod and pollock catch 1981-1998.* Published electronically on the Alaska Fisheries Science website (24th October), 2000: http://www.refm.noaa.gov/stocks/Presentations/norfish/gfish98.htm

ICES (International Council for the Exploration of the Sea). 2000. http://www.ices.dk/committe/acfm/comwork/report/2000/contents.html

ICES (International Council for the Exploration of the Sea). 1999. *Advisory Committee on Fishery Management Report 1999.* Online at: http://www.ices.dk/committe/acfm/acfm.htm

ICES (International Council for the Exploration of the Sea). 2000. *ICES Advisory Committee on Fishery Management Report 2000.* Available online (24th October 2000): http://www.ices.dk/committe/acfm/acfm.htm

IUCN Polar Bear Specialist Group. 1998. Worldwide Status of the Polar Bear. pp. 23-44, In: Proceedings of the 12th Working Meeting of the IUCN Polar Bear Specialists, January 1997. Oslo, Norway.

Kaftanovski, Yu.M. 1951. *The Alcids from the eastern part of the Atlantic. Moscow Society of Nature Explorer.* Moscow. 170p. (in Russian)

Kampp, K., D.N. Nettleship, and P.G.H. Evans. 1994 Thick-billed murres of Greenland: status and prospects. *BirdLife Conservation Series* 1:133-154.

Kojola, I. 1999. Suurpetojen lukumäärä ja lisääntyminen vuonna 1999 (In Finnish). *Riistantutkimuksen tiedote* 165:1-8.

Krasnov, Yu.V. 1995. Seabirds (retrospective analysis of population development). In: The environment and ecosystems of Novaya Zemlya archipelago and shelf. p. 138-147. (in Russian).

Krasnov, Yu.V., and R.T. Barrett. 1995. *Large-scale interactions among seabirds, their prey and humans in the southern Barents Sea.* In: H.R. Skjodal, C. Hopkins, K.E. Erikstad, and H.P. Leinaas, eds. *Ecology of Fjords and Coastal Waters.* p. 443-456.

Madsen, J., A. Reed, and A. Andreev. 1996. Status and trends of geese (*Anser* sp.), Branta sp. in the world: a review, updating and evaluation. *Gibier Faune Sauvage, Game Wildlife* 13: 337-353.

Madsen, J., G. Cracknell, and A.D. Fox, eds. 1999. *Goose populations of the Western Palearctic. A review of status and distribution.* Wetlands International Publ. No. 48, Wageningen, The Netherlands. National Environmental Research Institute, Rönde, Denmark. 344p.

Mehlum, F., and V. Bakken. 1994. Seabirds in Svalbard (Norway): status, recent changes and management. In: D.N. Nettleship, J. Burger, and M. Gochfeld, eds. *Seabirds on islands threats, case studies and action plans.* BirdLife Conservation Series 1: 155-171.

Miyabayashi, Y., and T. Mundkur. 1999. Atlas of Key Sites for Anatidae in the East Asian Flyway. Wetlands International. 148p.

Morrison, R.I.G., Y. Aubry, R.W. Butler, G.W. Beyersbergen, G.W. Donaldson, P.W. Hicklin, V.H. Johnston, R.K. Ross, and C.L. Gratto-Trevor. 2000. Declines in North American shorebird populations. Ms. presented at the Waterbird Society Meeting, Plymouth, Massachusetts, November 2000.

Mosbech, A., and D. Boertmann. 1999. Distribution, abundance and reaction to aerial surveys of post-breeding King Eiders (*Somateria spectabilis*) in western Greenland. *Arctic* 52: 188-203.

NAFO (Northwest Atlantic Fisheries Organization) *STATLANT 21A database.* Online at: http://www.nafo.ca/ (last update 24 August 2000).

NAFO (Northwest Atlantic Fisheries Organization). 1999. *Standing Committee on Fisheries Science Report 1999.* http://www.nafo.ca/

NAFO (Northwest Atlantic Fisheries Organization). 2000. *Standing Committee on Fisheries Science Report 2000.* http://www.nafo.ca/

NAMMCO (North Atlantic Marine Mammal Commission). 2000. *NAMMCO Annual Report 1999.* Trosmø: NAMMCO. 265p.

Pasitschniak-Arts, M. 1993. *Ursus arctos. Mammalian Species* 439:1-10.

Piatt, J.F., and P. Anderson. 1996. Response of common murres to the Exxon Valdez oil spill and long-term changes in the Gulf of Alaska marine ecosystem. Pages 720-737. In: S. D. Rice, R. B. Spies, D. A. Wolfe and B. A. Wright, eds. Proceedings of the Exxon Valdez oil spill symposium. American Fisheries Society Symposium 18, Anchorage, Alaska.

Pihl, S. 1997. European Species Action Plan: Steller's eider *Polysticta stelleri.* In: Final Technical Report and Claim. March 1996 – November 1997. Species Action Plans for eight European Threatened Bird Species. Sandy, U.K.: RSPB.

Pavlov, B.M., L.A. Kalpashchikov, and V.A. Zyryanov 1996. Population dynamics of the Taimyr reindeer population. *Rangifer* 9:381-384.

Servheen, C., S. Herrero, and B. Peyton. 1999. *Bears: status survey and conservation action plan.* IUCN/SSC Bear and Polar Bear Specialist Groups. Cambridge, U.K.: IUCN. 309p.

Smit, C.J., and A. Van Wijngaarden. 1981. *Threatened mammals in Europe.* Wiesbaden, Germany: Akademische Verlagsgesellschaft. 259p.

Solovieva, D.V. 1997. Steller's eider: national report from Russia. Proc. From Steller's Eider Workshop. *Wetlands International Seaduck Specialist Group Bull.* 7:7 –12.

Strøm, H., I.J. Øien, J. Opheim, E.A. Kuznetsov, and G.V. Khakhin. 1994. *Seabird censuses on Novaya Zemlya 1994.* Norwegian Ornithological Society. Report No. 2-1994. 38 pp.

Suydam, R.S., D.L. Dickson, J.B. Fadely, and L.T. Quakenbush. 2000. Population declines of King and Common eiders of the Beaufort Sea. *The Condor* 102:219-222.

Syroechkovski, E.E. Jr., C. Zöckler, and E.Lappo. 1998. Status of Brent Goose (*Branta b. bernicla* and *B. b. nigricans*) interbreeding in NW Yakutia. *British Birds* 91: 565-572.

Thomas, V.G. and S.D. MacDonald. 1987. The breeding distribution and current population status of the Ivory Gull in Canada. *Arctic* 40: 211-218.

Tomkovich, P., Syroechkovski, E.E. Jr., C. Zöckler, and E. Lappo. 2000. Spoon-billed sandpiper (*Eurynorhynchus pygmeus*) - probable population crash. http://www.surfbirds.com/Features/Spoonbsand.html

Trites, A.W., P.A. Livingston, S. Mackinson, M.C. Vasconcellos, A.M. Springer, and D. Pauly. 1999. Ecosystem change and the decline of marine mammals in the Eastern Bering Sea. *Fisheries Centre Research Reports*, Vol. 7.

U.S. Fish and Wildlife Service 2000. *Waterfowl Populations Estimates, 2000.* Washington, D.C.: U.S. Department of the Interior. 33p.+ appendices.

Wetlands International. 1999. Report on the Conservation of Migratory Waterbirds in the AEWA Agreement Area. Report prepared for the first meeting of the parties in Capetown 1999. 153p.

Witherell, D. 2000. *Groundfish of the Bering Sea and Aleutian Islands area: species profiles 2000.* Anchorage, Alaska: North Pacific Management Council. http://www.fakr.noaa.gov/npfmc/Reports/STATUS00.pdf

Witherell, D., and N. Kimball. 2000. *Status and trends of principal groundfish and shellfish stocks in the Alaska EEZ, 2000.* Anchorage, Alaska: North Pacific Management Council.

Wolf Specialist Group. 2000. Minutes of the 2000 Wolf Specialist Group (IUCN/SSC) meeting. Duluth, MN (23 February. http://www.wolf.org

Zeh, J.E., C.W. Clark, J.C. George, D. Withrow, G.M. Carroll and W.R. Koski 1993. Current population size and dynamics. In: J.J. Burns, J.J. Montague and C.J. Cowles (eds.). The Bowhead Whale. Special Publication number 2. The Society for Marine Mammalogy, 209-489.

10. Conclusions

> *The future of the Arctic environment and its living resources depends on effective conservation and management. Such efforts require an understanding of Arctic ecology, the role of humans in northern regions, and the status of flora, fauna, and habitats today. This report has covered each of these topics in summary form. In conclusion, a few points are worth emphasizing.*

Ecology

■ Much of the Arctic remains in a natural state, its ecological processes governed by climate, geology, and biology. Such processes are distinctive in the Arctic for several reasons. First, productivity in the Arctic is typically low, with the exception of the waters above some continental shelves. On land, most soils are frozen or waterlogged. The growing season is brief, though it may be intense. Recovery from disturbance, whether natural or caused by humans, is slow.

Second, Arctic systems can be highly variable and dynamic, most notably in the population cycles of some animals. Examining population trends, especially in the short term, can be misleading because cycles can mask real changes.

Third, although the Arctic has relatively few species, the genetic diversity of those species is typically very high, creating a mosaic of distinct populations and subspecies. On an evolutionary time scale, the Arctic and its flora and fauna are young. Some areas are still recovering from the last Ice Age or from glaciation events since then. Ecologically, the Arctic is rapidly evolving and developing.

Fourth, although there are many species endemic to the Arctic region that are well-adapted to its harsh and variable conditions, the Arctic is a marginal zone for a number of its inhabitants. Plants and animals at the limits of their ranges typically undergo wide fluctuations in abundance and distribution as a result of changing conditions. This pattern makes such species useful indicators of environmental change, but does not necessarily shed much light on ecosystem health.

Humans

The Arctic is home to many people and many distinctive cultures. Through time, they have been intimately connected to the land and sea. They have used local resources as the foundation for their communities and societies, in the process affecting plants and animals around them. These practices continue, their cultural significance perhaps greater as northern peoples establish their identity in the modern world. The human presence has been an integral part of ecosystems across much of the Arctic for millennia, a fact that must be recognized in efforts to understand and manage the region's "natural" systems and landscapes.

Over-exploitation

Many species and habitats in the Arctic have been harmed by over-exploitation. While some stocks have recovered, other populations remain precariously low.

The removal or severe reduction of one species may affect other plants and animals as ecological relationships are changed. Over-exploitation remains a problem, particularly in marine fisheries. In addition, certain harvest methods pose a serious threat to non-target species. By-catch mortality of non-target fish, seabirds, and marine mammals is high in some fisheries. Similarly, overgrazing has historically contributed to extensive degradation and loss of vegetated areas – a problem that continues today in parts of the Arctic. As human populations increase in many areas and as commercial harvesters seek new stocks, the exploitation pressure on Arctic species is likely to increase. Sustainable use must be the goal of resource management, both for commercial and other uses of Arctic flora and fauna.

Habitat disturbance

Worldwide, the disturbance, loss, and fragmentation of habitat is considered the most serious threat to biodiversity. Habitat fragmentation creates smaller, more isolated populations that are at greater risk from catastrophic events, genetic deterioration, or behavioral dysfunction. Fires, floods, storms, erosion, and volcanoes can be catastrophic, but are typically natural processes that shape landscapes. Disturbances by humans can also damage habitats, but these can hardly be considered natural. They include oil and gas exploration and development, roads and other infrastructure, off-road use of motorized vehicles, deforestation, hydropower development, bottom trawling, and dredging, each of which has affected Arctic habitats in various areas. A great deal of conservation effort will be required to minimize disturbance from increasing industrial development, acriculture and forestry in many Arctic areas. This point is especially true for assessing and addressing the cumulative impacts of many smaller projects and activities.

Biological disturbance

The reduction of genetic diversity, the introduction of alien species, and disruptions to the balance of ecosystems pose serious threats to the long-term health of Arctic ecosystems. In many places, species have been introduced intentionally or otherwise from other regions within the Arctic or from outside the Arctic entirely. Some perish quickly but others have thrived, displacing existing species and causing the demise of local flora and fauna. Diseases and parasites are particularly hard to get rid of once they have been introduced to a population or an ecosystem. Practices such as fish farming and the stocking of streams threaten the genetic diversity of highly varied species such as salmon. Genetic diversity enhances resistance to disease and may reflect distinctive local adaptations necessary for the long-term health of a population. Human efforts to shift the balance of ecosystems, for example by controlling or eliminating predators, can cause long-term effects throughout the system. All forms of biological disturbance must be avoided where possible, and their impact minimized when they cannot be prevented.

Contamination and pollution

Heavily polluting activities in the North such as mining and oil development have damaged or devastated surrounding areas. Long-range pollution, in the form of heavy metals, persistent organic pollutants, radionuclides, and other substances, has reached every part of the Arctic. Some freshwater systems are stressed as a consequence of acid rain. A number of environmental contaminants are found at relatively high levels in some species but their long-term biological and ecological effects are as yet unknown. In part, this is due to the complexity of the interactions among the many contaminants that have been found. In part, it is due to the uncertainties inherent in detecting and predicting what may be subtle effects on an individual animal, and then extrapolating those effects to the overall population of that species and to the ecosystem of which it is a part. Because the sources of contaminants are spread around the world, reducing pollution in the Arctic will require international action.

Climate change

Climate change is perhaps the greatest threat to the Arctic environment as it exists today. Observed changes in species abundance and distribution, and to permafrost and sea ice, are evidence that climate change is already having an impact. Predictions of climate change indicate substantial warming in much of the Arctic, accompanied by substantial cooling in some areas. Although the predictions cannot tell what will happen, they do indicate the types of ecological impacts that may occur. Ecosystems are likely to respond in complex ways that may reduce or magnify the impacts of changes in temperature and precipitation. Current knowledge about the relationships between climate and the functioning of individuals, populations, and ecosystems is limited. Increased ultraviolet radiation, resulting from reduced atmospheric ozone, is known to cause skin and eye maladies, and poses a serious threat to Arctic species and ecosystems. Monitoring to detect the impacts of climate change and ultraviolet radiation on Arctic ecosystems is vital, as is further research to understand the dynamics of the systems that will be affected.

Knowledge gaps

Current understanding of many aspects of Arctic biology and ecology is poor. The identification and classification of Arctic species, especially invertebrates and microorganisms, is at a rudimentary stage. Ecological processes that govern life in Arctic soils, in the Arctic Ocean, and at the ice edge are only beginning to be understood. Information on the status and trends of Arctic fauna is fragmentary at best, and almost entirely non-existent for flora. Advances in understanding of Arctic ecology often lead to more and more difficult questions. Much more monitoring is required to track trends in basic environmental parameters in the Arctic. More research is needed to document the species found in the Arctic and their genetic diversity, the ways that species interact locally and across ecosystems, and the ways that species interact with and respond to the physical environment. Access to and logistics in remote Arctic areas are a major challenge to researchers. New techniques, including remote sensing, and further studies will shed light on these critical areas. Nonetheless, the limits of current knowledge must be recognized before potentially harmful development is undertaken. Humans must tread carefully where the consequences of their actions cannot be foreseen.

Conservation

Conservation in the Arctic has achieved many notable successes, both in protecting the environment and in raising public awareness. It must build on such accomplishments while also developing more sophisticated conservation measures. The emphasis should be on preventing ecological problems, rather than less effective and more costly attempts to restore what has been damaged. Cooperative and collaborative conservation measures will be especially important, bringing various local, regional, and international interests together to identify and address common concerns. Managing ecosystems, rather than single species, needs more attention both in concept and in practice.

The future of Arctic biodiversity depends greatly on what happens in the rest of the world. Migratory birds and marine mammals link the Arctic to nearly every region of the globe, and are affected by conditions throughout their annual travels. Even in the Arctic, species face pollution from sources largely outside the region. International efforts to protect the Arctic must extend beyond the eight Arctic countries to encompass the entire global community.

This report began with the question, "What is the overall state of the Arctic environment?" An easy answer might be that the Arctic is generally in good condition. As this report makes clear, such a reply is inadequate. Much of the Arctic is indeed in its natural state, the impacts of human activity relatively minor. But development and pollution have devastated some areas. Individuals, species, and ecosystems throughout the Arctic face threats from many causes, and the long-term consequences of human impacts remain unknown. The challenges to conservation in the Arctic will always be present as humans continue to define their relationship with the natural world.

One thing is clear, though. While current approaches to conserving the Arctic environment and managing its living resources have had some successes, they cannot protect the Arctic forever. In the past, most threats were limited to small areas or to single species. Today, the Arctic also faces fragmentation of habitat, contamination, climate change, the cumulative impacts of many small projects, and human population growth – all of which can undermine the basic ecological processes of the region. The Arctic can no longer be considered remote and invulnerable. Instead, humans must recognize their capacity for causing damage, subtle and otherwise. The Arctic must be considered as a unit, its flora and fauna inextricably linked to each other and to the future of humans in the North.

The Arctic has many special and unique features. Many of its stark and beautiful landscapes and seascapes are pristine. Its flora and fauna are distinctive and highly adapted to the cold and dryness that characterize much of the region for most of the year. The many indigenous communities remain linked to their lands and waters, a connection between humans and the environment that is increasingly rare elsewhere. In much of the world, conservation is a matter of protecting what is left, or trying to restore what has been damaged. The Arctic offers a rare opportunity to demonstrate that humans can conserve a region, not as an afterthought, but as a priority.

The spoon-billed sandpiper
Eurynorhynchus pygmaeus, *a globally threatened Arctic species, is found only on the coastal tundra of Chukotka and Kamchatka, Russia.*

LIST OF ENGLISH AND RUSSIAN COMMON NAMES AND LATIN SCIENTIFIC NAMES OF SPECIES MENTIONED IN THIS BOOK

Note: Only species with an English common name are included.

ENGLISH NAME	LATIN NAME	RUSSIAN NAME
Mammals	*Mammalia*	Млекопитающие
Alpine marmot	*Marmota marmota*	Альпийский сурок
American mink	*Mustela vison*	Американская норка
Arctic hare	*Lepus arcticus*	Полярный заяц
Arctic fox	*Alopex lagopus*	Песец
Arctic shrew	*Sorex arcticus*	Арктическая бурозубка
Atlantic humpback whale	*Megaptera novaeangliae*	Горбач
Bearded seal	*Erignathus barbatus*	Морской заяц
Beaver	*Castor canadensis*	Канадский бобр
Beluga whale	*Delphinapterus leucas*	Белуга
Black bear	*Ursus americanus*	Черный медведь
Black-capped marmot	*Marmota camtschatica*	Черношапочный сурок
Black-tailed prairie dog	*Cynomys ludovicianus*	Чернохвостая луговая собачка
Blue whale	*Balaenoptera musculus*	Синий кит
Bowhead whale	*Balaena mysticetus*	Гренландский кит
Brown bear	*Ursus arctos*	Бурый медведь
Canadian river otter	*Lutra canadensis*	Канадская выдра
Caribou	*Rangifer tarandus*	Карибу, северный олень
Collared lemming	*Dicrostonyx groenlandicus*	Копытный лемминг
Coyote	*Canis latrans*	Койот
Dall's porpoise	*Phocoenoides dalli*	Белокрылая морская свинья
Elk/moose	*Alces alces*	Лось
Ermine/stoat	*Mustela erminea*	Горностай
Eurasian beaver	*Castor fiber*	Европейский бобр
Eurasian otter	*Lutra lutra*	Речная выдра
Fin whale	*Balaenoptera physalus*	Финвал
Giant bottle-nosed whale	*Berardius bairdii*	Дальневосточный бутылконос
Giant sperm whale	*Physeter catodon*	Кашалот
Gray whale	*Eschrichtius robustus*	Серый кит
Gray wolf/wolf	*Canis lupus*	Серый волк
Gray seal	*Halichoerus grypus*	Серый тюлень
Harbor seal	*Phoca vitulina*	Обыкновенный тюлень
Harbour porpoise	*Phocoena phocoena*	Морская свинья
Harp seal	*Phoca groenlandica*	Гренландский тюлень
Hooded seal	*Cystophora cristata*	Хохлач
Humpback whale	*Megaptera novaeangliae*	Горбач (мадагаскарская популяция)
Killer whale	*Orcinus orca*	Косатка
Laptev walrus	*Odobenus rosmarus laptevi*	Морж (лаптевский подвид)
Lynx	*Lynx lynx*	Рысь
Minke whale	*Balaenoptera acutorostrata*	Малый полосатик
Moose/elk	*Alces alces*	Лось
Mountain hare	*Lepus timidus*	Заяц-беляк
Muskox	*Ovibos moschatus*	Мускусный бык, овцебык
Narwhal	*Monodon monoceros*	Нарвал (единорог)
Northern birch mouse	*Sicista betulina*	Лесная мышовка
Northern bog lemming	*Synaptomys borealis sphagnicola*	Северный мышевидный лемминг
Northern fur seal	*Callorhinus ursinus*	Северный морской котик
Northern right whale	*Eubalaena glacialis*	Южный (гладкий) кит
Norwegian lemming	*Lemmus lemmus*	Норвежский лемминг

Norwegian rat/rat	*Rattus norvegicus*	Серая крыса
Pacific walrus	*Odobenus rosmarus divergens*	Морж (тихоокеанский подвид)
Peary caribou	*Rangifer tarandus pearyi*	Карибу Пири
Pine marten	*Martes martes*	Лесная куница
Polar bear	*Ursus maritimus*	Белый медведь
Putorean snow sheep	*Ovis nivicola borealis*	Путоранский снежный баран
Red fox	*Vulpes vulpes*	Обыкновенная лисица
Red squirrel	*Sciurus vulgaris*	Обыкновенная белка
Redbacked vole	*Clethrionomys gapperi*	Полевка Гаппера
Reindeer	*Rangifer tarandus tarandus*	Европейский северный олень
Ribbon seal	*Phoca fasciata*	Полосатый тюлень
Ringed seal	*Phoca hispida*	Кольчатая нерпа
Root vole	*Microtus oeconomus*	Полевка-экономка
Russian flying squirrel	*Pteromys volans*	Белка-летяга
Sakhalin vole	*Clethrionomys sikotanensis*	Шикотанская полевка
Sea otter	*Enhydra lutris*	Калан
Sei whale	*Balaenoptera borealis*	Сейвал
Siberian musk deer	*Moschus moschiferus*	Сибирская кабарга
Sibling vole	*Microtus epiroticus*	Полевка
Snow sheep	*Ovis nivicola*	Снежный баран
Snowshoe hare	*Lepus americanus*	Американский заяц
Sperm whale	*Physeter macrocephalus*	Кашалот
Spotted seal	*Phoca largha*	Пятнистый тюлень, ларга
Stejneger's beaked whale	*Mesoplodon stejnegeri*	Командорский ремнезуб
Steller's sea lion	*Eumetopias jubatus*	Сивуч
Stoat/ermine	*Mustela erminea*	Горностай
Svalbard reindeer	*Rangifer tarandus platyrhynchus*	Шпицбергенский северный олень
Triangle island vole	*Microtus townsendii cowanii*	Полевка Таунсенда
Vinogradov's lemming	*Dicrostonyx vinogradovi*	Лемминг Виноградова
Walrus	*Odobenus rosmarus*	Морж
Western barbastelle	*Barbastella barbastellus*	Европейская обыкновенная широкоушка
Wolverine	*Gulo gulo*	Росомаха

Birds / *Aves* / Птицы

Aleutian Canada goose	*Branta canadensis leucopareia*	Средняя канадская казарка
American golden plover	*Pluvialis dominica*	Бурокрылая ржанка
American widgeon	*Anas americana*	Американская свиязь
Arctic tern	*Sterna paradisaea*	Полярная крачка
Arctic warbler	*Phylloscopus borealis*	Таловка
Baikal teal	*Anas formosa*	Клоктун
Baird's sandpiper	*Calidris bairdii*	Бэрдов песочник
Barnacle goose	*Branta leucopsis*	Белощекая казарка
Barrow's goldeneye	*Bucephala islandica*	Исландский гоголь
Black-bellied plover	*Pluvialis squatarola*	Тулес
Black guillemot	*Cepphus grylle*	Чистик
Black-capped chickadee	*Parus atricapillus*	Черноголовая гаичка
Black-tailed godwit	*Limosa lapponica*	Малый веретенник
Black-winged stilt	*Himantopus himantopus*	Ходулочник
Brambling	*Fringilla montifringilla*	Зяблик
Brent goose	*Branta bernicla*	Черная казарка
Bristle-thighed curlew	*Numenius tahitiensis*	Таитийский кроншнеп
Broad-billed sandpiper	*Limicola falcinellus*	Грязовик
Buff-breasted sandpiper	*Tryngites subruficollis*	Канадский песочник
Cackling Canada goose	*Branta canadensis minima*	Малая канадская казарка
Capercaillie	*Tetrao urogallus*	Глухарь
Common crane	*Grus grus*	Серый журавль

Common eider	*Somateria mollissima*	Обыкновенная гага
Common guillemot/Common murre	*Uria aalge*	Тонкоклювая кайра
Common scoter	*Melanitta nigra*	Синьга
Common snipe	*Gallinago gallinago*	Обыкновенный бекас
Crested auklet	*Aethia cristatella*	Большая конюга
Curlew sandpiper	*Calidris ferruginea*	Краснозобик
Dark-bellied brent goose	*Branta bernicla bernicla*	Евразийская черная казарка
Dotterel	*Charadrius morinellus*	Хрустан
Dunlin	*Calidris alpina*	Чернозобик
Emperor goose	*Chen canagica*	Гусь-белошей
Eskimo curlew	*Numenius borealis*	Эскимосский кроншнеп
Eurasian wigeon	*Anas penelope*	Свиязь
Far eastern curlew	*Numenius madagascariensis*	Дальневосточный кроншнеп
Fulmar	*Fulmarus glacialis*	Глупыш
Gadwall	*Anas strepera*	Серая утка
Gannet	*Sula bassana*	Северная олуша
Golden eagle	*Aquila chrysaetos*	Беркут
Golden plover	*Pluvialis apricaria*	Золотистая ржанка
Goosander	*Mergus merganser*	Большой крохаль
Gray phalarope	*Phalaropus fulicarius*	Плосконосый плавунчик
Great gray owl	*Strix nebulosa*	Бородатая неясыть
Great knot	*Calidris tenuirostris*	Большой песочник
Great skua	*Stercorarius skua*	Большой поморник
Greater scaup	*Aythya marila*	Морская чернеть
Greater snow goose	*Chen caerulescens atlantica*	Голубой атлантический гусь
Greater white-fronted goose	*Anser albifrons frontalis*	Большой белолобый гусь
Greenland white-fronted goose	*Anser albifrons flavirostris*	Гренландский белолобый гусь
Grey plover	*Pluvialis squatarola*	Тулес
Greylag goose	*Anser anser*	Серый гусь
Gyrfalcon	*Falco rusticolus*	Кречет
Harlequin duck	*Histrionicus histrionicus*	Каменушка
Hazel grouse	*Tetrastes bonasia*	Рябчик
Herring gull	*Larus argentatus*	Серебрисая чайка
Ivory gull	*Pagophila eburnea*	Белая чайка
King eider	*Somateria spectabilis*	Гага-гребенушка
Kittiwake	*Rissa tridactyla*	Моевка
Knot/red knot	*Calidris canutus*	Исландкий песочник
Least sandpiper	*Calidris minutilla*	Песочник-крошка
Lesser snow goose	*Chen caerulescens caerulescens*	Малый белый гусь
Lesser white-fronted goose	*Anser erythropus*	Пискулька
Lesser-black backed gull	*Larus fuscus*	Клуша
Light-bellied brent goose	*Branta bernicla hrota*	Атлантическая черная казарка
Little auk	*Alle alle*	Люрик
Little stint	*Calidris minuta*	Кулик-воробей
Long-tailed duck	*Clangula hyemalis*	Морянка
Long-tailed jaeger	*Stercorarius longicaudus*	Длиннохвостый поморник
Long-toed stint	*Calidris subminuta*	Длиннопалый песочник
Mallard	*Anas platyrhynchos*	Кряква
Marbled godwit	*Limosa fedoa*	Пятнистый веретенник
Marbled murrelet	*Brachyramphus marmoratus*	Длинноклювый пыжик
Merlin	*Falco columbarius*	Дербник
Naumann's thrush	*Turdus naumanni*	Дрозд Науманна
Osprey	*Pandion haliaetus*	Скопа
Pacific/black brant	*Branta bernicla nigricans*	Американская черная казарка
Parakeet auklet	*Cyclorhynchus psittacula*	Белобрюшка
Pectoral sandpiper	*Calidris melanotos*	Кулик-дутыш

Peregrine falcon/falcon	*Falco peregrinus*	Сапсан
Pine grosbeak	*Pinicola enucleator*	Щур
Pink-footed goose	*Anser brachyrhynchus*	Короткоклювый гуменник
Pintail	*Anas acuta*	Средний поморник
Pochard	*Aythya ferina*	Шилохвость
Pomarine skua	*Stercorarius pomarinus*	Тундряная куропатка
Purple sandpiper	*Calidris maritima*	Морской песочник
Raven	*Corvus corax*	Ворон
Razorbill	*Alca torda*	Гагарка
Red-breasted goose	*Branta ruficollis*	Краснозобая казарка
Red-breasted merganser	*Mergus serrator*	Средний (длинноносый) крохаль
Red-legged kittiwake	*Rissa brevirostris*	Красноногая моевка
Red-necked stint	*Calidris ruficollis*	Кулик-красношейка
Red phalarope	*Phalaropus fulicarius*	Плоскоиосый плавунчик
Red-necked phalarope	*Phalaropus lobatus*	Круглоносый плавунчик
Red-throated loon/diver	*Gavia stellata*	Средний поморник
Reed bunting	*Emberiza schoeniclus*	Краснозобая гагара
Ringed plover	*Charadrius hiaticula*	Камышевая овсянка
Rock ptarmigan/ptarmigan	*Lagopus mutus*	Галстучник
Rock sandpiper	*Calidris ptilocnemis*	Берингийский песочник
Ross's gull	*Rhodosthestia rosea*	Розовая чайка
Rough-legged buzzard	*Philomachus pugnax*	Зимняк
Ruff	*Buteo lagopus*	Турухтан
Sanderling	*Calidris alba*	Песчанка
Sandpiper	*Totanus totanus*	Травник, красноножка
Scaup	*Aythya marila*	Морская чернеть
Sedge warbler	*Acrocephalus schoenobaenus*	Камышевка-барсучок
Semi-palmated plover	*Charadrius semipalmatus*	Перепончатопалый песочник
Semi-palmated sandpiper	*Calidris pusilla*	Малый (перепончатопалый) песочник
Shag	*Phalacrocorax aristotelis*	Хохлатый баклан
Sharp-tailed sandpiper	*Calidris acuminata*	Остроxвостый песочник
Short-billed dowitcher	*Limnodromus griseus*	Короткоклювый американский бекасовидый веретенник
Short-tailed albatross	*Phoebastria (Diomedea) albatrus*	Белоспинный альбатрос
Shoveler	*Anas clypeata*	Широконоска
Siberian jay	*Perisoreus infaustus*	Кукша
Siberian tit	*Parus cinctus*	Сероголовая гаичка
Slender-billed curlew	*Numenius tenuirostris*	Тонкоклювый кроншнеп
Slint sandpiper	*Micropalama himantopus*	Ходулочниковый зуек
Snow crane	*Grus leucogeranus*	Стерх, Белый журавль
Snow goose	*Chen caerulescens*	Белый гусь
Snowy owl	*Nyctea scandiaca*	Белая сова
Spectacled eider	*Somateria fischeri*	Очковая гага
Spoon-billed sandpiper	*Eurynorhynchus pygmaeus*	Кулик-лопатень
Steller's eider	*Polysticta stelleri*	Сибирская гага
Steller's sea-eagle	*Haliaeetus pelagicus*	Белоплечий орлан
Stilt sandpiper	*Micropalama himantopus*	Ходулочниковы зуек
Surfbird	*Aphriza virgata*	Бурунный кулик
Svalbard ptarmigan	*Lagopus mutus hyperboreus*	Шпицбергенская тундряная куропатка
Taverner's canada goose	*Branta canadensis taverneri*	Канадская казарка (подвид)
Teal	*Anas crecca*	Чирок-свистунок
Temminck's stint	*Calidris temminckii*	Белохвостый песочник
Thick-billed murre/Brünnich's guillemot	*Uria lomvia*	Толстоклювая кайра
Three-toed woodpecker	*Picoides tridactylus*	Трехпалый дятел
Tufted duck	*Aythya fuligula*	Хохлатая чернеть
Tundra bean goose	*Anser fabalis rossicus*	Западный тундровый гуменник

Tundra swan	*Cygnus columbianus*	Малый лебедь
Turnstone	*Arenaria interpres*	Камнешарка
Western sandpiper	*Calidris mauri*	Перепончатопалый песочник
White wagtail	*Motacilla alba*	Белая трясогузка
White-fronted goose	*Anser albifrons*	Белолобый гусь
White-rumped sandpiper	*Calidris fuscicollis*	Бонапартов песочник
White-tailed sea eagle	*Haliaeetus albicilla*	Орлан-белохвост
Whooper swan	*Cygnus cygnus*	Лебедь-кликун
Whooping Crane	*Grus americana*	Американский журавль
Whimbrel	*Numenius phaeopus*	Средний кроншнеп
Willow ptarmigan/Willow grouse	*Lagopus lagopus lagopus*	Лапландская белая куропатка
Willow tit	*Parus montanus*	Пухляк
Wood thrush	*Hylocichla mustelina*	Американский (лесной) дрозд
Wren	*Troglodytes troglodytes*	Крапивник
Yellow-billed loon/White billed diver	*Gavia adamsii*	Белоклювая (полярная) гагара

Fishes / Рыбы

Arctic char	*Salvelinus alpinus*	Арктический голец
Atlantic cod	*Gadus morhua*	Треска (атлантическая)
Arctic grayling	*Thymallus arcticus*	Сибирский хариус
Atlantic salmon	*Salmo salar*	Атлантический лосось, семга
Atlantic wolffish	*Anarhichas lupus*	Обыкновенная зубатка
Barndoor skate	*Raja (Dipturus) laevis*	Острорылый скат, скат-левис
Big skate	*Dipturus binoculata*	Бурый калифорнийский скат
Brook trout	*Salvelinus fontinalis*	Американская палия
Brown trout	*Salmo trutta*	Кумжа
Bull trout	*Salvelinus confluentus*	Большеголовый голец
Burbot	*Lota lota*	Налим
Capelin	*Mallotus villosus (arcticus)*	Мойва
Common skate	*Raja batis*	Гладкий скат
Deepwater redfish	*Sebastes mentella*	Клюворылый морской окунь
Dolly Varden char	*Salvelinus malma*	Мальма, голец
Greenland cod	*Gadus ogac*	Гренландская треска
Greenland halibut	*Reinhardtius hippoglossoides*	Черный палтус
Gulf sturgeon	*Acipenser oxyrinchus*	Американский осетр
Halibut	*Hippoglossus hippoglossus*	Обыкновенный, белокорый палтус
Herring	*Clupea harengus*	Сельдь
Inconnu sheefish	*Stenodus leucicthys nelma*	Нельма
Lamprey	*Lampetra fluviatilis*	Речная минога
Lake sturgeon	*Acipenser fulvescens*	Озерный осетр
Lake trout	*Salvelinus namaycush*	Кристивомер
Lake whitefish	*Coregonus clupeaformis*	Сельдевидный сиг
Lena river sturgeon	*Acipenser baerii stenorrhynchus*	Восточносибирский осетр
Minnow/European minnow	*Phoxinus phoxinus*	Европейский гольян
Navaga	*Eleginus navaga*	Навага
Northern pike	*Esox lucius*	Обыкновенная щука
Pacific cod	*Gadus macrocephalus*	Треска (тихоокеанская)
Polar/Arctic cod	*Boreogadus saida*	Сайка
Rainbow trout	*Oncorhyncus mykiss*	Микижа
River lamprey	*Lampetra fluviatilis*	Речная (невская) минога
Round whitefish	*Prosopium cylindraceum*	Валёк
Saffron cod	*Eleginus gracilis*	Дальневосточная навага
Scalebelly eelpout	*Lycodes squamiventer*	Чешуебрюхий ликод
Sea lamprey	*Petromyzon marinus*	Морская минога
Sea-trout	*Salmo trutta trutta*	Кумжа
Shortjaw cisco	*Coregonus zenithicus*	Широкоротая ряпушка

Shortspine thornyhead	*Sebastolobus alascanus*	Аляскинский щипощёк
Siberian sturgeon	*Acipenser baerii*	Сибирский осетр
Slimy sculpin	*Cottus cognatus*	Бычок-подкаменшик
Spotted wolffish	*Anarhichas minor*	Пятнистая зубатка
Sturgeon	*Acipenser baerii baerii*	Байкальский осетр
Vendace	*Coregonus albula*	Европейская ряпушка
Walleye pollock	*Theragra chalcogramma*	Минтай
Weatherfish	*Misgurnus fossilis*	Вьюн
Whitefish spp.	*Coreogonus spp.*	Сиги
Wolffish	*Anarhichas spp.*	р. Зубатка
Yellowfin sole	*Limanda aspera*	Желтоперая камбала
Yellowtail flounder	*Pleuronectes ferrugineus*	Желтохвостая камбала
Svetovidov's long-finned char	*Salvethymus svetovidovi*	Длиннопёрая палия Световидова

Reptiles and amphibians / Рептилии, амфибии

Common lizard	*Lacerta vivipara*	Живородящая ящерица
Siberian salamander	*Ranodon sibiricus*	Сибирский углозуб
Wood frog	*Rana sylvatica*	Лесная лягушка

Insects / Насекомые

Autumn(al) moth	*Epirrita autumnata (Oporinia) autumnata (Bkh.)*	Осенняя пяденица
Biting midge	*Ceratopogonidae* spp.	Мокрецы
Black flies	*Simulium* spp.	Мошки
Bumble bee	*Bombus* spp.	Шмель
Caddis flies	*Limnephilus* spp.	Ручейники
Chironomid midge	*Chironomidae* spp.	Комары-звонцы
Common furniture beetle	*Anobium punctatum*	Обыкновенный точильщик
Fox tapeworm	*Echinococcus multilocularis*	Эхинококк
Freija's Fritillary	*Clossiana freija*	Перламутровка Фрейя
Geometrid moths	*Epirrita autumnata* and *Operophtera* spp.	Осенняя пяденица и Пяденицы
Malarial mosquitoes	*Anopheles* spp.	Малярийные комары
Mayflies	*Ephemeroptera*	Поденки
Mosquitoes	*Aedes* spp, *Culex* spp.	Настоящие комары
Mountain burnet	*Zygaena exulans*	Бабочка пестрянка
Scale insect	*Coccoidea* spp.	Червецы, щитовки, кокциды
Spruce bark beetles	*Dendroctonus rufipennis*	Короед
Woolly-bear caterpillar	*Gynaephora groenlandica*	Мохнатая гусеница

Marine invertebrates / Морские беспозвоночные

Atlantic sea scallop	*Placopecten magellanicus*	Гребешок Магеллана
Baltic clam	*Macoma balthica*	Балтийская макома
Gorgonheaded brittle star	*Gorgonocephalus caputmedusae*	Офиура Голова Горгоны (Горгоноцефал)
Common sea urchin	*Echinus esculentus*	Съедобный морской еж
Copepods	*Copepoda*	Веслоногие ракообразные
Deep-sea coral	*Lophelia pertusa*	Коралл
Edible mussel	*Mytilus edulis*	Обыкновенная мидия
Krill	*Euphausiida*	Эуфазиевые рачки (криль)
Opossum shrimp	*Mysis relicta*	Мизида
Red gorgonian	*Paragordia arborea*	Горгонарий
Red king crab	*Paralithodes camtschatika*	Камчатский краб

Vascular plants / Сосудистые растения

Alpine azalea	*Loiseleuria procumbens*	Лаузелеурия лежачая
Alpine bearberry (black bearberry)	*Arctostaphylos alpina*	(Толокнянка) Арктоус альпийский
Androsace-like wormwood	*Artemisia senjavinensis*	Полынь сенявинская
Arctic paintbrush	*Castilleja arctica vorkutensis*	Кастиллея арктическая воркутинская

Arctic poppy	*Papaver radicatum*	Мак полярный
Arctic willow	*Salix polaris*	Ива полярная
Arctic wormood	*Artemisia aleutica (arctica)*	Полынь алеутская (арктическая)
Aspen (American)	*Populus tremuloides*	Тополь осинообразный
Aspen (European)	*Populus tremula*	Осина
Balsam poplar	*Populus balsamifera*	Тополь бальзамический
Bering Sea douglasia	*Douglasia beringensis*	Дугласия берингийская
Big mouse-ear chickweed/ common mouse-ear	*Cerastium fontanum*	Ясколка ключевая
Bilberry	*Vaccinium myrtillus*	Черника обыкновенная
Black spruce	*Picea mariana*	Ель черная
Blue flax	*Linum lewisii*	Лен Льюиса
Bog bilberry	*Vaccinium uliginosum*	Голубика (гонобобель)
Bog cranberry	*Oxycoccus microcarpus*	Клюква мелкоплодная
Bramble	*Rubus arcticus*	Княженика арктическая, поляника
Cloudberry	*Rubus chamaemorus*	Морошка приземистая
Common chickweed	*Stellaria media*	Звездчатка средняя, мокрица
Cotton-grasses	*Eriophorum* spp.	Пушица (осоковые)
Cowberry	*Vaccinium vitis-idaea*	Брусника
Creeping azalea	*Loiseleuria procumbens*	Луазелеурия лежачая
Crowberry/black crowberry	*Empetrum nigrum* spp. *hermaphroditum*	Морошка приземистая, водяника черная
Dahurian larch	*Larix gmelinii*	Лиственница даурская
Diapensia	*Diapensia lapponica*	Диапенсия лапландская
Drummond's bluebell	*Mertensia drummondii*	Мертензия Друммонда
Dwarf birch	*Betula nana*	Береза карликовая
Dwarf or shrubby alder	*Alnus fruticosa*	Ольха кустарниковая
Dwarf willow	*Salix polaris*	Ива полярная
European birch	*Betula pendula*	Береза белая
Europen larch	*Larix decidua*	Лиственница европейская
Fireweed	*Chamaeneriun (Epilobium) angustifolium*	Иван-чай
Forget-me-not	*Myosotis arvense*	Незабудка полевая
Goat willow	*Salix caprea*	Ива козья
Grey alder	*Alnus incana*	Ольха серая
Heather	*Calluna* or *Erica* spp.	Вересковые
Labrador tea	*Ledum palustre*	Багульник болотный
Lodgepole pine	*Pinus contorta*	Сосна скрученная
Lyme grass	*Elymus arenarius*	Колосняк песчаный
Meadow-grass	*Poa annua*	Мятлик однолетний
Mossberry	*Vaccinium oxycoccus*	Клюква обыкновенная (болотная)
Moss campion	*Silene acaulis*	Смолевка бесстебельная
Mountain birch	*Betula pubescens*	Береза пушистая Черепанова
Muir's fleabane	*Erigeron (Aster) muirii*	Мелколепестник Мюра
Nootka lupine	*Lupinus nootkatensis*	Люпин нуткинский
Norway spruce	*Picea abies*	Ель обыкновенная
Polunin's reedgrass	*Calamagrostis poluninii*	Вейник Полунина
Poppy species	*Papaver leucotrichum*	Мак беловолосковый
Ptarmigan wormwood	*Artemisia lagopus (abbreviata)*	Полынь куропаточья
Purple saxifrage	*Saxifraga oppositifolia*	Камнеломка супротивнолистная
Purple wormwood	*Artemisia globularia* var. *lutea*	Полынь шаровницевая (разновидность Желтая)
Reflexed saltmarsh-grass	*Puccinellia distance*	Бескильница расставленная
Rowan/mountain ash	*Sorbus aucuparia*	Рябина обыкновенная
Russet sedge	*Carex saxatilis*	Осока скальная
Scots pine	*Pinus sylvestris*	Сосна обыкновенная
Scurvy-grass	*Cochlearia officinalis*	Ложечница лекарственная
Sea lungwort/oysterplant	*Mertensia maritima*	Мертензия приморская

Sea sandwort	*Honckenya peploides*	Гонкения бутерлаковидная
Siberian dwarf pine	*Pinus pumila*	Кедровый стланик
Siberian fir	*Abies sibirica*	Пихта сибирская
Siberian larch	*Larix sibirica*	Сибирская лиственница
Siberian stone pine	*Pinus sibirica*	Кедр сибирский
Sitka alder	*Alnus crispata* spp. *sinuata*	Ольха кудрявая, п/вид вырезанная
Sitka spruce	*Picea sitchensis*	Ель ситхинская
Snow goose grass	*Phippsia algida*	Фиппсия холодная
Stone birch	*Betula ermanii*	Береза каменная
Stone bramble	*Rubus saxatilis*	Костяника обыкновенная
Subalpine fir	*Abies lasiocarpa*	Пихта субальпийская
Twin-flower	*Linnea borealis*	Линнея северная
Viviparous alpine meadow-grass	*Poa alpina* var. *vivipara*	Мятлик альпийский (вариация живородящий)
White birch	*Betula alba*	Береза пушистая
White mountain avens	*Dryas octopetala*	Дриада восьмилепестная
White spruce	*Picea glauca*	Ель канадская
Wild azalea	*Loiseleuria procumbens*	Луазелеурия лежачая
Wild rose	*Rosa* spp.	Роза (Шиповник)
Willow	*Salix* spp.	Ива

Non-vascular plants, lichens, and fungi / Лишайники и грибы

Birch fungus	*Polyporous betulinus*	Березовый трутовик
Common brown sphagnum/rusty peat moss	*Sphagnum fuscum*	Сфагнум бурый
Fly agaric mushroom	*Amanita muscaria*	Красный мухомор
Marimo algal ball	*Cladophora aegagropila*	Кладофора эгагропила
Reindeer lichen	*Cladonia rangiferina*	Олений мох, ягель
Sea kale	*Laminaria* spp.	Морская капуста
Sphagnum peat moss	*Sphagnum* spp.	Торфяной мох, сфагнум
Stairstep moss	*Hylocomium splendens*	Гилокомий блестящий
Wooly fringe-moss	*Racomitrium lanuginosum*	Ракомитрий шерстистый

LIST OF FIGURES

Figure 1	The Arctic	10
Figure 2	The limits of the Arctic according to various definitions	14
Figure 3	Major biomes (vegetation zones) of the Arctic	18
Figure 4	Surface ocean currents and the minimum extent of sea ice in the Arctic Ocean	19
Figure 5	Average atmospheric sea-level air pressure in winter and summer	19
Figure 6	Metamorphoses of snow	20
Figure 7	Average air temperatures in January and July	21
Figure 8	Deglaciation of North America	22
Figure 9	Reindeer calf/female ration in relation to annual snow accumulation	25
Figure 10	The North Atlantic Oscillation (NAO) index	25
Figure 11	The average length of the growing season across the Arctic	28
Figure 12	The Normalized Difference Vegetation Index (NDV), or Greenness Index	28
Figure 13	Number of vascular plant species found on Surtsey between 1965 and 2000	40
Figure 14	Migration routes of the humpback whale **Megaptera novaeangliae**	42
Figure 15	Migration routes of the arctic tern **Sterna paradisaea**	42
Figure 16	Breeding areas, migration routes, and staging areas of different subspecies of knot **Calidris canutus**	42
Figure 17	Places of origin and routes of transport for Arctic driftwood	53
Figure 18	The seasonal cycle of hunting, fishing and gathering for Polar Inuit in 1980	53
Figure 19	Approximate distribution of indigenous groups of the Arctic	58
Figure 20	The proportion of indigenous and non-indigenous residents in different areas around the Arctic	58
Figure 21	Industrial activities and oil and gas reserves in the Arctic	62
Figure 22	The Inuvialuit Settlement Region (ISR)	64
Figure 23	Organization of renewable resource management in the Inuvialuit Settlement Region (ISR)	65
Figure 24	Diamond mines on the range of Barthurst caribou herd in NW Territories, Canada	66
Figure 25	Percentage of the total protected area in the Arctic that is located in each biome	79
Figure 26	Percentage of the territory of each Arctic biome that is protected	79
Figure 27	Protected areas (>500 hectares) in the Arctic by IUCN Categories I-VI	80
Figure 28	Areas of high population densities of different groups of marine vertebrates in the Beringia Region	81
Figure 29	Protected areas (>500 hectares) in Fennoscandia by IUCN Categories I-VI	82
Figure 30	Ramsar Sites, World Heritage Sites, and Important Bird Areas in the Arctic	83
Figure 31	Decline of the thick-billed murre **Uria lomvia** population in central West Greenland	89
Figure 32	Murre colonies in the Arctic	92
Figure 33	Concentrations of DDT (an insecticide) and PCBs (industrial oils) in the blubber of ringed seals **Phoca hispida** in the 1980s and early 1990s	94
Figure 34	Oil drilling activities and related infrastructure on the North Slope of Alaska	96
Figure 35	Decrease of wilderness areas in Norway north of the CAFF boundary	98
Figure 36	Human impact on Arctic biodiversity and ecosystems in 1980 and 2050 using three rates (50-100-200%) of increase in resource development and infrastructure growth	99
Figure 37	The Arctic treeline, with characteristic species for various areas.	112
Figure 38	Plant successional changes in Nootka lupine **Lupinus nootkatensis** patch in southwestern Iceland	123
Figure 39	Classification of northern ecozones according to North American and Eurasian usage	132
Figure 40	The carbon cycle	135
Figure 41	Development of a palsa mound	137
Figure 42	Mire zones and provinces in Russia, showing the change of mire types from south to north	138
Figure 43	Distribution of **Lemmus** lemmings in the Arctic	142
Figure 44	Distribution of **Dicrostonyx** lemmings in the Arctic	143
Figure 45	The number of **Calidris** species breeding in each Arctic country	146

Figure 46	Distribution of wild reindeer, caribou, and domesticated reindeer in the Arctic	155
Figure 47	Die-offs of Peary caribou **Rangifer tarandus pearyi** from the 1960s to 1990s	157
Figure 48	The Peary caribou **Rangifer tarandus pearyi** population on Bathurst Island from the 1960s to 1990s	157
Figure 49	The distribution of rare endemic vascular plants of the Arctic	159
Figure 50	Circumpolar distribution of arctic char complex, **Salvelinus alpinus**	167
Figure 51	The upper food web of Toolik Lake, Alaska	168
Figure 52	The microbial food web of Toolik Lake, Alaska	168
Figure 53	Catchment area of the Arctic Ocean, showing the annual discharge (in cubic kilometers) of major rivers.	169
Figure 54	River channel fragmentation and major river dams in the Arctic catchment area	174
Figure 55	Mercury accumulation in food chains of subarctic lakes in Quebec, Canada	177
Figure 56	Trends in mercury level in fish muscle after damming of rivers in Quebec, Canada	177
Figure 57	The decline of Barrow's goldeneye **Bucephala islandica** at Lake Myvatn, Iceland	179
Figure 58	The distribution of four closely related benthic species of the genus **Pyrgo** (Foraminiferida) around Iceland	184
Figure 59	Annual mean temperature of water at the sea floor around Iceland	185
Figure 60	The Great Siberian Polynya	192
Figure 61	Generalised pattern of primary production in the Bering Sea	194
Figure 62	Distribution of oceanic zooplankton on the northern Bering-Chukchi continental shelf	194
Figure 63	Simplified Arctic marine food web	195
Figure 64	Distribution of bearded seal **Erignathus barbatus**	196
Figure 65	Distribution of narwhal **Monodon monoceros**	197
Figure 66	Traditional ecological knowledge of beluga whale **Delphinapterus leucas**, Uelen, Chukotka, Russia	200
Figure 67	Distribution of walrus **Odobenus rosmarus**	201
Figure 68	Distribution and migration of Barents Sea capelin **Mallotus villotus** and Atlantic herring **Clupea harengus** stocks	202
Figure 69	Cannibalism of cod in relation to deviations from long-term average sea temperature in the southwestern Barents Sea	202
Figure 70	Recovery of the Aleutian Canada goose **Branta canadensis** population	207
Figure 71	Changes in the black guillemot **Cepphus grylle** population on Flatey island, Breidafjördur, from 1974 to 1999.	213
Figure 72	Distribution of Steller's sea lion **Eumetopias jubatus** in the Arctic	214
Figure 73	Estimated numbers of non-pup of Steller's sea lions **Eumetopias jubatus** in different areas of the Arctic	214
Figure 74	Distribution and population status of polar bear **Ursus maritimus** in the Arctic	221
Figure 75	Breeding and wintering distribution of spectacled eider **Somateria fischeri**	223
Figure 76	Population trends in colonies of Arctic murres **Uria** spp.	224
Figure 77	Total catch of Atlantic cod **Gadus morhua** in the northeast Atlantic, Iceland, Davis Strait, and Labrador Sea	229
Figure 78	The fishable stock and proportion caught of Icelandic cod from 1925 to 1999	229
Figure 79	Spawning stock biomass and total landings of Pacific and Atlantic cod (**Gadus macrocephalus** and **G. morhua**) from 1970 to 1999	230
Figure 80	Population fluctuations of Svalbard reindeer **Rangifer tarandus platyrhynchus** from 1978 to 1999	232
Figure 81	Distribution of wolf **Canis lupus** in the Northern Hemisphere	235
Figure 82	Distribution of wolverine **Gulo gulo** in the Northern Hemisphere	237
Figure 83	Distribution of muskox **Ovibos moschatus** in the Arctic	241
Figure 84	Trends of Arctic goose populations	244

LIST OF TABLES

Table 1	*Effects of variation in snow cover*	*31*
Table 2	*Insect adaptations in the High Arctic*	*36*
Table 3	*Estimated numbers of species in selected groups worldwide and in the Arctic*	*48*
Table 4	*Protected areas in the Arctic as of 2000*	*78*
Table 5	*Population size and distribution of sandpipers in the Arctic*	*147*
Table 6	*The number of male ducks (various species) in the Myvatn area in spring from 1975 to 1989*	*179*
Table 7	*Estimated population size and trends for marine mammal stocks in the Arctic*	*219*
Table 8	*Estimated population size and trends for eiders in the Arctic region*	*222*
Table 9	*Trends for commercial fish stocks in the Arctic*	*228*
Table 10	*Estimated population size and trends of Arctic terrestrial mammals*	*242*
Table 11	*Estimated population size and trends of geese in the Arctic region*	*245*
Table 12	*Estimated population size and trends of shorebirds breeding in the Arctic*	*246*
Table 13	*Globally threatened species and populations within the Arctic Region*	*248*

INDEX

A

acid rain, acidification, acidity 86, 95, 126, 152, 176, 177, 254
Agenda 106
Agreement on Fishing Vessels on the High Seas 106
Agreement on Straddling Fish Stocks 106
Agreement on the Conservation of Polar Bears 107, 108
Alaska Nanuuq Commission 108
alder 131, 263, 264
Aleutian Canada goose 206, 207, 245, 258
algae 35, 41, 45, 165, 167, 168, 169, 170, 180, 181, 184, 185, 187, 188, 190, 191, 207
alkalinity 152
American golden plover 149, 258
Anadyr Current 194, 195
Aquificiales 45
archaea 45
Arctic Basin 17, 18, 29, 52, 53, 183, 184, 187, 207
arctic char 12, 44, 46, 47, 48, 71, 166, 167, 176, 178, 250
arctic cod 185, 208, 228
Arctic Council 9, 11, 12, 13, 107
Arctic Environmental Protection Strategy (AEPS) 9, 13
Arctic fox 20, 144, 206
arctic hare 145, 161
Arctic National Wildlife Refuge 240
arctic tern 42, 43, 44, 89, 226, 258
Ary-Mas 121
aspen 117, 263
Athabascans 115
Atlantic cod 202, 227, 228, 229, 230, 261
Atlantic herring 202, 228, 231
Atlantic salmon 73, 166, 172, 178, 203, 261
Atlantic wolffish 231
Atlas of Rare Endemic Vascular Plants of the Arctic 85, 147, 158, 159
autumn moth 112, 120

B

bacteria 40, 41, 45, 134, 145, 165, 167, 168, 170, 180, 181, 183, 186, 188, 191, 212
Baffin Bay 23, 57, 219, 220, 248
Baffin Island 27, 53, 146, 160, 226, 242
Baltic clam 189, 262
Barents Sea 59, 73, 86, 183, 184, 185, 186, 187, 193, 198, 202, 208, 219, 220, 223, 225, 226, 227, 228, 229, 231, 250, 251
barnacle goose 93, 144, 239, 245, 258
barndoor skate 249, 250, 261
Barrow's goldeneye 179
Bathurst Caribou Herd 66, 73, 101
Bear Island 223
bearded seal 170, 187, 188, 192, 196, 218, 220, 257
Beaufort Sea 22, 23, 59, 65, 183, 208, 219, 251
beaver 33, 167, 257
beluga whale 12, 29, 170, 181, 198, 200, 219, 257
Bering Sea 17, 73, 108, 109, 132, 146, 158, 181, 192, 193, 194, 195, 198, 200, 203, 206, 207, 208, 209, 215, 219, 220, 226, 227, 228, 229, 230, 251
Bering Strait 17, 22, 57, 59, 183, 193, 194, 195, 200, 201, 209
Beringia 22, 81, 132
Berne Convention on European Wildlife and Natural Habitats 107
big skate 250, 261
bilberry 74, 263

biosphere reserves 81
birch 52, 74, 111, 113, 115, 116, 117, 118, 120, 121, 129, 131, 132, 153, 263, 264
black brant, see also Pacific brant 167, 239, 245
black guillemot 189, 203, 212, 213, 258
black spruce 117, 263
black-capped marmot 85, 257
black-legged kittiwake 192
blackfly 168, 170, 178
bog 35, 116, 121, 128, 132, 136, 163
boreal forest 12, 26, 52, 77, 78, 110, 111, 115, 116, 120, 121, 125, 128, 129, 138, 143
bowhead whale 23, 57, 59, 189, 195, 196, 198, 208, 217, 220, 248, 257
brittle star 204, 262
broad-billed sandpiper 146, 247
brook trout 261
brown bear 32, 39, 82, 133, 167, 171, 232, 238, 242, 257
brown fat 38
brown trout 171, 176, 178, 261
brucellosis 145
Brünnich's guillemot, see also thick-billed murre 88, 193
burbot 168, 170, 176, 261
butterfly 36, 124, 140
bycatch 87, 90, 198, 203, 205
Bylot Island Bird Sanctuary 93

C

cadmium 152
Caloplaga elegans 141
Canadian archipelago 13, 17, 131
capelin 86, 185, 198, 202, 228, 231, 261
capercaillie 33, 82, 124, 125, 258
Carex saxatilis 165, 263
caribou 12, 190
Cerastium fontanum 40, 263
choanoflagellates 191
Chrysophyte algae 190
Chukchi 59, 74, 108, 181, 183, 194, 195, 220
Chukchi Sea 183, 192, 193, 195, 200, 219
Chukotka 56, 59, 74, 108, 146, 199, 200, 242, 245, 247, 256
ciliates 168, 191
Circumpolar Protected Areas Network 12, 77, 78, 79
Circumpolar Seabird Working Group 90, 203, 222, 250
clams 23, 186, 195
climate change 13, 20, 24, 49, 60, 81, 86, 95, 101, 109, 126, 127, 129, 144, 145, 154, 157, 160, 161, 166, 180, 181, 198, 207, 208, 209, 238, 241, 255, 256
cloudberry 74, 134
co-management, cooperative management 61, 64, 65, 87, 101, 208, 239
Cochlearia officinalis 40, 263
collembola 134
common crane 258
common eider 89, 203, 215, 222, 251, 259
common guillemot 203, 223, 259
common murre 89, 90, 203, 223, 251, 259
common scoter 179, 259
Conservation of Arctic Flora and Fauna 9, 11, 12, 13, 79, 85, 90, 109, 159, 203, 208
Convention Concerning the Protection of World Cultural and Natural Heritage 81

Convention for the Protection of the Marine Environment of the North East Atlantic 107
Convention on Biological Diversity 106
Convention on Environmental Impact Assessment in a Transboundary Context 107
Convention on Prevention of Pollution from Ships 106
Convention on the Conservation of Migratory Species of Wild Animals 106
Convention on Trade in Endangered Species 106
Convention on Wetlands of International Importance 81, 106
copepod 188, 262
coral 205
Cortinarius 41
Corydalis arctica 85
Cree 75, 180, 181, 209
cryoturbation 24
Cryptonatica clausa 193
curlew sandpiper 49, 146, 148, 247, 259
cushion plants 30, 132, 138, 152
cyanobacteria 41, 167, 170, 181

D

Dahurian larch 121, 263
dam, damming 66, 86, 87, 94, 121, 136, 165, 172, 174, 175, 176
deepwater redfish, see also ocean perch 228, 231, 261
deforestation 26, 118, 129, 254
Dene 66, 115
detritus 194
diatoms 178, 190, 191
Dictyoha speculum 190
dinoflagellates 190, 191
dipterans 167
dissolved organic carbon 165, 180
dissolved organic matter 176, 180, 181, 191
Dolgan 54, 55, 74
Dolly Varden char 46, 47, 166, 261
dotterel 49, 146, 259
Drummond's bluebell 158
dunlin 148, 239, 247, 259

E

earthworm 134
East Siberian Sea 183
eelpout 186, 187, 261
eider 12, 89, 90, 93, 141, 144, 167, 188, 189, 192, 193, 203, 215, 222, 223, 250, 251, 259, 260
El Niño-Southern Oscillation 17
endemic plant 85, 158
environmental impact assessments 94, 160
ermine (stoat) 20, 139, 141, 142, 144, 257, 258
erosion 40, 52, 57, 60, 86, 98, 123, 126, 140, 144, 150, 151, 153, 163, 165, 171, 175, 208, 254
estuaries 35, 170, 171, 183, 191
EU Bird and Habitat Directives 107
Eurasian wigeon 179, 259
Even 74
Evenk 51, 70, 74
extinction 44, 60, 82, 93, 198, 206, 233

F

falcon 9, 87, 121, 127, 191, 259, 260
fern 48, 49, 132
fire 26, 40, 98, 112, 115, 117, 118, 124, 127, 140, 254
fireweed 74, 263

Fisheries Joint Management Committee (FJMC) 64
Fomitopsis rosea 82
Foraminiferida 184
forest-line 110, 112, 116, 127
forestry 52, 66, 82, 98, 124, 125, 129
fox tapeworm 145, 262
fragmentation 12, 65, 86, 87, 98, 102, 121, 124, 174, 237, 238, 254, 256
freeze-thaw cycles 24, 29, 101, 134, 165
fritillary butterflies 140
frog 167, 262
fulmar 88, 189, 203, 259
fungi, fungus 39, 40, 41, 45, 48, 49, 74, 82, 115, 124, 126, 132, 264

G

gannet 226, 259
Gastropoda 193
genes, genetics 44, 82, 152, 159
glaciation, glaciers 21, 23, 24, 51, 78, 118, 131, 147, 163, 167, 253
golden eagle 121, 126, 259
goosander 179, 259
goose 12, 42, 49, 93, 137, 141, 144, 146, 160, 167, 172, 181, 206, 207, 211, 232, 233, 239, 243, 244, 245, 248, 251, 258, 259, 260, 261
gray phalarope 259
gray whale 188, 191, 195, 199, 219, 257
grayling 168, 173, 261
Great Arctic Reserve 84
great black-backed gull 225
great grey owl 119
Great Siberian Polynya 35, 185, 192, 193
greater scaup 167, 179, 259
greater snow goose 233, 245
greenhouse gases 180
Greenland cod 228, 229
Greenland halibut, see also Greenland turbot 228, 230
Greenland ice cap 163
Greenland Sea 183, 190, 220
Greenland shark 187
Greenland turbot, see Greenland halibut 230
Greenness Index, see also Normalized Difference Vegetation Index (NDVI) 28
grey plover 149, 247, 259
guillemot, see also murre 12, 88, 189, 193, 212, 213, 223, 225, 258, 259
Gulf Stream 17, 24, 25, 132
gull 29, 40, 188, 193, 225, 226, 227, 250, 251, 259, 260
gyrfalcon 9, 121, 259
Gyrodactylus salaris 176

H

harlequin duck 165, 179, 259
harp seal 38, 218, 221, 257
hazel grouse 82, 259
heavy metals 18, 86, 95, 126, 152, 177, 180, 201, 254
helminth 145
herbivore 30, 39, 115, 116, 140, 161, 187
herder, herding 24, 25, 51, 55, 60, 65, 66, 70, 71, 73, 74, 75, 98, 120, 126, 127, 150, 151, 152, 161, 173, 234, 235, 237
herring 40, 87, 185, 198, 202, 228, 231, 259, 261
herring gulls 40
hibernating, hibernation 20, 32, 38, 85, 133
Honckenya peploides 40, 263
hot springs 45
Humane Trapping Standards 107
humpback whale 42, 217, 219, 248, 257

humus 124
hunters, hunting 24, 51, 53, 55, 57, 60, 61, 64, 66, 67, 68, 69, 70, 72, 73, 74, 77, 81, 86, 87, 88, 90, 94, 98, 101, 102, 109, 118, 120, 121, 124, 144, 150, 151, 152, 156, 160, 184, 193, 198, 199, 200, 207, 208, 209, 214, 223, 232, 233, 234, 235, 237, 238, 239, 240, 241

I

Ice Age 22, 23, 24, 51, 55, 118, 126, 131, 147, 161, 193, 253
International Murre Conservation Strategy and Action Plan 90, 250
International Whaling Commission 208
Inuit 12, 13, 24, 51, 53, 57, 59, 60, 61, 63, 65, 66, 68, 75, 89, 101, 115, 156, 180, 181, 207, 209, 226
Iñupiat 65, 115
Inuvialuit 64, 65, 109, 115
ivory gull 225, 226, 227, 250, 251, 259

J

jaeger, see also skua 139, 141, 143, 144, 192, 259

K

kelp 39, 49
Khanty 72
killer whale 188, 216, 257
king eider 192, 193, 215, 222, 251, 259
kittiwake 88, 259
knot 42, 43, 49, 239, 259

L

Lake Myvatn 170, 178, 179
lake trout 168, 170, 173, 177, 180, 261
lake whitefish 177
lamprey 250, 261
land claims 61, 65
Lapland diapensia 30
Laptev walrus 192, 257
larch 52, 74, 111, 113, 121, 128, 263, 264
least weasel 139, 142
lemming 20, 29, 39, 49, 115, 116, 139, 141, 142, 143, 144, 150, 161, 257
Lena Delta, Lena Delta Nature Reserve 52, 77, 84, 85, 240
Leonucula bellotii 193
lesser snow goose 245, 259
lesser white-fronted Goose 93, 233, 243, 245, 248, 259
lesser-black backed gulls 40
Leymus arenarius 40
lichen 24, 30, 31, 40, 41, 45, 48, 49, 74, 123, 124, 126, 132, 139, 141, 152, 160, 161, 264
little auks 88, 211, 259
little stint 148, 247, 259
lodgepole pine 128, 263
London Dumping Convention 106
long-tailed duck 179, 192, 259
long-tailed jaeger 144, 259
Lophelia pertusa 204, 205, 262
lynx 33, 82, 115, 120, 147, 257

M

mallard 179, 259
Man and the Biosphere Program 81
Marimo algal balls 178, 264
marmot 85, 133, 257
medicinal herbs 74
Mertensia maritima 40, 263
methane 186, 187
microbes 24, 39, 45, 152, 163, 168

microclimate 21, 121, 135
microorganism 255
microparasites 145
microrganisms 135
migration, migratory 14, 39, 42, 43, 44, 46, 49, 51, 54, 55, 57, 66, 78, 81, 86, 87, 90, 95, 101, 106, 107, 116, 120, 128, 133, 137, 138, 139, 140, 141, 143, 144, 146, 150, 152, 154, 160, 163, 171, 175, 178, 187, 188, 190, 191, 192, 193, 195, 199, 200, 202, 207, 208, 232, 239, 250, 251, 255
mines, mining 59, 66, 86, 87, 94, 95, 98, 103, 121, 126, 131, 137, 150, 152, 158, 176, 178, 179, 198, 254
mire 27, 132, 135, 136, 137, 138
mite 48, 134
mollusks 22, 185, 188, 204
moose 82, 114, 127, 145, 167, 234, 238, 257
morphological adaptations 32
mosquitoe 37, 139, 168, 262
moss 24, 30, 40, 48, 49, 95, 115, 117, 123, 124, 132, 134, 136, 139, 142, 152, 170, 181, 264
moss campion 138
moth 37, 112, 116, 118, 120, 124, 262
mountain birch 113, 116, 118, 263
mountain burnet 140, 262
mountain hare 30, 33, 82, 257
mud volcanoes 184, 186
murre, see also guillemot 12, 88, 89, 90, 91, 92, 259, 260
mushroom 41, 72, 74, 264
muskox 38, 49, 51, 73, 87, 133, 140, 145, 150, 152, 157, 161, 231, 240, 241, 242, 250, 257
muskox lungworm 145
mycorrhiza, mycorrhizae, mycorrhizal 41, 115

N

nanoflagellates 168
narwhal 197, 216, 219, 257
national park 14, 47, 63, 77, 101, 117, 123, 128, 129, 164, 240
National Petroleum Reserve-Alaska (NPRA) 150
Naumann's thrush 121, 259
navaga 227, 228
nematodes 145, 167
Nenets 55, 74, 100, 115
Nganasan 74
Nootka lupine 86, 122, 123, 263
Nordic Environmental Protection Convention 107
Normalized Difference Vegetation Index (NDVI) 28
North American Agreement on Environmental Co-operation 107
North American Waterfowl Management Plan 107
North Atlantic Oscillation (NAO) 17, 24, 25, 195
North East Atlantic Fisheries Agreement 107
Northeast Greenland National Park 63, 77, 240
northern fulmars 189
northern fur seals 207
northern pike 177, 261
Norway rat 207
Norwegian lemming 142, 143, 257
Norwegian Sea 183, 202
Norwegian spruce 121

O

ocean perch, see also deepwater redfish 231
oil 59, 62, 63, 86, 87, 90, 94, 95, 96, 97, 98, 100, 121, 150, 151, 199, 241, 251, 254
Olenek Delta 239
Operophtera spp. 262

opossum shrimp 86, 176, 262
osprey 175, 259
Oxycoccus microcarpus 263
ozone, ozone depletion 49, 86, 95, 152, 180, 181, 203, 255

P

Pacific brant, see also black brant 239
Pacific cod 188, 227, 251, 261
Pacific salmon 166, 170, 173, 207, 208
Pacific walrus 108, 258
pack ice 29, 188, 193
Palsa, palsa mire 137, 138
paludification 117
Papaver leucotrichum 85, 263
parasites, parasitism 39, 41, 139, 145, 157, 161, 172, 176, 212, 241, 254
passerine 33, 239
Peary caribou 87, 156, 157, 190, 232, 242, 248, 258
peat, peat bogs 131, 134, 136, 137, 163, 264
Pectoral sandpiper 148, 247, 259
peregrine falcon 87, 121, 127, 260
periphyton 170
permafrost 12, 19, 24, 26, 35, 49, 71, 86, 95, 98, 102, 111, 112, 115, 117, 121, 127, 128, 131, 134, 135, 140, 142, 150, 154, 160, 171, 172, 255
Persistent organic pollutants (POPs) 152, 201
pesticides 177
phalarope 48, 146, 149, 247, 259
phytoplankton 29, 35, 168, 185, 187, 195
pine 41, 52, 72, 74, 111, 116, 118, 126, 128, 263, 264
pine marten 82, 258
pingo 24, 171
pintail 179, 260
plankton, plankton blooms 29, 35, 86, 139, 168, 176, 177, 180, 185, 187, 188, 189, 190, 192, 193, 195, 202, 203, 207, 231
Poa annua 40, 263
polar bear 12, 29, 32, 33, 54, 59, 65, 73, 86, 107, 108, 145, 150, 160, 184, 188, 189, 190, 192, 203, 207, 209, 215, 221, 238, 251, 258
polar front 17, 187, 202
polychaetes 185, 205
polygons, polygon tundra 21, 24, 137, 138
polymorphism 44
polynya 35, 88, 185, 189, 192, 193, 209
pomarine jaegers, pomarine skuas 192, 260
poplar 23, 34, 52, 111, 115, 263
Porcupine Caribou Herd 160
pounikko 137
primary production 45, 169, 170, 180, 187, 194, 195, 203
protozoa 45, 145
ptarmigan 30, 32, 33, 88, 133, 140, 260, 261, 263
Puccinellia distans 40
puffin 212
purple Sandpiper 148, 247
purple saxifrage 34
purple wormwood 158, 263

R

rabies 145
radionuclide 71, 106, 152, 180, 201, 254
rainbow trout 176, 261
Ramsar sites, Ramsar Convention 77, 78, 81, 83, 106, 239
rat 206, 207, 258
raven 39, 260
razorbill 203, 226, 260
red fox 145, 258

red gorgonian 204, 262
red squirrels 127
red-breasted merganser 179, 260
red-legged kittiwake 248, 260
red-necked stint 148, 260
red-throated loon 260
redbacked vole 121, 258
refugia, refugium 22, 49, 81
reindeer, domestic reindeer, wild reindeer 12, 20, 25, 30, 33, 38, 41, 42, 54, 55, 56, 57, 59, 60, 65, 66, 67, 70, 71, 73, 74, 75, 98, 100, 109, 112, 116, 120, 121, 126, 127, 133, 139, 145, 150, 151, 152, 154, 155, 161, 173, 175, 231, 232, 233, 234, 235, 237, 241, 242, 251, 258
Rhodothermus 45
right whale 59, 198, 216, 221, 248, 257
ringed plover 149, 239, 260
ringed seal 12, 94, 109, 192, 218, 221, 258
ross´s gull 227
rotifer 167, 168

S

Saami 13, 57, 60, 61, 71, 73, 75, 98, 161
sacred sites 12, 77, 100
salinity 42, 171, 176, 185, 187, 191
salmon 42, 57, 73, 75, 166, 170, 171, 172, 173, 175, 176, 178, 191, 193, 203, 207, 208, 254, 261
sanderling 146, 148, 239, 247, 260
sandpiper 45, 48, 49, 146, 147, 148, 149, 167, 239, 247, 248, 251, 256, 258, 259, 260, 261
sarcocystis 145
sawfly 37, 45
scallop 86, 262
Scolopacidae 48, 147
Scotch pine 128
Scots pine 41, 126, 263
sculpin 168
sea kale 74
sea level rise 22
sea otter 39, 49, 57, 248, 258
sea urchin 39, 204, 262
seal 12, 29, 38, 54, 60, 68, 73, 86, 94, 95, 109, 145, 170, 184, 185, 187, 188, 189, 192, 196, 198, 207, 208, 218, 221, 248, 257, 258
sedge 29, 132, 136, 142, 161, 260, 263
Senecio monstrosus 34
sewage 106, 172
shag 226, 260
shearwaters 193
sheep 24, 57, 60, 86, 116, 120, 123, 151, 153, 232, 237, 248, 258
shoveler 179, 260
Siberian jay 82, 124, 125, 260
Siberian larch 128, 264
Siberian newt 167
Siberian pine 72, 264
Siberian shelf 192
Siberian tit 82, 124, 125, 260
sites of scientific interest 77
Sitka spruce 128, 264
Skaftafell National Park 123
skate 187, 249, 250, 261
skua, see also jaeger 139, 143, 192, 259, 260
snail 168
snow 17, 19, 20, 21, 25, 29, 30, 31, 32, 33, 36, 38, 67, 73, 75, 101, 102, 115, 133, 136, 139, 140, 143, 154, 156, 157, 163, 165, 172, 177, 180, 181, 190, 192, 240

snow bunting 192
snow goose 49, 93, 144, 146, 233, 245, 259, 260
snowshoe hare 115, 145, 258
snowy owl 139, 141, 143, 191, 260
species diversity 48, 132, 191
spectacled eider 215, 222, 223, 250, 260
Sphagnum moss 136
sponge 205
spoon-billed sandpiper 146, 247, 251, 256, 260
spotted wolffish 231
springtail 37, 48, 134
spruce bark beetles 118, 262
Steller's eider 144, 222, 223, 250, 251, 260
Steller's sea cow 198
Steller's sea lion 214, 218, 221, 248, 258
stilt sandpiper 146, 247
stone flies 44
sturgeon 173, 249, 261, 262
sulfur bacteria 186
Surtsey 40
sustainable development 13, 65, 106, 129
sustainable use 13, 106, 254
Svalbard ptarmigan 33, 260
Svalbard reindeer 33, 145, 232, 233, 258

T

tardigrades 167
teal 179, 248, 258, 260
Temminck's stint 148, 260
The Global Program of Action for Protection of the Marine Environment from Land-Based Activities 106
The Regional Program of Action for Protection of the Arctic Marine Environment against Land-Based Activities 107
thermokarst 26
thick-billed murre, see also Brünnich´s guillemot 88, 89, 90, 91, 189, 193, 203, 225, 250, 251, 260
Thingvallavatn 46, 47
threatened species 124, 211, 247, 248, 250, 251, 267
three-toed woodpecker 82, 260
threespined stickleback 44
tourism, tourists 61, 66, 75, 86, 95, 98, 102, 103, 104, 109, 116, 121, 150, 172, 179, 232
toxoplasma 145
Tracheophyta 250
traditional ecological knowledge 12, 61, 75, 108, 109, 181, 200, 209
Transpolar Drift 52
trapping 52, 59, 61, 64, 66, 70, 73, 77, 107, 120, 150, 198, 207, 234
trawling 86, 198, 204, 205, 206, 254
treaties 160, 208
treeline 11, 12, 13, 14, 20, 23, 27, 34, 41, 49, 55, 66, 86, 95, 110, 111, 112, 113, 115, 116, 118, 121, 126, 127, 129, 131, 132, 139, 146, 147, 150, 154, 171, 239
trichinosis 145
trichoptera 44
tufted duck 179, 260
tundra swan 167
turnstone 149, 239, 247, 261
tussock tundra 132

U

U.S./Russia Bilateral Agreement on the Conservation and Management of the Alaska-Chukotka Polar Bear Population 108
Uelen 200

ultraviolet radiation 13, 49, 86, 95, 126, 152, 165, 176, 180, 203, 255
United Nations Convention on Law of the Sea 106

V

vendace 176, 262
viruses 139, 145, 167, 168, 212
vole 29, 116, 119, 121, 145, 258

W

walleye pollock 228
walrus 23, 54, 59, 95, 108, 145, 187, 188, 189, 192, 193, 195, 198, 201, 208, 218, 221, 257, 258
warbler 111, 258, 260
weasel 33, 139, 142, 143
western sandpiper 148, 261
whale 12, 23, 29, 42, 53, 55, 57, 59, 73, 95, 106, 109, 170, 181, 188, 189, 191, 195, 198, 199, 200, 203, 208, 216, 217, 219, 221, 248, 257, 258
white fat 38
white mountain-avens 138
white spruce 117, 264
white-tailed eagle 175
whitefish 44, 85, 166, 167, 170, 173, 176, 177, 262
Whooper swan 261
wild rose 74, 264
wilderness 98, 102, 103, 128, 235, 237
willow 30, 41, 45, 74, 131, 132, 142, 153, 263, 264
wolf 82, 120, 121, 157, 234, 235, 236, 243, 250, 251, 257
wolffish 228, 231, 262
Wolverine 82, 120, 121, 152, 232, 236, 237, 243, 248, 250, 251, 258
woolly-bear caterpillar 138, 262
World Charter for Nature 106
World Conservation Union (IUCN) 236
world heritage sites 81, 83
World Wide Fund for Nature (WWF) 103

Y

yellowfin sole 228
Yukon-Kuskokwim Delta Goose Management Plan 239
Yupiit 115

Z

zapovedniks 84
zooplankton 29, 86, 168, 176, 177, 185, 187, 188, 190, 193, 195, 207